The City in Need

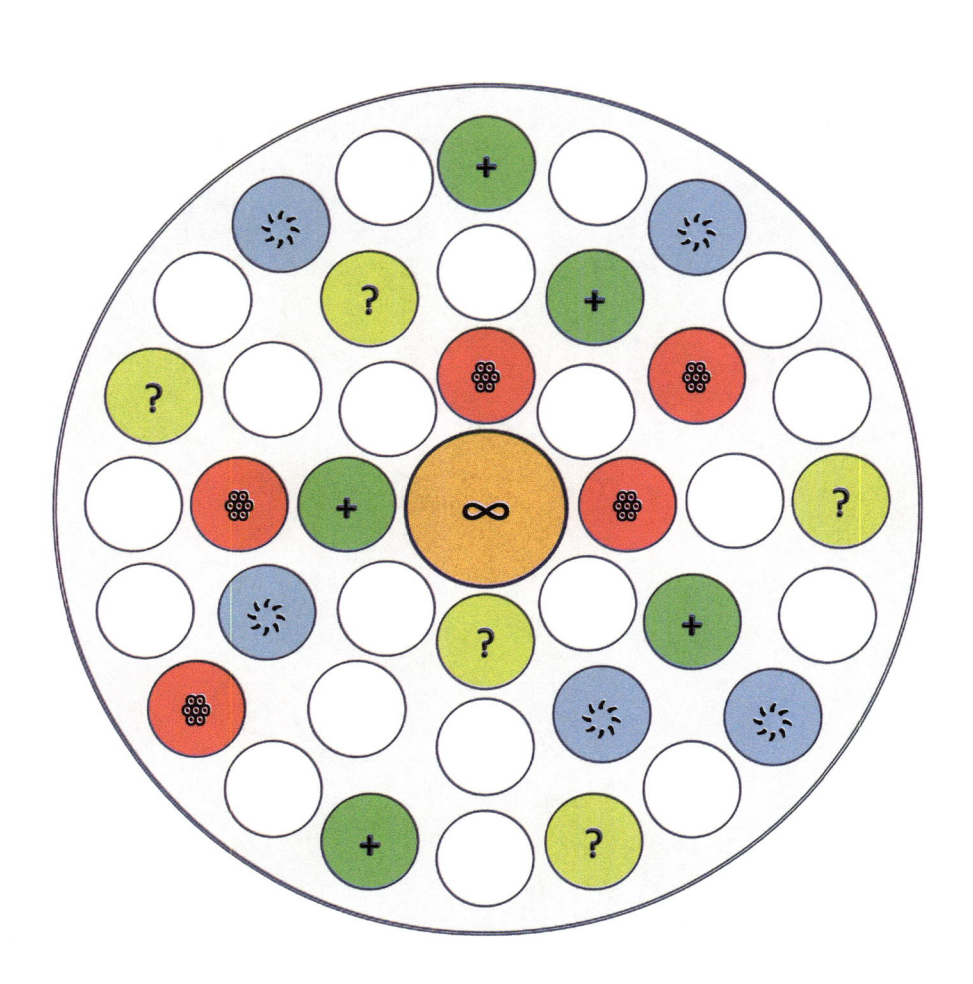

Ali Cheshmehzangi

The City in Need

Urban Resilience and City Management
in Disruptive Disease Outbreak Events

 Springer

Ali Cheshmehzangi
University of Nottingham Ningbo China
Ningbo, China

ISBN 978-981-15-5486-5   ISBN 978-981-15-5487-2   (eBook)
https://doi.org/10.1007/978-981-15-5487-2

This Springer imprint is published by the registered company Springer Nature Singapore Pte Ltd.
The registered company address is: 152 Beach Road, #21-01/04 Gateway East, Singapore 189721,
Singapore

To my family—they know I will arrive home in another 10 minutes! Thanks for your patience. I know I am not so easy.

I dedicate this book to the City of Wuhan and all victims of the COVID-19 outbreak. In particular, I dedicate this book to Dr. Li Wenliang (1986–2020). His efforts shall be remembered.

Preface

This is not an ordinary academic book. This timely book has ascended from the lockdown period of many cities and provinces across China and the globe, which was caused by the spread of the novel coronavirus disease (also known as COVID-19). This period shall be remembered for decades to come, as the world's big suffer and the global moment of instability and health emergency. However, there are lessons to be learned, and through a reflection, we can utilise them more thoughtfully for a more prepared future. Apart from the professional academic overview of the topic, this book reflects on practical examples and recommendations as well as the author's own experience of this dreadful outbreak event. The book, shortened for *The City in Need*, is an immediate and comprehensive response to the city's vulnerabilities.

In facing the many disease outbreak events, we note how cities and communities can become more vulnerable. We note the impacts and highlight how adversities are transmitted to the city. We also note how we should enhance our resilient thinking. We also suggest several forces that can help to respond more robustly to the situation of the outbreak, including governance, society, urban resilience measures, and city management practices. The latter two are the focus of this book, for us to suggest preparedness through comprehensive urban resilience, and responsiveness through reflective city management. In parallel to the fast-tracking vaccine research and its development, the race against what could be longer adversity, many nations and their societies may possibly face new challenges. These involve public health, societal wellbeing, socio-economic values, and an array of vulnerabilities.

The disease outbreak is a case of emergency and often regarded as a crisis or even a wartime situation. The impacts are very severe and healthcare systems cannot cope with populated conditions of emergencies. Currently, no city has the capacity to handle thousands of infected cases, let alone millions that could be a possibility. During and after the outbreak, the economic crisis is very serious, but so are the impacts on humanity and the wellbeing of our societies. Given the rising number of outbreak events globally, this is also a matter of governance—and hopefully virtuous governance. Hence, we require to have a holistic understanding of multiple sectors, multiple levels, multiple objectives, and multiple stakeholders.

And this book addresses these at the scale of the city, and through the powers of urban resilience and city management in disruptive outbreak events. These are meant to save the city in need.

This book will be a useful guide specifically for city authorities, urban planning teams, emergency units, urban managerial teams, urban bureaus, and research centers. The practical guidelines should provide support to practitioners, specific emergency task forces, prevention and protection teams, disease outbreak management teams, policymakers, and decision-makers. The practical recommendations should support some of the city-level and even national strategies and action plans. The extended knowledge on the topics of urban resilience and city management should also serve the needs of those relevant stakeholders, businesses/enterprises, researchers, academics, and communities.

In the first days of the outbreak, I developed a comprehensive urban resilience framework, which was passed to the municipal government of the City of Ningbo, East China. I believe some of the work helped to improve the city's strategic and action plans at the city management level. But later, I realised there is scope for more. In this uncertain situation, I thought about the main difference between the person who filters his own drinking water at home, and the one who works much harder to filter the lake. What is similar between the two is the act of filtering, but the scale and the results are not comparable. I was certain that I could not clean the oceans, but I could lend a hand to filter the lake. Hopefully, this helps the future lakes too.

It is useful to remember that sometimes we have to be vigilant in a different way, simply to make things better. Finally, I strongly believe we usually forget how resilient we can be.

Ningbo, China

Ali Cheshmehzangi
Head of Department of Architecture
and Built Environment
Director of Urban Innovation Lab

In the midst of the COVID-19 outbreak
(which was declared a pandemic on 11 March 2020)
January–March 2020

Acknowledgements

The idea of this book came to life under the influence of one dramatic moment. It was at around 7.20 am on the 28th of January 2020, when we had to go over a bridge on our way to Changsha airport, in central-south China. This was only 5 days after the lockdown of Wuhan City, and just before the many impending lockdowns. On a cold winter day, the bridge was frozen and the roads were slippery. We witnessed a movie-like scene on the frozen bridge with cars crashing onto both sides and leaving their passengers in shock. It was then that I realised how vulnerable a city can be in every other sector when it has to deal with the adversities of the outbreak. The first word came to my mind was 'resilient'. There was my moment of light bulb effect!

Last time, we experienced an outbreak by traveling into it (MERS outbreak in South Korea), and this time we lived through it with all its adversities. The experiences we gained this time are just invaluable!

I owe a special debt of gratitude to my family for their resilience and positive attitude during the outbreak. This was quite a ride. I acknowledge the National Natural Science Foundation of China (NSFC), for funding project number 71950410760, which was used for the provision of materials for the book. I humbly thank the Ningbo local government, and specifically Ningbo Urban Infrastructure Development Center, for their heartening response to my urban resilience framework document, which was sent to them on 10 February 2020. My sole intention was to help the city at this very unprecedented time. Their appreciation was simply rewarding, and I thank them for referring to me as the New Ningbo Citizen (*Xin Ningbo Ren* in pinyin).

I thank Ayra for her critical—but inspiring—feedback. Her comment was a turning point for me. It was then that I decided this book must be a full monograph and not a short book. I also highly appreciate her (somewhat) superpower skills and sense of humor. She once said we washed our hands so much that she mistakenly put soap on her toothbrush. She learned how to be more resilient and positive even in the hardest times. And she managed to create a new board game called 'Beat-the-Corona'. Ayra, you are just the best.

Above all, I show my appreciation to those individuals and groups who selflessly and tirelessly sacrificed their lives and time to maintain community safety and security throughout the outbreak. Those are the ones who received and treated the infected patients, dealt with so many emergencies, and sustained the city's operations and services through high levels of security and continuous monitoring. You are the real heroes.

About This Book

The sons of Adam are limbs of each other, having been created of one essence.

When the calamity of time affects one limb, the other limbs cannot remain at rest.

If you have no sympathy for the troubles of others, you are unworthy to be called a Human.

Saadi Shirazi, Persian Poet

This book is written at the time of a recurring outbreak, but just with a novel disease. 'Recurring' as we knowingly or unknowingly deal with many disease outbreaks regularly, and 'novel' because this disease was never seen or recorded before. This time the disease was unknown, it was prevalent and fast, and it could last in one's body for about 14 days before the person could show any symptoms. This means a much longer incubation period, twice longer than the two previous coronavirus diseases (SARS-CoV and MERS-CoV) and almost four times longer than the common influenza virus. More importantly, this was a novel disease with a dangerous combination of mysterious, vigorous, and surreptitious characteristics.

In facing the outbreak, there was a range of actions and inactions across the globe. Despite the fact that we deal with disease outbreaks very regularly, this time it felt unprecedented and experiences were surreal. There were signs of growing fear and anxiety due to the mass of uncertainties. But same as the previous times, we could either wait or innovate in various ways.

In this outbreak, there seemed to be a clear message from our mother earth that says, "If you play with my nature then I will play with yours". This was a breathing time for the planet and our environments, with less travel and minimised negative impacts on our environments. The outbreak truly became a test for humanity, and the adversities may just last for much longer than anticipated initially. Globally, there were profound measures and approaches against the disease and many research institutes and companies were in the not-so-fast—but intense—race of vaccine development. There were also signs of growing hatred and fear in between the societies, such as the records of many racism acts against certain groups of people and wrong reactions against those who wore facial masks at the early stages. There were signs of negligence, reckless manners, misbehaviors, maladministration, and many other issues that raised our concerns. In a simple response to the UK's health secretary who boldly said, *It is important that the country keeps moving as much as possibly can, within the limits of the advice we have been given*, I have to remind us all that it is actually the power of people who keeps things moving forward. Without it, our humanity is meaningless. Temporary closure of regular operations, with the aim to enhance the resilience and public health, would mean much more to any country that to just keep moving in a misty path. It is perhaps time for us to reflect on and revise our political will.

All of a sudden, the unknown City of Wuhan (at the global level) became a well-known epicenter of the disease as it struggled the battle more than any other city in China. Many cities struggled in different ways and many economies were worsened during this impactful event. There were signs of disruption in every single country/region and in every single sector. This was not an ordinary outbreak event. The governmental officials referred to this disease outbreak as the time of war, an unprecedented battle, a once-in-a-generation event, etc. We are convinced the impacts would last for long, but we remain uncertain how long they would continue and how far they would go. Hence, this book is a comprehensive guide for the city in need—may be slightly late for this time, but early for the future. The available knowledge about the topic had to be collected in such form, allowing a larger group of stakeholders and actors to gain more knowledge and act more responsively and responsibly in future occasions.

I am certain about one thing that after finishing this book, my eyesight is worsened because of much longer hours of reading and writing. The whole process felt like doing another Ph.D. studies, but much tenser and much faster. This book became my priority at the time I thought cities and communities (as well as their governments and healthcare systems) are struggling the most. From a novel framework development, the ideas were soon developed further to a short book and then to a full monograph. This decision was under the influence of this much-longer-than-expected outbreak (COVID-19). It simply made this book happen. The book then addresses a very important topic that requires immediate attention and awareness, and I hope it fulfills this need.

In six chapters, this book puts together knowledge from various disciplines, from scientific knowledge on epidemiology to details of urban planning and management. By putting together two valuable approaches to 'urban resilience' and 'city management', the book is a genuine reflection on the needs of the city. On the day, I started the conclusion chapter of the book, the event was officially declared 'pandemic'. The situation continued and was much worse as the book was getting completed.

As the framework was aimed to help the city in need, this book is aimed to help us more in the future. We have to find the right ways of facing outbreaks and we have to find the most effective ways to prepare and respond. We have to know where humanity stands and overcome the surreal feeling of the outbreak. We have to find answers to 'how humane is our humanity?' And we have to learn where we simply need to do better. As Karen Salmansohn puts in well, *When you can't control what's happening, challenge yourself to control the way you respond to what's happening*. Finally, let the efforts be remembered, and let humanity revive!

Contents

About the Author

Dr Ali Cheshmehzangi holds a Ph.D. degree in Architecture and Urban Design, a Masters degree in Urban Design, a Graduate Certificate in Professional Studies in Architecture, and a Bachelor degree in Architecture—all from the UK. He is an urbanist and urban designer by profession and by heart. His research is generally about cities and city transitions, sustainable urbanism directions, and integrated urban design strategies.

Ali is currently Head of Department of Architecture and Built Environment and Director of the Centre for Sustainable Energy Technologies (CSET), at the University of Nottingham Ningbo China. He is also an Associate Professor of Architecture and Urban Design and Director of Urban Innovation Lab (UIL). More recently, he works on two research projects on 'Sponge City Programme in China' and 'Integrated Urban Modelling Framework'. Ali has previously worked in several UK universities and practices, and has worked on several practice and research projects on eco-cities in China (Caofeidian, Meixi Lake, Chongming Islands, etc.), low-carbon town planning, urban modelling of residential neighbourhoods in several countries, green infrastructure of cities, toolkit for resilient cities (with Arup and Siemens), sponge city programme and green development in Ningbo City, and other projects related to urban transitions in various contexts. He has developed a comprehensive planning toolkit, called 'Integrated Assessment of City Enhancement (iACE)'. More recently, he has developed a *Comprehensive Urban Resilience Framework for the City of Ningbo* (January–

Febuary 2020). So far, he has four other published books, titled *Designing Cooler Cities: Energy, Cooling and Building Form—The Asian Perspective* (2017), the award-winning *Eco-development in China: Cities, Communities and Buildings* (2018)—awarded by Springer Nature on 22 August 2019, *Sustainable Urban Development in the Age of Climate Change—People: The Cure or Curse* (2018), and *Identity of Cities and City of Identities* (2020).

Chapter 1
Introduction: The City During Outbreak Events

I can be changed by what happens to me. But I refuse to be reduced by it.
—Maya Angelou.

1.1 Introduction: A General Overview

As the largest quarantine in human history, the City of Wuhan, China, with more than 11 million people went under a complete lockdown situation on 23 Jan 2020. An unprecedented situation that lasted longer than ever imagined. This occurred solely due to the spread of the novel coronavirus disease (later renamed as "COVID-19"), just one day before the celebration of the Chinese New Year. The situation became more intensified and new measures were gradually introduced and implemented. The lockdown measures were later practiced at a larger scale as it covered a larger area at the provincial level, taking into account a total population of more than 57 million people. Soon after, the spread of this novel disease reached other locations, pushing many cities and provinces to implement similar lockdown measures. By then, Wuhan's lockdown was no longer the largest quarantine. The situation of the lockdown was applied to more cities, regions, provinces, states, and also at the country level (Cheshmehzangi 2020). Only two months after Wuhan's lockdown, more than a quarter of the world's population was living under some form of lockdown. Despite all our technological advancement and economic progressions of the last two centuries, a novel virus managed to put everything at a halt.

The idea of this book was developed during this specific outbreak event, while the author experienced the situation from inception to later stages. The combined knowledge from this unique experience and a range of professional research studies provided a unique opportunity to not only portray what has occurred in this particular event but also what should be done in the future occasions of such kind. As the disease became widely-spread both nationally and internationally, it became more evident that this was a major test for the city's resilience against disruptive disease outbreak events. From the beginning, as the cities prepared themselves to control the spread of the disease, their resilience for other aspects became more vulnerable. Hence, the city authorities needed to pay more attention and have better preparedness in place, as it is only a matter of days or weeks before a city could literally collapse.

In their report on 'Health Emergency and Disaster Risk Management Framework', the World Health Organisation (WHO) (2019) announced that from 2012 to 2017, a total number of 1,200 outbreaks were recorded across 168 countries. This was later increased with a further 352 cases in 2018 before we experienced the newest outbreaks in China and Nigeria in late 2019 and early 2020. The numbers are not something that we see in our daily news reports but are certainly very alarming in many ways. It also proves how frequent we face outbreak events, and how important it is for us to study various methods of tackling the outbreak impacts on our cities and communities.

In an outbreak event, whether we deal with epidemic and pandemic events, cities suffer tremendously, societies can become extremely vulnerable, and economies can fail. Till the time normalisation can happen, there may be a lot of losses—from human life losses to impactful economic losses. Globally, there is no single city with an adequate healthcare capacity that can accommodate thousands of infected patients, set aside a scenario in which millions of people are infected in relatively small proximity of the city level. It has been proven that even the least deadly disease can spread at a gradual pace and have a gradual increase in mortality rates. This can occur mainly due to a lack of healthcare infrastructure (comprised of both capacity and provision) to accommodate larger groups of patients at the same time. In a similar situation, the City of Wuhan struggled to maintain the needs of patients in a timely manner. An example of which was the attempt to build a hospital in less than a week, indicating the importance of disaster management strategies and the need for urgently increasing the capacity of the healthcare infrastructure and facilities. In such events, time is exceptionally scarce and responses need to be fast. Unfortunately, not all responses will be accurate in this process and not all can respond effectively. Nevertheless, the process needs to happen at a large scale, and covering a range of factors; particularly if there were no or little preparedness in place. In this process, decisions are made rapidly but carefully. They are often short-lived or temporary, while the situation remains uncertain during multiple phases of the outbreak event. These phases will be discussed more thoroughly in the next chapter.

In general, the second generation of human-to-human cases in an outbreak event may raise new concerns about the handling of the situation. As such, containing the disease is no longer the only priority. More than ever, the city would then need to focus on enhancing the resilience of multiple aspects and manage the situation promptly. Hence, urban resilience and city management measures are highly important, as they can essentially save 'the city in need'. Such an approach requires holistic planning with its three primary characteristics of (1) predictive, (2) prescriptive, and (3) preventive. In an attempt to strengthen urban health resilience, such outbreaks require urgent attention to maintain and support the needs of residents, visitors, and those who are directly and indirectly affected. This requires the city's preparedness to maintain and manage multiple and essential city systems, such as health, food and clean water (including both supply and delivery), assets, medical support, safety and security, and social stability. Most importantly, this requires people's resilience and how society can handle the situation effectively. Through many global examples of outbreak events, we verify that (urban) resilience should be strongly backed up by regional management and national strategies. It requires support both internally and externally. This needs to be planned and implemented in a network of multiple stakeholders, enabling the resilience to be at multiple levels and considering multiple health aspects. Throughout the whole process of an outbreak, we need to have careful measures for urban resilience; and these should be holistic and inclusive to better contain people, health, infrastructure, and management of the situation.

This book is an immediate response to a major pandemic outbreak event at the dawn of this new decade. It started as an outbreak in the central part of China. It did not take long for it to be an epidemic event. It remained as a case of the epidemic for about two months, before it was characterised and declared a pandemic. It took more than two months for the infected cases to reach 100,000 cases globally. But then it took less than two weeks to double the number of cases, and only a few days till it passed 300,000 cases. The numbers were then much higher than initially expected. This declaration of change from epidemic to pandemic was due to deep concerns about the irrepressible spread of the disease and its severity, lack of resilience, and alarming levels of inaction at the global level (WHO 2020b). This decision was believed to be delayed already (announced on 11th of March 2020), which marks a difference between reality and realisation. However, we are rest assured that there are lessons to be learned from what has been done in the past, what has been experienced, and what can be done in the future. In all cases, cities play a major part in managing the situation as well as to avoid the widespread disease, and to contain the situation as promptly as possible. All these require careful planning. But before we delve into the details, this chapter will serve as an introduction to first explain what is 'outbreak event'? And how it differs from other events and patterns of large scale outbreaks, such as an epidemic, pandemic, and endemic? Afterward, the following sub-sections will explain the issues associated with city vulnerability and urban resilience. This is narrated through a broader understanding of theories, literature review, and current practices. This chapter will conclude with an overview of three R's in the practice of urban resilience, namely 'reflections', 'readiness', and 'responses'. The eventual discussions of this chapter will set a good foundation for the following chapters in order to assess the multiple stages of outbreak events and provide a range of theoretical and practical suggestions. The later suggestions are shaped around the idea of a comprehensive urban resilience framework in outbreak events, which is novel in the field of resilient cities and health-related city management scenarios (e.g. health emergency and health crisis). But before we do so, it is important to understand the definition of outbreak event and how it differs from other definitions in the field.

1.2 What Is "Outbreak Event"?

There are common misinterpretations between different health-oriented and disease events. In the field of epidemiology, a typical outbreak event is defined as the sudden spread of a contagious disease that occurs at a particular spatial scale in a certain period. Epidemiology itself is an interdisciplinary field, bridging between scientific disciplines like biology, statistics for investigation and analysis, social sciences for multiple uses, and engineering for exposure assessment. It is a discipline that is now commonly used in research and practices of biological sciences, public health, and clinical research (Porta 2014). In epidemiology, there are a variety of research studies that stretch from outbreak investigation to clinical trials, covering a range of analytical, scientific, and comparative studies that investigate the cause and spread of disease, analyse the pattern and progress, as well as control and guidelines to support information dissemination, knowledge share, and decision making processes.

The World Health Organisation (WHO) (WHO webpage, on *environmental health in emergencies*, sub-section 'Disease Outbreaks' 2020a) define outbreak diseases as the ones usually:

> "...*causes by an infection, transmitted through person-to-person contact, animal-to-person contact, or from the environment or other media. Outbreaks may also occur following exposure to chemicals or to radioactive materials*".

While there is an immediate need to investigate the actual cause or source of an outbreak, the investigation can take a long time, and often result in the development of potential scenarios, hypothesis, or continuing scientific research. WHO (ibid) also categorises the outbreak events into three distinct categories of: (1) Communicable disease outbreaks, (2) Disease outbreak events caused by Chemicals, and (3) Disease outbreaks of unknown etiology. The first category generally includes particular environmental factors as the main source of the outbreak, this can be caused by humans, but the source itself comes from the environmental factors influencing the spread of disease; such as from air quality, food, water, and sanitation as four common examples. The second category occurs less frequently and is mostly due to exposure to chemicals or toxins in a particular area. The third category occurs more regularly around the globe, and usually the cause if not clear from inception and it may remain undetected for a while. This category is the focus of this book, through which we try to address resilience and management measures to overcome the urban challenges and diverse disruptions of disease outbreak events.

One factor to note is that there are different ways of dealing with different outbreak events, e.g. there are differences between different categories (i.e. shown in the above three categories) and our responses to them, clear differences between natural disasters and disease outbreaks (Alwidyan et al. 2020), as well as differences between different stages of a particular event (see Chap. 2). Some studies refer to disease outbreaks as disaster scenarios (Sandi and Kangbai 2019) or include them in the same category (Lee et al. 2012). However, even though there are some overlapping factors and measures between the two, the author suggests refraining from categorising outbreak events as disaster events or scenarios. The rationale behind this is due to the apparent differences in the nature, progression, and multi-stage characteristics of outbreak events. Also, not only that disease outbreak events are different themselves, but some of them may be defined differently at different times and in different contexts, such as the reoccurring case of Dengue disease outbreak (Brady et al. 2015). Nevertheless, the stages of disease outbreaks are very similar in how they develop over a period, and only differ in terms of how they can progress, spread, and eventually become contained. In each event, the city resilience and management measures and methods are not the same, but they are generally similar in terms of how we should respond to the impacts and vulnerabilities caused by the situation.

In addition, disease outbreaks are different in scale and have patterns of occurrence, recognised as 'epidemic', 'pandemic', and 'endemic' situations. These are different to the general outbreak categories of 'common source' (both continuous source and point source), 'propagated' that is generally transmitted between person to person, 'behavioural risk related', and 'zoonotic' that is normally transmitted from animals to humans (extracted from 'Glossary of Epidemiology Terms'). There are common misunderstandings between these patterns of occurrence, and some research studies confuse one with another. Each of these categories represents a different stage at different scales of a disease spread. In most cases of outbreak events, there exists a later or immediate epidemic situation; hence, it is usually regarded as an 'epidemic disease outbreak'. If the spread is contained in just one location, then it can be regarded as just a disease outbreak. However, as this is generally unlikely, it is often regarded as an 'epidemic outbreak'. This is a common case as the spread can occur only in a few days and can easily go beyond the boundaries of a particular region. In most cases, this is almost inevitable as we frequently commute and mobilise, and as we are constantly in contact with multiple groups of people who do the same, too. This cycle of mobility provides an opportunity to increase the probabilities of disease spread or transmission and helps to transfer it from one location to another in a blink of an eye. On the other hand, Green et al. (2002) acknowledge the distinction between the two terms, namely "outbreak" and "epidemic", and argue that the difference is indeed related to the size of the event, referring in particular to the scale a disease eventually spreads. In this regard, an epidemic situation is defined as the further expansion of the outbreak event, normally including a larger number of cities and communities, beyond just a particular contained region. This is very common for novel diseases. Also as Brady et al. (2015) conclude in their studies, there is still scope to understand a practical definition of an outbreak event; one that is unconventional, holistic, and clear.

The case of pandemic becomes more momentous as the spread becomes a near-global or global situation, meaning that it includes not only multiple regions but also multiple countries across the globe. The actual occurrence of a pandemic event mainly depends on how fast and how efficient an epidemic event is managed. In some cases, a pandemic can last for a much longer period (such as HIV AIDS pandemic, recognised from 1966 onwards). In reality, it can last until it is completely cured. Finally, endemic is defined as an infection spread that is *constantly maintained at a baseline level in a particular geographic area without external inputs*" (extracted from 'Centre of Disease Control and Prevention (CDC)' webpage, Division of Scientific Education and Professional Development (DSEPD) 2020). This is not necessarily defined as an event but is recognised as a continuing situation of the disease spread.

Henceforth, we mainly use the terms 'outbreak' or 'outbreak event', as we focus mostly on the defined scale of the city. This book addresses this scale as it requires further attention from the perspective of resilience and management, two factors that will be assessed and discussed throughout. However, the suggestions are common for both epidemic and pandemic situations, but more closely to a more common case of 'epidemic'; an outbreak event that we can say is not completely disastrous but is exceedingly impactful on the societies.

1.3 City Vulnerability in Disease Outbreak Events

Before reaching the *"vaccine effectiveness period"* (Pezzoti et al. 2018), the city and its communities suffer from the invasive disease outbreak. In a situation like this, cities are more vulnerable. In a common scenario, the situation is always unexpected; hence, preparedness is not exactly adequate to respond to earlier stages of the outbreak and is not as prompt as it should be. In the first few days, or even in the first few weeks of the outbreak, the situation appears to be uncertain and difficult to handle. This is mainly because of the outbreak changes drastically, the broad-spectrum life patterns change in a sudden, and our daily routines and operations become completely disrupted. The multiplicity of impacts is sensed across multiple sectors, affecting the most: our society. While the long term vulnerabilities can be reduced in a more progressive way (Lim et al. 2013), the short term treatments may take longer than initially expected.

As suggested by Brady et al. (2015) with appropriate and timely control, the disease outbreak burden can be minimised. Yet, this requires preparedness as early as possible and it requires ready-made planning to reduce the city's vulnerability in crucial areas/factors, wherein need the most. As addressed by other studies, fluctuations in case numbers and regular surges can frequently disrupt and slow down the progress of treatment and containment (ibid), which can drive *"already-stretched healthcare resources to breaking point"* (also see Hay et al. 2003a; Hay et al. 2003b, Garg et al. 2008, Cotter et al. 2013; Brady et al. 2015). More importantly, the city's vulnerability increases as the spread continue to affect the primary services and systems of the city, such as healthcare, food systems, transportation services, etc. In general, disease outbreaks are usually fast-developing situations with indeterminate progress and constantly changing updates. These effects or factors put pressure on the society as much as they cause an excessively *"high burden due to the lack of response capabilities"* (Garg et al. 2008, Grais et al. 2007, Najera 1999, WHO Ebola Response Team 2014; Brady et al. 2015). Also, there is an urgent need to reduce vulnerabilities, by enhancing emergency risk management for health, which is believed to be multi-sectoral (Emergency Risk Management for Health Fact Sheets 2013, p. 1), including:

"The systematic analysis and management of health risks, posed by emergencies and disasters, through a combination of (i) hazard and vulnerability reduction to prevent and mitigate risks, (ii) preparedness, (ii) response and (iv) recovery measures".

This includes a range of factors for individuals, larger groups of people populations, infrastructure, services, and other community factors. In this regard, the need for primary health care at multiple levels is essential to reduce any *"underlying vulnerability, protect health facilities and services, and scale-up the response to meet the wide-ranging health needs"* (ibid). Hence, during an outbreak event, as the city becomes more vulnerable, we should ensure that supporting measures and the role of urban resilience is not reduced. To name a few, for instance, we have patients with other health issues who require attention, we have the elderly whom are more vulnerable than the others, we consistently need essential daily supplies and services, such as food, water, energy, etc., as well as other factors that should be taken into full consideration. Therefore, we cannot just avoid all those dynamism or else it cannot last long for a city to fall apart.

In the outbreak events, the vulnerability of the city goes beyond just vulnerable communities. The situation is dissimilar to those examples that only target a particular group. Hence, outbreak events are usually widespread. This is also one of the reasons why outbreaks are different from those examples of disasters and are instead more related to health emergency conditions (examples by WHO 2019). Generally, in the outbreak events, cities and communities with weak(er) or no institutional structure suffer the most. For the case of the COVID-19 pandemic, this was evident from inception. This was tracked from the earlier records and updates from WHO (2020b): *"The international community has asked for US$675 million to help protect states with weaker health systems as part of its Strategic Preparedness and Response Plan"*. The numbers were later increased and included more countries and regions. Two months after, the United Nations requested for a total of US$2 billion cash contributions for nations that will struggle to contain the outbreak. Those countries/regions that are not well prepared are likely to experience a shock before they could cope with the unexpected situation. Municipal actions, from managerial and decision-making bodies, try to assess the situation as any response is sensitive and require careful processing and monitory. Society gets the biggest hit as they experience panic attacks, anxiety, and uncertain conditions. In such events, our voracious nature vivacities as we are alarmed to prepare for survival. In general, people often rush to store more necessary food and supplies, and create a self-imbalance in the equilibrium of regular production and production trends. We do not buy what we need; but instead, we buy what we need and what we think we may need. This is caused by the uncertain circumstances that can change in any direction at any time. The vulnerability does not end there, and becomes more severe if the society is not reassured (Schoch-Spana et al. 2020), and if immediate measures are not in place. Hence, there are major debates around society management challenges during outbreak events, such as community/public engagement processes (Biehler et al. 2018; Jamrozik et al. 2018), public information policies (Maxwell 2003), health and risk communication (Miller 2017), community values (Schoch-Spana et al. 2020), public health ethics (Kenny et al. 2010; Lee 2012; Marckmann et al. 2015, Spike 2018), etc. In addition, an extended communication plan is required to effectively respond to those societal needs to ensure vulnerabilities are minimised at multiple stages and throughout the outbreak event.

Furthermore, the contagious disease could affect many people directly and indirectly, and this potentially increases the vulnerability of society from multiple dimensions (American Psychological Association 2018), more than just the disease itself:

> *"The threats to psychological well-being that outbreak pose often can be overcome with the skills of resilience, which can serve as a kind of emotional vaccine. We all can develop resilience. It involves behaviours, thoughts, and actions that can be learned over time".*

Therefore, as a response to vulnerability increase, there is a major need to increase and maintain the urban resilience measures; those interventions that boost the city's management, and those strategies that eventually save the city from a looming disaster.

1.4 Urban Resilience in Disease Outbreak Events

There is more to urban resilience in the disease outbreak events than just a typical example of a resilient city strategic plan. As discussed earlier, outbreak events are inimitable situations; and indeed, they require exclusive responses, too. This was evident from the time when we developed the first resilient city toolkit several years ago (Siemens, Arup, and RPA 2013). Through a variety of studies on resilient cities and urban resilience measures, we can verify a range of direct impacts on health, emergency medical services, communities, infrastructure, economy and businesses, profitability, production systems, social well-being, and quality of life. As many foundations of the city deteriorate at a rapid pace, we have to ensure the city is prepared enough to handle the situation before making progress.

More recent published books on urban resilience planning are mostly related to natural disasters (to name a few: Clarke and Dercon 2016; Sanderson et al. 2016; Lamond et al. 2017; Baldwin and King 2018; Miyata et al. 2019) or other disaster scenarios (to name a few: Pasteur 2010; Masterson et al. 2014; Matsuoka and Shaw 2014; Tierney 2014; Coaffee and Lee 2016; Shaw et al. 2016; Fekete and Fiedrich 2017; Lucini 2017; Borsekova and Nijkamp 2019; Lindell 2019; Pelz et al. 2019) that address factors of preparedness, resilience, responses, and action plans; those that then respond to immediate, gradual, and long-term transformations. There are fewer examples that include disease and contagion as part of those resilience planning (Rodin 2014; Jones 2016; Yang 2017; Singh et al. 2020; Yang Chan and Shaw 2020), which include a range of measures for risk management or bring together related reflections and initiatives (Burayidi et al. 2019). Hence, in order to be effective in practice, urban resilience needs pioneering state-of-the-art thinking. In addition to this, the need for supporting guidelines (Ihekweazu, et al. 2010) and frameworks are certainly essential, too.

As our cities grow more in numbers and size, and as they face more adversities and challenges (from the webpage of 100 Resilient Cities 2020), we may not always be ready to mitigate particular events; sometimes, we have to adapt, and often we have to enhance what we may have or develop what we may not have (Cheshmehzangi 2016; Cheshmehzangi and Dawodu 2018). In this regard, we need to develop a set of strategies to combat those particular events that can cause significant disruptions or could progressively jeopardise our cities and societies. Urban resilience measures are ever needed to respond to those adversities and challenges, those we foresee and not foresee. An outbreak event is one of those examples of specific events, which can be damaging in multiple ways and can increase the burden on the overall city management. In such incidences, the vulnerabilities are extensive and affect the multiple operations of cities. The situation of an outbreak event of any kind suggests an insalubrious city status; it is unhealthy for the government, governance, institutions, economy, health, and on top of all, the society.

To date, there is little literature or specific research associated with urban resilience in outbreak events. On the other hand, there are generic examples of practical measures, frameworks, tools, and guidelines that enable us to support those cities in need. Yet, in real practice, the city authorities often require to make decisions fast and precisely. Those decisions need to be context-specific and should address cultural factors, social needs, and economic concerns of that specific place/city. The process is so fierce that it can cause significant disruptions in any direction. Any decision needs to be carefully crafted before it is released to the public, or else it can turn into playful games of multiple means of the media, from social media to a more monstrous international media. They can generate false news, increase anxiety and fear, and they can simply make a pandemonium. In the outbreak events, there are many issues associated with the overarching public health factors that require a new (or revised) perspective; hence, vulnerabilities are at a much higher rate when outbreak events hit the city and its communities. Undoubtedly, resilience is needed for any occasion for any community. More importantly, we also need to acquire those resilience skills, both individually and collectively as part of society. The probabilities of failures or failed occasions are high, and reversing them can take longer and interrupt the progress of containment and recovery at any time. Any minimal disruption is perceived as a major issue, and the impacts are felt event much greater. Many sectors come into an absolute halt, even if temporary, but they suffer significantly. They usually look forward to novel solutions, innovations, and findings in medical research—in other words, something that can save the situation sooner than later.

In general, urban resilience must be realised as the backbone of how cities can be managed, both effectively and efficiently, especially during the outbreak events. There are only a handful of preparedness (i.e. including but not limited to services, supplies, and facilities) that can be in place before the actual start of the outbreak; hence, the majority of the work is not necessarily related to preparedness but is indeed the immediate and strategic responses that should develop, shape and get implemented during the event itself. This is exactly why, similar to some of those disaster events, vulnerabilities are high and cities and communities are at a high-risk level. In their report, WHO (2019, p. 15) suggest that while the *"leadership in managing infectious risks and responding to outbreaks is clear, the health sector also has a critical role in preventing and minimizing the health consequences of emergencies due to natural, technological and societal hazards"*. In addition to this statement, it is important to note that from the city management perspective, multiple sectors must experience similar—if not the same—situation. Other sectors associated with such incidents aim to promptly deal with emergencies and meritoriously cope with the disruptions caused by the event. In other words, the city as a whole becomes a new entity that requires to deal with emergencies at multiple levels and in multiple sectors. By having a resilience plan (i.e. in any practice-based or practical form), the city can work more effectively in managing the event and its negative impacts on the society. Hence, it is suggested to enhance the city's resilience where we can and where it seems feasible to do so in the specific context (i.e. in terms of capacity, capabilities, economic background, social issues, etc.). With such a planning approach, we can speed up the containment and recovery processes of the outbreak—i.e. to better contain the spread of disease, and avoid the event to evolve from an emergency status into a disaster situation.

Finally, what has to be addressed is the way we prepare and respond in a process. This requires a framework that could reflect on the ever-changing situation(s) of outbreak events, which will be addressed in later chapters in more detail (see Chaps. 3 and 4). Therefore, there is an essential need to act reflectively and responsibly by multiple actors of the government (of multiple departments), emergency units and emergency medical and health services, and other associated stakeholders of the public sector, private organsiations, non-governmental organisations (NGOs), community organisation groups, and the general public. In reality, the situation creates a new ecosystem of management and operations, one that requires to have resilience measures and adaptive capabilities. In their report on 'Communicable diseases following natural disasters', WHO (2006) proposed a set of risk assessment and priority interventions, to ensure the needs of the society are addressed promptly throughout the events and adequate planning is operational for both therapeutic and preventive interventions. These factors, apart from having adequate planning measures, requires a tangible resilience capacity in order to reflect quickly, be ready, and respond to those situations in the best possible way. These would be the concluding remarks of this introduction chapter in the following section.

1.5 Reflections, Readiness, and Responses

As an introduction to the book, this chapter has summarised a general overview of outbreak events, and how they are progressively more important in the fields of 'urban resilience' and 'city management'. The aim of this book is then to see how we can save the city before it becomes too vulnerable, and how we can respond to those unexpected and unfortunate events that could cost us many human lives and many other pressures. In a way, the analysis so far indicates that the extent to which we see outbreak events in a city boundary needs more scholarly attention. As mentioned earlier, there are many cases of outbreak events that we do not even hear about; as we only learn about the ones of the global importance or higher contagious nature. Some are only become more visible because of their political importance, or the economic impacts they may have at the much larger global scale (such as the recent COVID-19 outbreak). Yet, as it appears from the recorded data of global outbreak events (WHO 2019), it is indicated that we constantly deal with various outbreak events in communities all around the globe. How we may reflect on those events are something that we should take into full consideration for better future preparedness and a much-enhanced resilience (Fig. 1.1). It only makes a logical sense that we should plan this ahead and not wait for the unforeseen impacts. As we continue to neglect the importance of outbreak events, we may continue to neglect the importance of resilience measures we should develop for our cities and communities around the world. This is not a simple task, but without a doubt, it is one very important task that can save the lives of many people who could live if we take action either in advance or as early as possible.

So far, we highlighted the basic knowledge, the existing literature on outbreak events, and the perspectives that refer to the situation of a typical outbreak event. We learned outbreak events are not universal; as they differ from one disease to another, and from one context to another. We realised the same disease can also be different

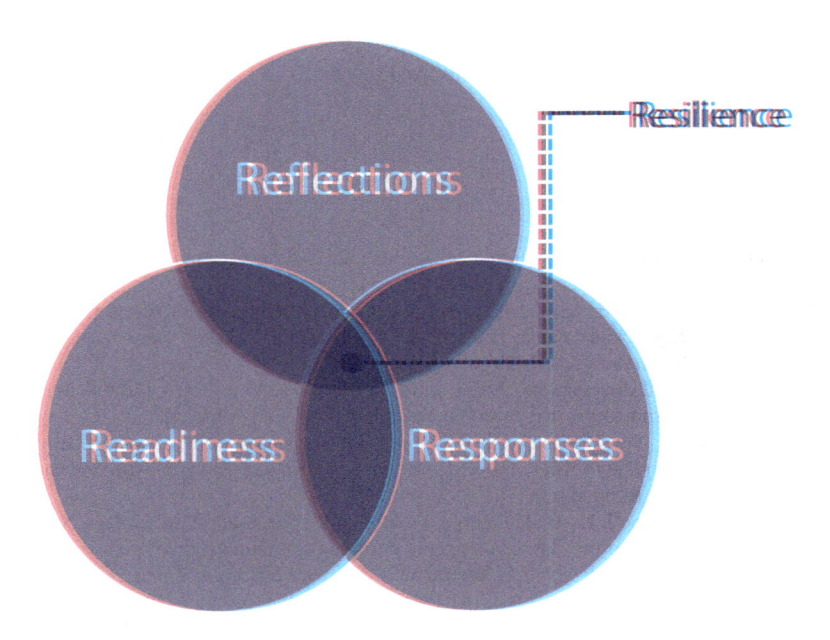

Fig. 1.1 The three primary R's in the practice of urban resilience. *Source* The Author's Own

from one location to another. Our cities are also very different, with primary differences in how they operate, their size, infrastructure, capabilities, economic conditions, social values, services, networks, etc. Therefore, our cities and communities would certainly reflect on outbreak events very differently, too. This means their readiness will be different, and how they may respond will ultimately be different. However, certain commonalities cannot be simply avoided in case of an outbreak event. For instance, the commonalities are mostly related to those institutional and societal needs, meaning how we plan to keep up the operations of our systems and services, how we may provide access to clean water, daily food supplies, energy, hygiene, amenities, and how our emergency units are supported and enhanced, as well as many other factors. In sum, cities, although different in many ways, will face similar difficulties/challenges in the case of outbreak events. Careful and comprehensive planning cannot be anything less than an assurance to overcome those difficulties that can simply threaten any community that exists in our world. In reality, we have to understand there is no immune community; at least, there is none that we know about.

The next few chapters of the book focus purely on key factors of urban resilience and city management to address their practicalities in a probable case of an outbreak event. Henceforth, the book addresses methods and strategies to enhance urban resilience during outbreak events. These ideas are generated through existing literature, practices, available tools and frameworks, dialogues with multiple experts of different disciplines, continuous discussions with local governmental authorities and global organisations, and the invaluable experience gained from standing with the community in a particular pandemic outbreak event.

References

100 Resilient Cities webpage.(2020). Retrieved February 2, 2020, from http://www.100resilientcit ies.org/.

Alwidyan, M. T., Traunor, J. E., & Bissell, R., A. (2020). Responding to natural disasters vs. disease outbreaks: Do emergency medical service providers have different views?, *International Journal of Disaster Risk Reduction*, *44*, 101440.

American Psychological Association (2018). *Resilience and Disease*, Psych Central, Retrieved February 18, 2020, from https://psychcentral.com/lib/resilience-and-disease/.

Baldwin, C., & King, R. (2018). *Social Sustainability, Climate Resilience and Community-Based Urban Development: What About the People? (Routledge Focus on Environment and Sustainability)*. Oxon: Routledge.

Biehler, D., Baker, J., Pitas, J. H., Bode-George, Y., Jordan, R., Sorensen, …, & Ladeau, S. (2018). Beyond "the Mosquito people": The challenges of engaging community for environmental justice in infested urban spaces. In R. Lave, C. Biermann, & S. Lane (Eds.), *The Palgrave Handbook of Critical Physical Geography* (pp. 295–318). Palgrave Macmillan: Cham.

Borsekova, K., & Nijkamp, P. (Eds.). (2019). *Resilience and Urban Disasters: Surviving Cities (New Horizons In Regional Science Series)*. Cheltenham: Edward Edgar Publishing Ltd.

Brady, O. J., Smith, D. L., Scott, T. W., & Hay, S. I. (2015). Dengue disease outbreak definitions are implicitly variable. *Epidemics, 11*, 92–102. https://doi.org/10.1016/j.epidem.2015.03.002.

Burayidi, M. A., Allen, A., Twigg, J., & Wamsler, C. (Eds.). (2019). *The Routledge Handbook of Urban Resilience* (Routledge International Handbooks). Oxon: Routledge.

Centre of Disease Control and Prevention (CDC) webpage, Division of Scientific Education and Professional Development (DSEPD). (2020). Retrieved February 5, 2020, from https://www.cdc.gov/csels/dsepd.

Cheshmehzangi, A. (2016). City Enhancement beyond the notion of "Sustainable City": Introduction to integrated assessment for city enhancement (iACE) Toolkit. *Energy Procedia, 104,* 153–158.

Cheshmehzangi, A., & Dawodu, A. (2018). *Sustainable Urban Development In The Age of Climate Change–People: The Cure or Curse.* Singapore: Palgrave Macmillan.

Cheshmehzangi, A. (2020). *Blame game will not help contain outbreak.* Article on China Daily. Retrieved May 12, 2020, from http://global.chinadaily.com.cn/a/202005/12/WS5eb9e1d9a310a8b241154e3e.html.

Clarke, D. J., & Dercon, S. (2016). *Dull Disasters?: How Planning Ahead Will Make A Difference.* Oxford: OUP Oxford.

Coaffee, J., & Lee, P. (2016). *Urban resilience (planning, environment, cities).* Red Globe Press, part of Macmillan International Higher Education.

Cotter, G., Sturrock, H. J., Hsiang, M. S., Liu, J., Phillips, A. A., Hwang, J., …, & Feachem, R. G. (2013). The changing epidemiology of malaria elimination: new strategies for new challenges. *Lancet, 382,* 900–911.

Emergency Risk Management for Health Fact Sheets. (2013). *Emergency risk management for health: Overview,* global platform, May 2013, pp. 1–6.

Fekete, A., & Fiedrich, F. (Eds.). (2017). *Urban Disaster Resilience and Security: Addressing Risks in Societies (The Urban Book Series).* Singapore: Springer.

Garg, P., Nagpal, J., Khairnar, P., & Seneviratne, S. L. (2008). Economic burden of dengue infections in India. *Transactions of the Royal Society of Tropical Medicine and Hygiene, 102,* 570–577.

Grais, R. F., Dubray, C., Gerstl, S., Guthmann, J.-P., Djibo, A., Nargaye, K., …, & Ihekweazu, C. (2007). Unacceptably high mortality related to measles epidemics in Niger, Nigeria, and Chad, *PLoS Med.,* 4, p. e16.

Green, M. S., Swartz, T., Mayshar, E., Lev, B., Leventhal, A., Slater, P. E., et al. (2002). When is an epidemic an epidemic? *Israel Medical Association Journal, 4*(1), 3–6.

Hay, S.I., Renshaw, M., Ochola, S.A., Noor, A.M., & Snow, R.W. (2003a). Performance of forecasting, warning and detection of malaria epidemics in the highlands of western Kenya. *Trends Parasitol, 19,* 394–399.

Hay, S. I., Were, E. C., Renshaw, M., Noor, A. M., Ochola, S. A., Olusanmi, I., …, & Snow, R. W. (2003b). Forecasting, warning, and detection of malaria epidemics: a case study, *Lancet, 361,* 1705–1706.

Ihekweazu, C., Basarab, M., Wilson, D., Oliver, I., Dance, D., George, R., et al. (2010). Outbreaks of serious pneumococcal disease in closed settings in the post-antibiotic era: A systematic review. *Journal of Infection, 61*(1), 21–27.

Jamrozik, E., & Selgelid, M. J. (2018). Ethics, health policy, and Zika: from emergency to global epidemic? *Journal of Medical Ethics, 44,* 343–348.

Jones, G. (2016). *HIV and Young People: Risk and Resilience in The Urban Slum (Springerbriefs In Public Health).* Singapore: Springer.

Kenny, N. P., Sherwin, S. B., & Baylis, F. E. (2010). Re-visioning public health ethics: A relational perspective. *Canadian Journal of Public Health, 101*(1), 9–11.

Lamond, J., Booth, C., Hammond, F., & Proverbs, D. (Eds.). (2017). *Floor Hazards: Impacts and Responses for the Built Environment.* Taylor and Francis Group: CRC Press.

Lee, J. D., Healy, S., & Lee, M. (2012). *Prepare for Disaster: The One Book You Need To Plan For Emergencies.* Florence, Alabama: Whitman Publishing.

Lee, L. M. (2012). Public health ethics theory: review and path to convergence. *Journal of Law Med Ethics, 40,* 85–98.

Lim, S. S., et al. (2013). A comparative risk assessment of burden of disease and injury attributable to 67 risk factors and risk factor clusters in 21 regions, 1990–2010: a systematic analysis for the Global Burden of Disease Study 2010. *Lancet, 380,* 2224–2260.

Lindell, M. K. (Ed.). (2019). *The Routledge Handbook of Urban Disaster Resilience: Integrating Mitigation, Preparedness, and Recovery Planning.* Oxon: Routledge.

Lucini, B. (2017). *The Other Side of Resilience to Terrorism: A Portrait of a Resilient-Healthy City.* Singapore: Springer.

Marckmann, G., Schmidt, H., Sofaer, N., & Strech, D. (2015). Putting public health ethics into practice: a systematic framework Front. *Public Health, 3,* 23.

Masterson, J. H., Peacock, W. G., van Zandt, S. S., Grover, H., Schwarz, L. F., & Cooper, J., T. (2014). *Planning for Community Resilience: A Handbook for Reducing Vulnerability to Disasters.* Washington D.C.: Island Press.

Matsuoka, Y., & Shaw, R. (2014). *Hyogo Framework for Action and Urban Disaster Resilience (Community, Environment and Disaster Risk Management).* 16, Bingley, UK: Emerald Group Publishing Limited.

Maxwell, T. A. (2003). The public need to know: emergencies, government organizations, and public information policies. *Gov. Inf. Q., 20*(3), 233–258.

Miller, A. N. (2017). Appeals to morality in health and risk messaging. In J. F. Nussbaum (Ed.), *The Oxford Research Encyclopedias of Communication: Health and Risk, Communication.* Oxford University.

Miyata, Y., Shibusawa, H., Permana, I., & Wahyuni, A. (2019). *Environmental and Natural Disaster Resilience of Indonesia (New Frontiers in Regional Science: Asian Perspectives).* Singapore: Springer.

Najera, J. (1999). Prevention and control of malaria epidemics. *Parassitologia, 41,* 339–347.

Pasteur, K. (2010). *From Vulnerability to Resilience: A framework for analysis and action to build community resilience.* Rugby, Warwickshire: Practical Action Publishing.

Pelz, M. (Ed.). (2019). *Protecting Historic Coastal Cities: Case Studies in Resilience (Gulf Coast Books).* Texas: Texas A & M University Press.

Pezzotti, P., Miglietta, A., Neri, A., Fazio, C., Vacca, P., Voller, F., …, & Stefanelli, P. (2018). Meningococcal C conjugate vaccine effectiveness before and during an outbreak of invasive meningococcal disease due to *Neisseria meningitidis* serogroup C/cc11, Tuscany, Italy. *Vaccine. 36*(29), 4222–4227.

Porta, M. (2014). *A Dictionary of Epidemiology* (6th ed.). New York: Oxford University Press.

Rodin, J. (2014). *The Resilience Dividend: Managing disruption, Avoiding Disaster, and Growing Stronger in an Unpredictable World.* London: Profile Books.

Sanderson, D., Kayden, J. S., & Leis, J. (Eds.). (2016). *Design for Urban Disaster: Response. Resiience and Transformation,* Oxon: Routledge.

Sandi, A. A., & Kangbai, J. B. (2019). *Disaster Management and the West African Ebola Outbreak.* Independently published, 89 pages.

Shaw, R., Ur-Rahman, A., Surjan, A., & Parvin, G., R. (2016). *Urban Disasters and Resilience in Asia.* Oxford: Butterworth-Heinemann.

Schoch-Spana, M., Watson, C., Ravi, S., Meyer, D., Pechta, L. E., Rose, D., A., …, & Sell, T. K. (2020). Vector Control in Zika-Affected Communities: Local Views on Community Engagement and Public Health Ethics during Outbreaks. *Preventive Medicine Reports,* 101059 (in Press).

Siemens, Arup, & RPA. (2013). *Toolkit for Resilient Cities: Infrastructure, Technology and Urban Planning,* 60 pages document. Retrieved January 24, 2020, from https://assets.new.siemens.com/siemens/assets/public.1543066657.641ee2256c5a0d5919d1aa3094a701f6ec9c3f90.toolkit-for-resilient-cities.pdf.

Singh, R. B., Srinagesh, B., & Anand, S. (Eds.). (2020). *Urban Health Risk and Resilience in Asian Cities (Advances in Geographical and Environmental Sciences).* Singapore: Springer.

Spike, J. P. (2018). Principles for public health ethics. *Ethics Med Public Health, 4,* 13–20.

Tireney, K. (2014). *The Social Roots of Risk: Producing Disasters, Promoting Resilience (High Reliability and Crisis Management).* Stanford: Stanford Business Books.

World Health Organisation (WHO) Ebola Response Team. (2014). Ebola virus disease in West Africa—the first 9 months of the epidemic and forward projections. *New England Journal of Medicine, 371,* 1481–1495.

World Health Organisation (WHO). (2006). *Communicable diseases following natural disasters: Risk assessment and priority interventions.* Report number WHO/CDS/NTD/DCE/2006.4. Geneva, Switzerland (also see: http://www.who.int/diseasecontrol_emergencies/en/).

World Health Organisation (WHO). (2019). *Health Emergency and Disaster Risk Management Framework.* Geneva: Switzerland.

World Health Organisation (WHO). (2020a). Webpage source on the category of 'Environmental Health in Emergencies', sub-section 'Disease Outbreaks". Retrieved February 8, 2020, from https://www.who.int/environmental_health_emergencies/disease_outbreaks/en/.

World Health Organisation (WHO). (2020b). *Rolling updates on coronavirus disease (COVID-19),* as part of updates on 11 March 2020: WHO characterizes COVID-19 as a pandemic. Retrieved March 12, 2020, from https://www.who.int/emergencies/diseases/novel-coronavirus-2019/events-as-they-happen.

Yang Chan, E. Y., & Shaw, R. (Eds.). (2020). *Public Health and Disasters: Health Emergency and Disaster Risk Management in Asia (Disaster Risk Reduction).* Singapore: Springer.

Yang, W. (Ed.). (2017). *Early Warning for Infectious Disease Outbreak: Theory and Practice.* Elsevier: Academic Press.

Chapter 2
How Cities Cope in Outbreak Events?

Growth is inevitable and desirable, but destruction of community character is not. The question is not whether your part of the world is going to change. The question is how.
—Edward T. McMahon.

2.1 Understanding the Progress of Outbreak

An outbreak can cause more problems than just the spread of disease. It can be an antagonistic nemesis to our cities and communities, particularly if we lack preparedness and resilience. Its progress is usually unclear as it can be completely different from case to case, and can react differently in different contexts and with different groups of people. Such reactions may purely relate to climatic conditions, hygienic status, and environmental attributes of the context. Those reactions can also differ from one group of people to another, while the disease has to find its correct host as well the way it can transmit and evolve. Consequently, the magnitude of impacts would depend on many factors, of which the nature of the disease is very important during the whole outbreak progress. In most cases, we have little information about the speed of disease spread, its impacts, its fatality rate, how contagious it can be, how long the whole process could last, as well as many other factors that may take a while to get identified, analysed, verified, and reported. At all scales, the process should be recorded carefully and formally, to ensure all stages are verified accurately, and all measures are considered adequate. Unlike other emergencies and many disaster events, outbreaks can develop progressively in a longer period. Tracking their progress requires careful monitory, and decision making becomes more intense and sensitive. The outbreak's prolonged—yet stage by stage—progression adds to existing uncertainties, as it can reverse or deviate to earlier or later stages, or even intensify further if the particular disease advances, spreads faster, or even mutates. Hence, sudden jumps in infected cases always cause anxiety to a wider group of stakeholders (i.e. particularly medical scientists). And every step is carefully taken to ensure no mistakes happen that may eventually worsen the situation.

A. Cheshmehzangi, *The City in Need*, https://doi.org/10.1007/978-981-15-5487-2_2

The more recent and unexpected outbreak of the novel coronavirus (COVID-19) has tested the adaptability, responsiveness, and robustness of many cities as they rise to tackle this rapidly growing challenge. With no exceptions, any infected zone had to come up with new measures and practical changes to the overall management of the city. Evidenced through extended literature from multiple disciplines, we can see it was predominantly after the first Ebola outbreak in between 2014 and 2016 (with the first case in December 2013) that issues of resilience and management during outbreak events were taken into consideration (Note: some earlier research on MERS and SARS outbreaks were also found). These examples include multiple perspectives from various areas of research, such as resilient health system (Kieny and Dovlo 2015; Kruk et al. 2015; Dieleman et al. 2016; Haldane et al. 2017), Resilience and smart city framework (Liotine et al. 2016), public health impact and management (Alexandre et al. 2017; Schwerdtle et al. 2017), policy priority areas (Buseh et al. 2015), and community-based resilient health system (Siekmans et al. 2017). Hence, this can be recognised as a breaking point in scholarly research (on outbreak events) that developed new research pathways to also study those issues that are associated with various management and resilience measures. The impacts from Ebola's prolonged first outbreak were so significant that it made resilience research to play a part in studies of outbreak events; even though that some of those resilience studies (Emergency Preparedness 2017; Alexandre et al. 2017; Schwerdtle et al. 2017; Siekmans et al. 2017) only focus on a particular system or aspect. These mostly focused on enhancing health system resilience (Bloom et al. 2015; Kruk et al. 2017) rather than incorporating other systems or sectors of the city resilience. It is partially assumed that other systems are also studied individually, or at most, associated with the health system or emergency medical services/system (EMS) of specific communities and cities. Furthermore, for the first Ebola outbreak, primary reports from the World Health Organisation (2014–16) highlight three main objectives, namely: "*to interrupt all remaining chains of Ebola transmission; to respond to the consequences of residual risks;* [and] *to work on health systems recovery*". The latter is certainly a major highlight of this specific outbreak that helped to shape new recovery measures for infected areas. Again, from the perspective of the broader urban studies, such research innovation is important in the study of outbreak progress.

For the case of cities and communities, the main factors that come to one's mind are methods and approaches to practical 'resilience enhancement' in the progress of the outbreak. These have to be implementable, adaptable, and effective in practice. Hence, they include key factors—but not limited to—adaptability, robustness, responsiveness, and preparedness that should support the city's multi-sectoral management and governance. But where can get "*sources of resilience*" as Schwerdtle et al. (2017) suggest? And how can we guarantee those sources will be ready for the specific time of an outbreak event? To answer these, we have to delve into further studies of resilience management and assess how cities and communities cope in such outbreak events. In their studies about resilience management, Massaro et al. (2018) highlight the importance of 'resilience assessment and management' (Linkov et al. 2013), as an emerging field, particularly from the perspective of practical implementation (Gao et al. 2016; Ganin et al. 2016, 2017) during the time of need. Massaro et al. (2018, p. 7) then argue that such practical approach should *"evaluate cross-domain alternatives to identify a policy design that enhances the system's ability"*, which should be used for during the event itself, in the recovery period, as well as for the ensuing prediction and preparation for the future.

In urban studies, there is no scholarly research that focuses on outbreak events and issues of city resilience. Hence, this book aims to address this research gap. Before we delve into discussions of urban resilience and city management, we have to develop a broader understanding of outbreak events, as well as their impacts on cities and communities. This chapter serves to address this important topic. At first, we verify and study multiple phases (or stages) of outbreak events, before evaluating the outbreak event in each specific phase. By doing so, we will have a better overview of how the outbreak happens, how it is expected to develop, and how it progresses in specific phases. This chapter concludes with a perspective on the development of 'responsive city management' in the face of the outbreak event progression. This chapter's more elaborate discussion on phase-by-phase impacts on cities provides new knowledge to the field of research. The information here will be utilised in the next chapter, where we focus more on measures and practicalities of urban resilience and city management in disease outbreak events.

2.2 Six Phases of Outbreak Events

As important as global health security matters (CDC Global Health 2014; Heymann et al. 2015; Sands et al. 2016), it is vital to study and understand the progress of disease outbreak/epidemic event at its source, and in the known and unknown immediate locations. This may sound like, or even seem like a robbery chase at first, but it will progress very differently soon after. In an outbreak, we are dealing with a case of emergency, not a disaster (although with some overlaps). As described by Lisnyj and Dickson-Anderson (2018), the impacts of such emergencies on communities can be devastating and enduring; arguing in favour of having post-crisis resilience plans, too. This requires a good understanding of multiple phases so that we can develop a range of practical resilience measures at different phases of the disease outbreak event (Cheshmehzangi 2020a, Cheshmehzangi 2020b) and building them over a period of time (Chimusoro et al. 2018). This will eventually lead into the development of a framework (see Chap. 3) that is not only conceptual (DeRose and Long 2014) but is also relevant and applied (Gunawan et al. 2015; Cheshmehzangi 2020a) for the real-life situation.

On their study related to measuring resilience, Linkov et al. (2013, p. 10108) provide a viewpoint that resilience is defined as *"the ability of a system to perform four functions with respect to adverse events: (i) planning and preparation, (ii) absorption, (iii) recovery, and (iv) adaptation"* (also adapted from The U.S. National Academy of Sciences 2012). Linkov et al. (ibid) also argue that resilience is very much associated with the overall scenario of the event and its associated risks, by arguing that:

…resilience has a broader purview than risk and is essential when risk is incomputable, such as when hazardous conditions are a complete surprise or when the risk analytic paradigm has been proven ineffective. Therefore, resilience measurement must be advanced with novel analytic approaches that are complementary to, but readily distinguishable from, those already identified with risk analysis.

In their attempts to define and assess measuring resilience (ibid), they propose a follow up 'Resilience Matrix' consisted of time of progression and multiple domains affected from a particular event (as well as taken into consideration during the process). In another words, this systematic approach enables to map multiple system domains *"across an event management cycle of resilience functions"* (ibid). In this regard, a theoretical guideline can be generated for each system domain and help to measure the overall system resilience. This approach enables us to have a clearer understanding of progression, in four primary domains of (1) Physical, (2) Information, (3) Cognitive, and (4) Social (also Alberts 2003). And each domain would then need to respond according to those four stages of plan/prepare, absorb, recover, and adapt.

Based on Linkov et al. (2013, p. 10109), each of these domains are explained as below:

(1) **Physical Domain**—In the plan/prepare stage, to have state and capability of equipment and personnel, and network structure. In the absorb stage, to have event recognition and system performance to maintain function. In the recover stage, there is a need for system changes to recover previous functionality. Finally, in the adapt stage, there is a higher need for changes to improve system resilience.

(2) **Information Domain**—In the plan/prepare stage, it is vital to have data preparation, presentation, analysis, and storage. In the absorb stage, it is essential to conduct real-time assessment of functionality, and anticipation of cascading losses. In the recover stage, we anticipate the information domain to make good use of data to track recovery progress, and anticipate recovery scenarios. Finally, in the adapt stage, more support is required for the creation and improvement of data storage, and it is essential to use protocols.

(3) **Cognitive Domain**—In the plan/prepare stage, we have to proceed with system design and operation decisions, with anticipation of adverse events. In the absorb stage, it is essential to have and use contingency protocols and proactive event management. In the recover stage, there is a need for recovery decision-making and communication. Finally, in the adapt stage, we need the design of new system configuration, objectives and decision criteria.

(4) **Social Domain**—In the plan/prepare stage, we have to utilise as much as possible the extent of social network, social capital, institutional and cultural norms, and training. In the absorb stage, we have to be fully prepared to allocate resourceful and accessible personnel and social institutions for event response. In the recover stage, it is essential to promote teamwork and knowledge sharing

to enhance system recovery. Finally, in the adapt stage, we need to pay attention to addition of/changes of institutions, policies, training, programmes, and culture.

Consequently, this type of approach is identified to be closer to the principles of 'Network Centric Warfare (NSW)', which is developed by military scholars (ibid) and is generally focused on *"creating shared situational awareness and decentralized decision-making by distributing information across networks"* operating in those four mentioned domains. According to Alberts (2003, also in ibid), these details are recognised as: (1) a variety of sensors, facilities, equipment, system states, and capabilities for the 'Physical Domain'; (2) the combination of creation, manipulation, and storage of data for the 'Information Domain'; (3) careful consideration of understanding, mental models, preconceptions, biases, and values for the 'Cognitive Domain'; and (4) A range of interaction, collaboration and self-synchronisation between individuals and entities, for the 'Social Domain'. In an example of measuring city resilience, there are four key domains or dimensions of economy, society, governance, and environment (The Organisation for Economic Co-operation and Development (OECD) 2019). In their measurement, they include the following factors of all four domains:

(1) **Economy**—to include factors, such as 'GDP growth rate', 'Unemployment', 'Number of start-ups and business failures', and 'Age and gender of: (i) employed, and (ii) working population'.
(2) **Society**—to include factors, such as 'Migration age and gender', 'Poverty levels', 'Household income', 'Percentage of population', and 'Living 500 meters from services'.
(3) **Governance**—to include factors, such as 'revenue by source', and 'Number of: (i) Community organisations, (ii) Public sector officials, and (iii) Sub-national governments'.
(4) **Environment**—to include factors, such as 'Population density', 'Accessible green area levels, specifically for: (i) percentage of built up areas, (ii) percentage of brownfield sites, (iii) percentage of citizens near open space, and (iv) percentage of new development, and 'Near transit locations'.

The above puts 'city resilience' in the middle of these four primary domains/dimensions.

In order to reach a full state of resilience measures, the author believes that we should first understand how an outbreak progresses in multiple phases. In this regard, the whole progression should not be recognised as a cycle, but should indeed by reflective on how a particular disease outbreak develops from inception to post-recovery. This is mainly for the case of an outbreak and not the resilience that may be required to address the needs of each stage (see Chap. 3 for more details). By understanding those four mentioned stages of progression in resilience development and applications (Linkov et al. 2013), it is then more important to take the time factor into consideration of how an outbreak develops. Therefore, the author proposes six phases that are commonly experienced in outbreak events (Fig. 2.1). These phases are progressive, and can either prolong or reverse at certain points and should our

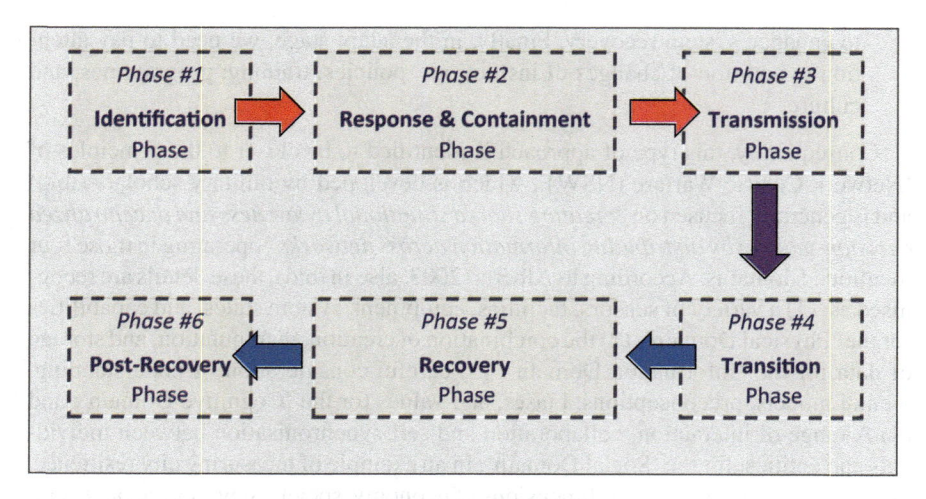

Fig. 2.1 Demonstration of 6 phases of outbreak event from inception to post-recovery, including first three alert/emergency phases, including 'identification', 'response and containment', and 'transmission' phases at the top (with red arrows in between), a 'transition' phase that can have multiple directions (e.g. positive or negative) at bottom right (with a Purple arrow in between), and two phases of 'recovery' and 'post-recovery (with blue arrows in between). Note: two phases of 'response and containment' and 'recovery' are intentionally drawn larger to represent both their longer time and importance throughout the outbreak progression. *Source* The Author's own

resilience and management fail to progress in the right direction. These six step-by-step phases are: (1) Identification phase, (2) Response and Containment phase, (3) Transmission phase, (4) Transition phase, (5) Recovery phase, and (6) Post-recovery phase. As shown in Fig. 2.1, the progress of the outbreak event is linear, but certainly includes a variety of parameters and factors that would be discussed in the next chapter. It is also important to note the difference between different phases of outbreak progression, and different phases of action plans against the outbreak. The former is the focus of what we cover in the following sub-sections.

While theoretically, we can define these six distinct phases of outbreak event progression, the actual boundaries in between all phases are somewhat fuzzy, reflecting on the ambiguous characteristics of the start and the closure of a disease outbreak event. In most cases, there is very little information about the know-about(s) of disease outbreak, its origin, when it started (Berthod et al. 2014; CDC 2014), how it emerged, and how it may progress. Hence, the progress itself is a source of more uncertainties, in terms of 'intensity', 'time', and 'location'. As described by CDC (2014, webpage source), "*a disease threat anywhere can mean a threat everywhere. It is defined by: (1) the emergence and spread of new microbes; (2) globalisation of travel and trade; (3) rise of drug resistance; and (4) potential use of laboratories to make and release—intentionally or not—dangerous microbes*". Hence, certain aspects can lead the path into new difficulties and challenges, or may simply intensify the situation from time to time. This certainly adds more complexity to the already complex situation of the disease outbreak. This may also alter in each phase, and can transmit between phases as the progress continues. In reality, each phase is intense in its way, and the impacts can differ depending on the nature of the disease itself. As a progressive event, a disease outbreak/epidemic travels from subject to subject, time to time, and location to location. In the latter, broader spatial understanding of outbreak events helps to respond more effectively to spatial attributes (Chanlekha and Collier 2010; Kramer et al. 2016). Furthermore, each phase needs to maintain a range of control strategies (Kawashima et al. 2016), resilience measures of multiple systems (Wallace and Wallace 2008; Yamagata and Murayama 2016), and planning. These factors require a careful assessment at multiple levels (i.e. beforehand, throughout, and afterward) in each step of this progression. In this regard, it is important to elaborate further on each phase, and discuss the characteristics of each phase, before we can assess how cities and communities respond to such adversities. The following six sub-section elaborate further on each phase of outbreak progression.

2.2.1 Identification Phase

The first phase of the outbreak event is defined as the 'identification phase'. The earlier this phase is discovered, the faster and more efficient the later phase(s) can be in practice. By identifying the causative agent of the disease, we can boost research and support to find suitable treatment and implement control measures (Hansen et al. 2018), when and where needed. Nevertheless, in most cases, it can take some time for a disease to show signs of spread from person to person; hence, this factor may delay its identification and usually delay the official announcement. Another factor that delays the official announcement of an outbreak is that the usually accountable authorities or sectors (associated with the public health) may assume they can contain the situation before it becomes a case of emergency or public health threat.

In this phase, as the whole situation remains obscured, there are potential chances of other disease development (Zaidi et al. 2018) that may even occur at a later stage. A careful monitory is essential as any information remains sensitive and any false step can escalate distress in the society. According to Cope and Ross (2020), the resource demands of surveillance are of much practical value in outbreak studies, particularly in the early stages. In this phase, the health system and emergency medical systems (EMS) are at high alert, while they assess further into the progress. The assessment must be comprehensive enough to shed a light on the situation, even if it does not clarify the origin/source of the disease (which is common in many outbreak cases). The focus should rather be on gaining knowledge about the disease itself, its symptoms, public health risks, and associated information that can support the medical and scientific sectors. In general, we can somewhat argue that the actual outbreak has not occurred yet, as it can take a while for the disease to show its presence and significance. However, since this is associated with the origin of the outbreak, it is therefore considered as the first phase of the overall outbreak progress. Although risks may seem to be at a lower rate, the uncertainties associated with this particular phase should be considered seriously. More importantly, in the identification phase, time and information are crucial factors; and both are needed for better handling of the outbreak event in its later phase(s).

2.2.2 Response and Containment Phase

The second phase of the outbreak event is defined as the '*Response and Containment phase*'. While there is little scholarly research that reflects on this particular phase, the response and containment phase is a very important part of the whole disease outbreak progress. The information provided from the earlier phase would certainly help to better "*inform the decision making processes for outbreak response*" (Cope and Ross 2020). If handled promptly, the whole situation may just turn into an endemic event, which can be regarded as early containment of a disease outbreak. While in most cases this may not seem possible, it is important to have a robust position in the response and containment phase. This requires speed, trustworthiness, surveillance, and precision (inspired from the editorial of Nature Microbiology, published on 22 January 2020) as to ensure enhancing the quality of response and its effectiveness (ibid, p. 228):

> The epidemiological characteristics of the pathogen will, of course, influence how quickly an outbreak can be contained. However, a response based on equipping local communities with surveillance and control capacity at the source, rapid and open communication of epidemiological and genetic information, and international community support, will increase our chances of controlling outbreaks earlier and thus potentially save lives.

In this regard, we can refer to the response and containment phase as a turning point; i.e. either the outbreak continues and worsens, or it can be contained as early as at this phase.

There are many examples of how our response rate can develop from media to practice. The latter is important as it requires enhancement of preparedness and support. A larger scale of such response mechanism is the WHO's main platform of the 'Global Outbreak Alert and Response Network (GOARN)', which also contributes towards 'global health security' by: "*(1) combating the international spread of outbreaks; (2) ensuring that appropriate technical assistance reaches affected states rapidly; and (3) contributing to long-term epidemic preparedness and capacity building*" (WHO webpage 2005, GOARN section). Developed in 2000, this particular response network reflects on the importance of this particular phase in the outbreak event. Through adequate planning, it is vital to have a robust response assessment mechanism that can suggest a series of practical priorities to inform decision making and response deliverables (Polio Eradication 2018; also National Institute for Communicable Diseases (NICD), an online source for 'outbreak response').

In this specific phase, careful planning is essential to develop an inclusive task force of multiple sectors and multiple stakeholders. An example of this is the systematic mechanism that is developed by Singapore in 2018. As a small nation, knowing that outbreaks can turn into catastrophic events in a short period, Singapore's Ministry of Health (MOH) developed the 'Disease Outbreak Response System Condition (DORSCON), which is a response-oriented guideline (MOH Website 2018). There is a range of prevention and reduction measures that suggest how the country should proceed with a risk assessment (ibid, also Russell et al. 2014) at an early stage (also implemented for the recent outbreak of COVID-19). This particular risk assessment includes four key factors of (1) the disease condition outside Singapore, (2) the disease's rate of transmissibility; (3) the likelihood of disease arrival into Singapore; and (4) its likely impact on the local community (ibid). By doing so, the assessments are conducted and responses are adjusted accordingly.

According to the guideline details of Singapore's 'Disease Outbreak Response System Condition (DORSCON)' developed in 2018, the three primary factors that should be addressed include: (1) nature of disease, (2) impacts on daily life; and (3) advise to public (Singapore's Ministry of Health 2018). The action plans of these three factors are then narrated based on four colour-coded scenarios, namely green, yellow, orange, and red. Here, we highlight details of all action plans under these scenarios:

- **GREEN**

  - **Nature of Disease**: *"Disease is mild OR Disease is severe but does not spread easily from person to person (e.g. MERS, H7N9)"*;
  - **Impacts on Daily Life**: *"Minimal disruption; e.g. border screening, travel advice"*;
  - **Advice to Public**: *"Be socially responsible: if you are sick, stay home; Maintain good personal hygiene; Look out for health advisories"*.

- **YELLOW**

  - **Nature of Disease**: *"Disease is severe and spreads easily from person to person, but is occurring outside Singapore. OR Disease is spreading in Singapore but is: (a) Typically mild, i.e. only slightly more severe than seasonal influenza. Could be severe in vulnerable groups (e.g. H1N1 pandemic). OR (b) being contained"*;
  - **Impacts on Daily Life**: *"Minimal disruption, e.g. additional measures at border and/or healthcare settings expected, higher work and school absenteeism likely"*;
  - **Advice to Public**: *"Be socially responsible: if you are sick, stay home; Maintain good personal hygiene; Look out for health advisories (same as GREEN)"*.

- **ORANGE**

  - **Nature of Disease**: *"Disease is severe AND spreads easily from person to person, but disease has not spread widely in Singapore and is being contained (e.g. SARS experience in Singapore)"*;
  - **Impacts on Daily Life**: *"Moderate disruption, e.g. quarantine, temperature screening, visitor restrictions at hospitals"*;

- **Advice to Public**: *"Be socially responsible: if you are sick, stay home; Maintain good personal hygiene; Look out for health advisories; and Comply with control measures"*.

- **RED**

  - **Nature of Disease**: *"Disease is severe <u>AND</u> is spreading widely"*;
  - **Impacts on Daily Life**: *"Major disruption, e.g. school closures, work from home orders, significant number of deaths"*;
  - **Advice to Public**: *"Be socially responsible: if you are sick, stay home; Maintain good personal hygiene; Look out for health advisories; Comply with control measures; Practise social distancing; and Avoid crowded areas"*.

In this guideline framework, we can that the authorities highlight examples of past epidemic/pandemic events, which shows their careful reflection on the earlier experiences as a vital tool for future responsiveness and readiness. Finally, we can argue the response rate plays a major part in informing the later transmission phase, in terms of what actions need to be elevated, what measures should be strengthened, and what practices should be enforced or reinforced. Hence, the response and containment phase is important from a multiple sectorial and multi-objective approach to help the development of guidelines and planning. It should be regarded as a prevention stage, through which containment is achievable. This can also be regarded as a decision making phase, as well as the period when our preparedness is increased according to the intensity of the situation.

2.2.3 Transmission Phase

The third phase of the outbreak event is defined as the *'transmission phase'*. Even though some of the primary characteristics of this phase are similar (or may seem overlapping) to the earlier phase, it should be regarded as a distinctive stage of its own during the outbreak progress. This phase is often regarded as a 'spreading' stage and is centered on the transmissibility and virulence of the disease. As demonstrated by Pigott et al. (2017), disease transmission is generally categorised into three stages that resemble the cases of viral transmission and how it develops into a case of human-to-human transmission from a smaller scale into a larger scale of multiple people, communities, and locations, including (1) index case potential, which is simply the transition from source, known as the reservoir host (e.g. animals, or non-human), to the first human case (i.e. index case); (2) outbreak potential, which is the start of human-to-human transition; and (3) epidemic potential when the disease starts spreading to multiple locations outside the boundaries of the original source (i.e. not in immediate proximity to the source). This is regarded as a conceptual progression of a disease spread, and how it develops in a transmissible process, from one source to another, and from one location to another.

Based on the analysis of the multistage pandemic study, conducted by Pigott et al. (2017), we can verify that transmission as a factor is elongated across multiple phases of outbreak event (i.e. from the inception of 'identification phases' to later phases, and even towards potential endemic or the 'post-recovery phase'). However, it only becomes a phase of its own, when the disease starts spreading at a faster pace, and on a wider scale; meaning that the spread progresses with a higher potential to develop an outbreak into a major pandemic event. Hence, in this phase of scaling up, transmissions are expected to be hostile with the rapid increase of infected cases or even upsurge in the number of deaths.

In this particular phase, scaling up prevention is essential. This is also visible in reports of the recent COVID-19 outbreak in China, suggesting to strengthen multiple aspects of 'disease prevention', 'control systems', and 'detection' (Xu 2020, in China Daily top news). Hence, the immediate plan should help to *"improve surveillance, diagnostic capabilities, and health systems in parallel with the design of policies for optimal responses"* (Pigott et al. 2017, p. 2662). Furthermore, there are also further risks for multi-transmissions, multiple critical conditions, and some unknown possibilities in this pulsating phase. In the case of having better control over the transmission rates and the number of cases, we may see the development of 'cluster of cases' (McCormick et al. 2012). Such a model of transmission is comparatively better than the individual spread of disease in multiple locations and communities. It is easier for speeding up the disease containment, as well as having a more efficient health system in place (ibid). In the case of multiple transmissions in multiple locations (often known as the epidemic event, if not any larger), we may end up with multiple hotspots and secondary zones of multiple cases, which can worsen the spread of disease. On the contrary, in a worse scenario, the progress may develop more aggressively as the outbreak can potentially escalate further through various modes of transmission. If this occurs, the outbreak may turn into a higher level emergency, a possible case of disaster, or a pandemic event. Hence, this can be verified in the fourth phase of outbreak progression; i.e. from the transmission to transition.

2.2.4 Transition Phase

The fourth phase of the outbreak event is defined as the *'transition phase'*. In this relatively longer phase outbreak progression, we expect to see changes in both disease activities and our actions. In some cases, and depending on the actual development of the disease, we expect to transit from reactive to proactive responses (Pigott et al. 2017). In this transition phase, such new responses may change our priorities, as well as our methods of detection, treatment, and control. If successful, we get closer to the recovery of the disease outbreak, and if not, it is likely we may remain in this phase for a longer period. If the former occurs, the transition period is relatively short as progress and control are expected to show signs of improvement. If in the case of the latter, we may require to review and revise our responses and reinforce some of the measures. In a longer transition phase, there are also potential risks for further mutation of the disease, out-of-control spread, and failures in resilience, health systems, and governance.

While resilience management reaches its peak at a larger scale (Massaro et al. 2018), the transition phase comes with some uncertainties until the steadiness of the situation is eventually reached. In this two-sided particular phase, the outbreak event can either intensity and escalate further or start to improve. Similar to transitions between multiple stages of transmission (Pigott et al. 2017), the transition can also happen for prevention and/or mitigation of the disease, as well as reduction of outbreak impacts. In their study of outbreak analysis, Chen et al. (2018, p. 396) suggest a potential "*tipping point pre-transition state, which is a critical state before the catastrophic event*". This reflects on their analysis of detecting early-warning signals of the outbreak (ibid) that may identify critical transitions (Liu et al. 2012) or show signs of abnormalities in the outbreak progression. If so, the already longer transition phase will be prolonged and further forces may be required to support the existing strategies and plans.

Furthermore, we can argue that the transition phase is often presumed as a steady phase towards early signs of either 'disaster' or 'improvement'—whichever the direction of the outbreak progression may be. If the latter occurs, the outbreak progression becomes closer to closure and reaching the crucial phase of recovery (inclusive of full containment). In this transition, should the disease mutates, the challenges will certainly escalate again. This will then further disrupt the systems as well as the progress itself. In such a scenario, this may break the pattern of linear disease progression and instead form into a cycle progression, resulting in a possibility of going back to earlier stages of the outbreak in a high-level emergency scenario. Nevertheless, in most cases, the transition phase is just extended to test and apply better reinforcement measures (if needed) and reach the ultimate results of control and recovery (if not absolute treatment). Since there is a slight chance of having the (new) vaccine to be tested on human trials in this phase, the transition should mainly indicate progress towards improvement, which resembles a good scenario in the right direction.

2.2.5 Recovery Phase

The fifth phase of the outbreak event is defined as the '*recovery phase*'. In this phase, a range of outbreak intervention strategies is developed and implemented towards outbreak containment and recovery, which are recognised to be context-specific and not universal (Kuhn and Calisher 2008). In this phase, treatment cases are expected to increase while the overall situation becomes relatively steady, with intermediate signs of slowing down and gradual decline in the number of cases and spread. If the containment can happen in a smaller cluster or community, then it may just be faster to proceed with treatment and healthcare support (Dwosh et al. 2003). As shown in their example case study, Neylon et al. (2010) highlight the role of the outbreak control team (OCT) in outbreak containment and recovery procedures. The team should be assembled in particular sites/locations to prevent further cases, to conduct extensive environmental screening, and to provide more meticulous isolation units.

At this particular phase, such underpinning measures will help to ensure the maximum effectiveness of management (van Hal et al. 2009), disinfection strategies (Neylon et al. 2010) and preventive procedures. By expanding on these measures, Lee et al. (2011) suggest a larger outbreak control interventions, including: "*(i) increased hand hygiene measures, (ii) enhanced disinfection practices, (iii) patient isolation, (iv) use of protective apparel, (v) staff exclusion policies, and (vi) ward closure*". While these control measures are applied for healthcare settings, with the potential application at smaller communities, there are also higher probabilities that we have to implement measures of control at different scales. For instance, in some cases, the approach may need to include a specific group of people (Dwosh et al. 2003; Calugar et al. 2006), a larger scale of people (van Hal et al. 2009), and even larger areas at a regional, sub-regional (CDC 2008), and country-wide scales (Leo et al. 2009; Agodi et al. 2011; Schwaber et al. 2011).

It is believed that with an outbreak, there should come a containment plan. This is regardless of how intense and time-consuming the process may be. Time is a crucial factor in the recovery phase, as it helps to have partial control (if not a full control) before full treatment of the disease (e.g. in most cases with a new vaccine). This requires a range of outbreak controllable measures that goes beyond the actions of isolation and tracing (Fraser et al. 2004) and towards enhanced treatment, planning, and management. By uncovering gaps in planning (Balicer et al. 2007) and reinforcing control measures, the containment can be successfully developed into a case of endemic (in less effective cases), and/or towards post-recovery status as the most effective result. As Heymann (2004) argues, the process may require the international response by the development of a response network and intense collaboration in multiple sectors of multiple disciplines, including virology, clinical medicine, and epidemiology, which can help to speed up the process. The recovery phase remains crucial as it is important to develop a steady pattern of outbreak improvement from multiple aspects/perspectives, and not just the decrease in the number of cases. If this occurs efficaciously, then the containment would succeed to put a gradual end and then an eventual closure to the outbreak momentum.

2.2.6 Post-recovery Phase

The final phase of the outbreak event is defined as the 'post-recovery phase'. While in smaller outbreak events, this may occur in a faster pace (White et al. 1986), in most cases it can take a much longer time than it requires to elaborate post-recovery strategies, long-term planning, proactive measures, and community-centered approach to better integrate social capital in the process (Thompson-Dyck and Mayer 2016; Lisnyj and Dickson-Anderson 2018). This phase is also realised as the normalisation phase, different to those normalisations of techniques in predictions and implementation

(Szomszor et al. 2010; Mustaffa and Yusof 2011; Yosuf and Mustaffa 2011), normalisation of early-stage disease detection and evaluation (Yang 2017), normalisation of the conditions of infected people (Kidd et al. 2004; Romantsov and Golofeevskii 2010), and normalisation of testing, the functionality of medical procedures, and treatment approaches (Brady et al. 2016). The term 'normalisation', can also be partially misleading as it can refer to the normalisation of the disease itself. Hence, this phase is named 'post-recovery phase' to avoid potential confusion with other normalisation factors of the outbreak.

It is important to note that post-recovery must only take place under the condition of reaching full outbreak recovery. In the case of an endemic situation, other measures should be taken into consideration before reaching the post-recovery phase. Hence, the system operations of health units and emergency medical services should eventually become more flexible as the situation eases. In this after event situation, it is important to continuously:

(1) Evaluate the recovery progress in every step to ensure no errors or mismanagement;
(2) Undertake a full monitory of any unexpected issues or uncertain conditions (e.g. that could even be from a new associated disease emerging from the outbreak);
(3) Shift to a normal status through a gradual pace, and with the full involvement of the community that should be practiced to compensate damages in the right way;
(4) Consider enhancement of certain units or sectors to ensure rapid recovery of those affected, and in particular, help to revitalise the governance of those sectors;
(5) Develop a set of (new) planning priorities that need implementation, monitory, and further adjustment to ensure they are feasible and practical in multiple sectors/domains; and
(6) Reflect on the whole outbreak progression, to capture and grasp lessons from the experience, and work towards enhancement of specific measures and supporting tools, where needed.

In a healthy process, a post-recovery may (possibly) take a longer time to ensure all units and domains are back to ordinary operation—hence, it is also a normalisation process of its own. As the society may continue to experience certain points of anxiety or mistrust, it is important to continue the outbreak monitory in a good capacity and allow for increasing the visibility of constructive progress and recovery management of them after the event. Therefore, a step-by-step progression is required to ensure support is provided to multiple stakeholders, including and at upmost, the general public. This phase must act as a reassurance period that signifies no further outbreak impacts on the society, or else it should not even commence and should instead remain at the earlier phase(s). As mentioned earlier, the pass to this phase is the full outbreak recovery that should occur in the previous phase of outbreak progression. Finally, public health should remain the priority while the community/city/country prepares to get back to the pre-outbreak operations. This phase must show evident signs of stability, even if prolonged. It also needs to immediately respond to the requirements of all affected sectors in a short period. The long-term planning would then need to reflect on some of those happenings during each phase, developing a set of enhanced preparedness measures and a more careful monitory for the future.

2.3 Cities During the Outbreak Progression

In all phases of the outbreak progression, cities can suffer tremendously and society may need to tolerate some of those negative impacts for a while. However, from a positive perspective, there come new opportunities for the enhancement of urban resilience and city management, which are the focus points of the book. Each phase of the outbreak progression offers a range of lessons (Hrudey et al. 2002) that should be used in a comparative assessment and should be utilised for the development and/or the enhancement of resilience measures, operational systems, and management factors. As highlighted by Monstadt and Schmidt (2019), we have to look into critical infrastructures of cities (also Carlson et al. 2012) and their governance to have a better overview of how we can enhance their urban resilience. A generalised structure of site assessment contribution to community resilience analysis is dependent on multiple factors. For instance, for critical infrastructures, we have supply chain, economic, governance, and civil society. In their order of importance, supply chain is the most essential and civil society is the least out of the four named factors. For instance, *"if we wanted to assess the civil society subsystem of a community at the same level as the critical infrastructure subsystem analysis, we would need to make use of additional data collection methods and tools outside of the site assessments"* (ibid).

This approach then helps to enhance the city's resilience. In doing so, we are then capable of increasing the *"institutional capacity of the local authorities and utility companies for risk mitigation and preparedness"*, which seems to be limited in many of their case study locations (ibid, p. 2353). Through their exploration of multiple cities, Monstadt and Schmidt (2019, p. 2366) also identify the existence of apparent silo city operations of multiple sectors, by addressing:

> Contrary to requests for cultures of inter-organisational preparedness, individual service providers focus at best on the vulnerability of their own system, while urban concepts for integrated emergency management are almost completely absent.

Hence, as it appears hard to break these separated operations, urban resilience urges to enhance management mechanisms that could help tremendously to overcome the side effects and difficulties of outbreak events. Through this, we can address 'infrastructural vulnerabilities' and develop the *"necessarily place-based solutions to urban and infrastructural vulnerabilities…[which]…could help to harmonise the local governance of infrastructures"* (ibid, p. 2369). As this is portrayed as a set of procedural standards, such measures could then operationalise those effective and protection strategies in the time of need, and can in return enhance the cooperative operation of multiple units and a larger body of stakeholders. The result of such planning will be the enhancement of city resilience while reducing an array of vulnerabilities on a larger scale.

In addition, it is clear that in outbreak events, we deal with a range of vulnerabilities (see Chap. 1), as well as a range of affected groups/locales, who could be even more vulnerable than the other groups/locales. In most disease outbreak events, depending on their mortality rate and the response quality for containment and treatment measures, we can identify three distinctive categories of vulnerable groups/locales, who may suffer the most:

- Those individuals who are unaware of the situation, and for whatever reason, they do not know they may be infected or may carry the disease/virus. Hence, they may delay their response by acting at a late or later stage in the disease progression in their bodies;
- Those vulnerable groups of the society who are less resilient to the disease/virus/infection, including the elderly group, the ones with previous health conditions, or the ones at high health risks;
- Those communities, cities, and even countries with a lack of health system infrastructure and weaker resilience OR those that have limited resources without enough equipment, emergency units, detection and verification tools, medical forces, and associated supplies.

These mentioned vulnerable categories represent a wide range of groups, from individuals to a larger scale of cities or countries (or clusters of them). Hence, in outbreak events, the alert level reaches a higher level in the case of an outbreak reaching or staring in poorer nations or those regions with minimal health infrastructure/system, and many other associated factors that represent a more vulnerable locale. Therefore, we have to consider vulnerable groups/locals as the primary point of our resilient planning.

2.3.1 Cities from a Phase-by-Phase Perspective

To summarise here, it is important to briefly highlight how cities cope in the event of an outbreak, and through the phase-by-phase progression that was illustrated earlier in Sect. 2.2. In the first phase of the 'identification phase', cities may seem to operate normally without any unexpected disorders or disruptions. As the search for more information increases, and as the official announcements of a disease outbreak are eventually made, the city operations start to change slightly. Usually, there are only minor impacts, with earlier signs of fewer people in outdoor public places, shopping malls, and busier areas of the city. As this progresses into a more alerting stage of the response phase, other sectors are affected too. The Health system and emergency medical services/system (EMS) become more alert and the cities' operational changes increase more rapidly. The public domain gets the biggest hit, as many public services would gradually lower their operational rate or may temporarily stop their operations. There seems to be more decline in using public places, retails sites of large scale, public transportation, and other public facilities. In the subsequent phase of the 'transmission phase', depending on the intensity and scale of disease spread, cities would gradually stop the secondary operations. Hospitals and health clinics may turn into hotspots, hence they require further support to avoid vast spread in those critical locations. In this phase, cities will face major difficulties as most businesses, industries, and retail units may stop (or may be asked to stop) their operations. If this occurs, apart from the existing impacts on society, the impacts on the economy will become more perceptible, too. This leads to a further decline of multiple systems as cities face further disruptions in their operations. In this phase, economic resilience and management are crucial as many critical infrastructures and key institutions are under significant pressure. The cities' economic foundation suffers while society becomes more worried about the rapid increase of cases and mortality rates/cases. In this phase, the city lockdown is not necessary but it may seem essential if the response rate is not so fast at the early stage. This lockdown approach can be partial to avoid the negative impacts on a larger number of sectors. The situation should eventually become steadier, and the peak may be reached towards the end of this phase, or throughout the next phase.

The two-sided 'transition phase' usually starts with a moment of halt. This is a sensitive time for any city, as the progress may change direction. The outbreak can worsen and this may hit the city the most as it can put all operations in halt, with only health systems and EMS to operate on the ground. Food systems and other industries become significantly vulnerable as transportation comes to a near halt. If the outbreak shows signs of improvement in the transition phase, then the city operations will only start improving towards the end of this period. This becomes a gradual process for the city to slowly shift from an alert/emergency situation towards a safer phase of recovery. In the transition phase, if the situation worsens, the outbreak can also shift to a case of disaster with higher risks than anticipated in the response phase. If this occurs, then the situation requires further support at multiple governmental levels, and it may turn into a full lock-down situation (if it did not happen earlier). This means significant impacts on nearly all primary and secondary sectors of the city. However, if the situation starts improving in a steady pattern, then the situation will get closer to the next phase of outbreak recovery.

In the 'recovery phase', cities must adopt and implement the maximum level of monitory and control. To enable early containment, all operations should be either on halt or should run under an all-inclusive and careful monitory and control. Cities should not take any further risks to increase the vulnerabilities and should rather reach a decision to accept some temporary economic losses. The impacts should be temporary and can be compensated, should the recovery becomes a success through full outbreak containment. Any adversities in this phase would potentially prolong the situation and intensify the outbreak, which can ultimately put more pressure on the city authorities and the overall city operations. It will then become harder to maintain adequate operations of multiple systems/services, hence it is important to experience a temporary difficulty than longer-term adversity. Once the situation is shaping its steading pattern (after a certain period), then the operations may start gradually from primary systems/services to secondary systems/services. This gradual shift should be carefully managed before recovery is reached and the situation is under full control. In this phase, the city should boost its health systems to ensure increasing the treatment probabilities and then supporting the overall procedure of treatment and recovery. Once this occurs, the situation will gradually move towards full outbreak recovery. The pathway to do so should be paved by high security, high monitory, high level of risk management, and high resilience to any unexpected change.

Succeeding from the recovery phase, society will start to become more relaxed. It will eventually move into the final phase of 'post-recovery'. This occurs once the regular operations of multiple sectors, services, and systems become widespread and visible. As highlighted earlier in Sect. 2.2.6, it is important to make all improvements to be as visible as possible. Such an approach enables to improve the sense of distress in the society and reassure businesses and industries for their planning of regular operations. In this phase, it is crucial to be inclusive by all means and in all sectors, as one system can help the other in a healthier ecosystem of resilience and city management. Disruptions are expected to become minimal throughout this phase, and society needs the highest level of support to ensure having an early recovery. In this stage, the more people-centered and community-oriented operation should shape and revitalise the overall operations of the city. Careful monitory should remain in place to avoid any unexpected adversities. In doing so, we will be able to bring all sectors back to regular operations and retain the conditions of all systems and services to their original regular patterns. Throughout the end of the post-recovery phase, the city should recharge its resilience and maintain and/or improve its main institutions. Finally, the most important of all is to develop 'responsive city management', which will be the main concluding point of this chapter.

2.4 Progress Through Responsive City Management

In a broader sense, there are many theories about resilience and its applications (Carlson et al. 2012; Barrett and Constas 2014), but there is a significant difference between theoretical aspects of resilience and how we can develop it in the planning practice (Coaffee 2013; Coaffee and Lee 2016; Cheshmehzangi 2020a), as well as through multi-sectoral city management—i.e. how resilience theories can then form into effective and efficient practices that are appropriate, implementable, and integrated into multiple systems. This chapter has highlighted the importance of outbreak progression and the phases it contains. Much of the focus here was to cover this important topic before we look further into the context of the city. This is the first time in any scholarly work that we can see such elaborate knowledge of multistage outbreak progression. This chapter offers valuable knowledge to those studies that would like to focus on the specific phase of outbreak progression, as well as those that aim to assess the progression of one disease in a particular location, or in between multiple locales. Moreover, the importance of resilience is partially covered in some of these discussions, but we have not fully addressed them from the city perspective. It is in the next chapter that we delve into this major factor, and will provide a comprehensive overview of 'urban resilience and city management in disruptive disease outbreak events'.

Now onwards, the book focuses more on the context of the city, in particular the ones in need of outbreak and adversities. There are many cases of disruptions, society failures, economic decline, and health system collapses. Our job, as urbanists, should then be to respond more effectively and responsibly to those significant urban pressures of extraordinary situations (e.g. disease outbreak events). Therefore, cities and communities are important subjects of the study as we see significant impacts of outbreak events on them, as we see their decline in such disruptive events, and as we witness their success and failures. Through a better understanding of urban resilience (including a holistic overview of measures, applications, and practices) and multi-sectoral city management, we can propose for better responsiveness and a more enhanced preparedness in the face of the disease outbreak events. Our highlights here will be mostly driven from existing practices and experiences of outbreak events, through which we propose for responsive city management (Cheshmehzangi 2020a), which is comprehensive, multi-dimensional, and practical.

We learn from many city examples, many tools, and many frameworks that address urban resilience and city management from various perspectives and/or in various conditions. As mentioned in Chap. 1, the adversities of outbreak events are exceptional, more of an emergency than a disaster situation. Thus, it is important to not only study what can be done but how they can be implemented in practice. The unfortunate recent outbreak of COVID-19 has been inspirational for us to understand resilience from various perspectives, of which most of them are associated with this very important situation of the disease outbreak event. More importantly, it is evident that resilience is key, and managing the city is the ultimate answer to many of those practicalities. Resilience is scenario-based, but it is essential for the city and how it can survive various adversities, of which, the outbreak is just one. It is then important to note that city operations cannot stop for long, and cities cannot be neglected. The more we understand how cities can cope in such events, the more we can improve their resilience and support their management. Hence, it is believed that only by bringing together a range of systems and thinking holistically, can a city be truly resilient.

References

Agodi, A., Voulgari, E., Barchitta, M., Politi, L., Koumaki, V., Spanakis, N., et al. (2011). Containment of an outbreak of KPC-3-producing Klebsiella pneumoniae in Italy. *Journal of Clinical Microbiology, 49*(11), 3986–3989.

Alberts, D. S. (2003). *Power to the edge: Command and control in the information age.* The Command and Control Research Program: CCRP Publications, Washington, DC: Library of Congress.

Alexandre, D., Delvaux, T., El Ayadi, A., Beavogui, A. H., Okumura, J., Van Damme, W., et al. (2017). Public health impact of the 2014–15 Ebola outbreak in West Africa: Seizing opportunities for the future. *British Medical Journal Global Health, 2*(2), e000202.

Balicer, R. D., Reznikovich, S., Berman, E., Pirak, M., Inbar, A., Pokamunski, S., et al. (2007). Multifocal Avian Influenza (H5N1) outbreak. *Emerging Infectious Diseases, 13*(10), 1601.

Barrett, C. B., & Constas, M. A. (2014). Toward a theory of resilience for international development applications. *Proceedings of National Academy of Sciences, 111*(40), 14625–14630.

Berthod, O., Müller-Seitz, G., & Sydow, J. (2014). Out of nowhere? Interorganizational assemblage as the answer to a food-borne disease outbreak. *Schmalenbach Business Review, 66*(4), 385–414.

Bloom, G., MacGregor, H., McKenzie, A., & Sokpo, E. (2015). *Strengthening health systems for resilience*. Report number: 18, IDS Practice Paper in Brief, project PRRINN-MNCH.

Brady, T. M., Pruette, C., Loeffler, L. F., Weidermann, D., Strouse, J. J., Gavriilaki, E., et al. (2016). Typical hus: Evidence of acute phase complement activation from a daycare outbreak. *Journal of Clinical & Experimental Nephrology, 1*(2).

Buseh, A. G., Stevens, P. E., Bromberg, M., & Kelber, S. T. (2015). The Ebola outbreak in West Africa: Challenges, opportunities, and policy priority areas. *Nursing Outlook, 63*(1), 30–40.

Calugar, A., Ortega-Sanchez, I. R., Tiwari, T., Oakes, L., Jahre, J. A., & Murphy, T. V. (2006). Nosocomial pertussis: Costs of an outbreak and benefits of vaccinating health care workers. *Clinical Infectious Diseases, 42*(7), 981–988.

Carlson, J. L., Haddenden, R. A., Bassett, G., Buehring, W. A., Collins, M. J., Folga, S., et al. (2012). *Resilience: Theory and applications*. Report Document, Report number ANL/DIS-12-1, Decision and Information Sciences Division.

Centers for Disease Control and Prevention (CDC). (2008). Multistate measles outbreak associated with an international youth event–Pennsylvania, Michigan, and Texas. *Morbidity and Mortality Weekly Report (MMWR), 57*(7), 169.

Centers for Disease Control and Prevention (CDC). (2014). *Why global health security matters*. Retrieved from February 18, 2020, from https://www.cdc.gov/globalhealth/security/why.htm.

Chanlekha, H., & Collier, N. (2010). A methodology to enhance spatial understanding of disease outbreak events reported in news articles. *International Journal of Medical Informatics, 79*(4), 284–296.

Chen, P., Chen, E., Chen, L., Zhou, X. J., & Liu, R. (2018). Detecting early-warning signals of influenza outbreak based on dynamic network marker. *Journal of Cellular and Molecular Medicine, 23*, 395–404.

Cheshmehzangi, A. (2020a). Comprehensive urban resilience for the city of Ningbo (in Chinese: 宁波市城市综合抗灾弹性框架). Report submitted to local government units in February 2020, Ningbo, China.

Cheshmehzangi, A. (2020b). 10 Adaptive measures for public places to face the COVID-19 pandemic outbreak. *City & Society*, Article ID: CISO_12282. https://doi.org/10.1111/CISO.12282.

Chimusoro, A., Maphosa, S., Manangazira, P., Phiri, I., Nhende, T., Danda, S., et al. (2018). Responding to cholera outbreaks in Zimbabwe: Building resilience over time. In D. Clarbon (Ed.), *Current Issues in global health*, Chapter 4. London: IntechOpen.

Coaffee, J. (2013). Towards next-generation urban resilience in planning practice: From securitization to integrated place making. *Planning Practice & Research, 28*(3), 323–339. Deconstructing planning and resilience: Lessons in translating theory to practice.

Coaffee, J., & Lee, P. (2016). *Urban resilience (planning, environment, cities)*. Red Globe Press. Part of Macmillan International Higher Education.

Cope, R. C., & Ross, J. V. (2020, in press). Identification of the relative timing of infectiousness and symptom onset for outbreak control. *Journal of Theoretical Biology, 486*, 110079.

DeRose, R. J., & Long, J. N. (2014). Resistance and Resilience: A conceptual framework for Silviculture. *Forest Science, 60*(6), 1205–1212.

Dieleman, J. L., Schneider, M. T., Haakenstad, A., et al. (2016). Development assistance for health: Past trends, associations, and the future of international financial flows for health. *Lancet, 387*, 2536–2544.

Dwosh, H. A., Hong, H. H. L., Austgarden, D., Herman, S., & Schabas, R. (2003). Identification and containment of an outbreak of SARS in a community hospital. *CMAJ, 168*(11), 1415–1420.

Editorial of Nature Microbiology. (2020). Rapid outbreak response requires trust. *Nature Microbiology, 5*, 227–228. Published on January 22, 2020. Retrieved January 25, 2020, from https://www.nature.com/articles/s41564-020-0670-8.

Emergency Preparedness. (2017). Ebola Preparedness Resources. Retrieved February 12, 2020, from https://www.calhospitalprepare.org/ebola-outbreak.

Fraser, C., Riley, S., Anderson, R. M., & Ferguson, N. M. (2004). Factors that make an infectious disease outbreak controllable. *Proceedings of the National Academy of Sciences, 101*(16), 6146–6151.

Gao, J., Barzel, B., & Barabási, A.-L. (2016). Universal resilience patterns in complex networks. *Nature, 530*, 307–312.

Ganin, A. A., Massaro, E., Gutfraind, A., Steen, N., Keisler, J. M., Kott, A., et al. (2016). Operational resilience: Concepts, design and analysis. *Scientific Reports, 6*, 33.

Ganin, A. A., Kitsak, M., Marchese, D., Keisler, J. M., Seager, T. P., & Lincov, I. (2017). Resilience and efficiency in transportation networks. *Scientific Advances, 3*(12), e1701079.

Gunawan, I., Sagala, S. Amin, S., Zawani, J., & Mangunsong, R., (2015). *City risk diagnostic for urban resilience in Indonesia.* Report number 102982, 84 pages document. February 15, 2020, from http://documents.worldbank.org/curated/en/699691468179059268/pdf/102982-WP-City-Risk-edited-12-01-2016-Box394848B-PUBLIC.pdf.

Haldane, V., Ong, S.-E., Chuah, F. L.-H., & Legido-Quigley, H. (2017). Health systems resilience: Meaningful construct or catchphrase? *Lancet Correspondence, 389*(10078), 1513.

Hansen, S., Faye, O., Sanabani, S. S., Faye, M., Böhlken-Fascher, S., Faye, O., et al. (2018). Combination random isothermal amplification and nanopore sequencing for rapid identification of the causative agent of an outbreak. *Journal of Clinical Virology, 106*, 23–27.

Heymann, D. L. (2004). The international response to the outbreak of SARS in 2003. *Philosophical Transactions of the Royal Society of London, Series B: Biological Sciences, 359*(1447), 1127–1129.

Heymann, D. L., Chen, L. C., Takemi, K., et al. (2015). Global health security: The wider lessons from the West African Ebola Virus Disease epidemic. *Lancet, 385*(9980), 1884–1901.

Hrudey, S. E., Huck, P. M., Payment, P., Gilham, R. W., & Hrudey, E. J. (2002). Walkerton: Lessons learned in comparison with waterborne outbreaks in the developed world. *Journal of Environmental Engineering and Science, 1*(6), 397–407.

Kawashima, K., Matsumoto, T., & Akashi, H. (2016). Disease outbreaks: Critical biological factors and control strategies. In Y. Yamagata & H. Murayama (Eds.), *Urban resilience: A transformative approach.* Part of the Advanced Sciences and Technologies for Security Applications book series (ASTSA) (pp. 173–204). Singapore: Springer.

Kidd, S. E., Hagen, F., Tscharke, R. L., Huynh, M., Bartlett, K. H., Fyfe, M., MacDougall, L., Boekhout, T., Kwon-Chung, K. J., & Meyer, W. (2004). *A rare genotype of Cryptococcus gattii caused the cryptococcosis outbreak on Vancouver Island (British Columbia, Canada).* Proceedings of the National Academy of Sciences Dec 2004, *101*(49), 17258–17263. https://doi.org/10.1073/pnas.0402981101.

Kieny, M. P., & Dovlo, D. (2015). Beyond Ebola: A new agenda for resilient health systems. *Lancet, 385*, 91–92.

Kramer, A. M., Pulliam, J. T., Alexander, L. W., Park, A. W., Rohani, P., & Drake, J. M. (2016). Spatial spread of the West Africa Ebola epidemic. *Royal Society Open Science, 3*, 160294.

Kruk, M. E., Myers, M., Varpilah, S. Y., & Dahn, B. T. (2015). What is a resilient health system? Lessons from Ebola. *Lancet, 385*, 1910–1912.

Kruk, M. E., et al. (2017). Building resilient health systems: A proposal for a resilience index. *BMJ, 357*, j2323.

Kuhn, J., & Calisher C. J. (Eds.). (2008). *Filoviruses: A compendium of 40 years of epidemiological, clinical, and laboratory studies.* Archives of Virology: Supplementa Book 20, Singapore: Springer. (Specifically on Chapter 13: Outbreak Containment).

Lee, B. Y., Wettstein, Z. S., McGlone, S. M., Bailey, R. R., Umscheid, C. A., Smith, K. J., et al. (2011). Economic value of norovirus outbreak control measures in healthcare settings. *Clinical Microbiology & Infection, 17*(4), 640–646.

Leo, Y. S., Chow, A. L. P., Tan, L. K., Lye, D. C., Lin, L., & Ng, L. C. (2009). Chikungunya outbreak, Singapore, 2008. *Emerging Infectious Diseases, 15*(5), 836.

Linkov, I., Eisenberg, D. A., Bates, M. E., Chang, D., Convertino, M., Allen, J. H., et al. (2013). Measurable resilience for actionable policy. *Environmental Science and Technology, 47*, 10108–10110.

Liotine, M., Ramaprasad, A., & Syn, T. (2016). Managing a smart city's resilience to Ebola: An ontological framework. In *Proceedings of the 49th Hawaii International Conference on System Sciences (HICSS 2016)*, Kauai, Hawaii.

Lisnyj, K. T., & Dickson-Anderson, S. E. (2018). Community resilience in Walkerton, Canada: Sixteen years post-outbreak. *International Journal of Disaster Risk Reduction, 31*, 196–202.

Liu, R., Li, M., Liu, Z. P., Wu, J., Chen, L., & Aihara, K. (2012). Identifying critical transitions and their leading biomolecular networks in complex diseases. *Scientific Reports*, 2. Article number 813.

Massaro, E., Ganin, A., Perra, N., Linkov, I., & Vespignani, A. (2018). Resilience management during large-scale epidemic outbreaks. *Scientific Reports*, 8. Article number 1859.

McCormick, D., Thorn, S., Milne, D., Evans, C., Stevenson, J., Llano, M., et al. (collective on behalf of the Incident Management Team). (2012). Public health response to an outbreak of Legionnaires' disease in Edinburgh, United Kingdom. *Eurosurveillance, 14*(28). Retrieved February 14, 2020, from https://www.eurosurveillance.org/content/10.2807/ese.17.28.20216-en.

Ministry of Health (MOH). (2018). Singapore Government webpage, on Being Prepared for a Pandemic: Learn more about how Singapore is prepared to prevent & respond to disease outbreaks. Retrieved February 21, 2020, from https://www.moh.gov.sg/diseases-updates/being-prepared-for-a-pandemic.

Monstadt, J., & Schmidt, M. (2019). Urban resilience in the making? The governance of critical infrastructures in German cities. *Urban Studies, 56*(11), 2353–2371.

Mustaffa, Z., & Yusof, Y. (2011). A comparison normalization techniques in predicting dengue outbreak. In *Proceedings of International Conference on Business and Economics Research* (Vol. 1, pp. 345–349).

National Institute for Communicable Diseases (NICD). Webpage on 'Division of the National Health Laboratory Service', section on 'Outbreak Response'. Retrieved February 21, 2020, from http://www.nicd.ac.za/our-services/outbreak-response/.

Neylon, O., O'Connell, N. H., Slevin, B., Powell, J., Monahan, R., Boyle, L., et al. (2010). Neonatal Staphylococcal scalded skin syndrome: Clinical and outbreak containment review. *European Journal of Paediatrics, 169*(12), 1503–1509.

Organisation for Economic Co-Operation and Development (OECD). (2019). Resilient Cities. Retrieved February 20, 2020, from http://www.oecd.org/cfe/regional-policy/resilient-cities.htm.

Pigott, D. M., et al. (2017). Local, national, regional viral haemorrhagic fever pandemic potential in Africa: A multistage analysis. *Lancet, 390*(10113), 2662–2672.

Polio Eradication. (2018). *Poliovirus outbreak response assessment (OBRA)*, published in December 2018. Retrieved February 20, 2020, from http://polioeradication.org/wp-content/uploads/2018/12/polio-outbreak-response-assessment-aide-memoire-dec-2018-20181220.pdf.

Romantsov, M. G., & Golofeevskii, S. V. (2010). Cycloferon efficacy in the treatment of acute respiratory tract viral infection and influenza during the morbidity outbreak in 2009–10. *Antibiotics and Chemotherapy (Antibiotiki i khimioterapiia) [sic], 55*(1–2), 30–35.

Russell, C. A., Kasson, P. M., Donis, R. O., et al. (2014). Improving pandemic influenza risk assessment. *Elife, 3*, e03883.

Sands, P., Mundaca-Shah, C., & Dzau, V. J. (2016). The neglected dimension of global security—a framework for countering infectious-disease crises. *New England Journal of Medicine, 374*, 1281–1287.

Siekmans, K., Sohani, S., Boima, T., Koffa, F., Basil, L., & Laaziz, S. (2017). Community-based health case is an essential component of a resilient health system: Evidence from Ebola outbreak in Liberia. *BMC Public Health, 17*. Article number 84.

Schwaber, M. J., Lev, B., Israeli, A., Solter, E., Smollan, G., Rubinovitch, B., et al. (2011). Containment of a country-wide outbreak of Carbapenem-Resistant Klebsiella pneumoniae in Israeli Hospiatals via a nationally implemented intervention. *Clinical Infectious Diseases, 52*(7), 848–855.

Schwerdtle, P. M., De Clerk, V., & Plummer, V. (2017). Experiences of Ebola survivors: Causes of distress and sources of resilience. *Prehospital and Disaster Medicine, 32*(3), 234–239.

Szomszor, M., Kostkova, P., & De Quincey, E. (2010). #Swineflu: Twitter predicts swine flu outbreak in 2009. In *International conference on electronic healthcare* (pp. 18–26).

The U.S. National Academy of Sciences. (2012). *Disaster resilience: A national imperative.* Washington, DC: The National Academies Press.

Thompson-Dyck, K., & Mayer, B. (2016). Bringing people back in: Crisis planning and response embedded in social contexts. In Y. Yamagata & H. Murayama (Eds.). *Urban resilience: A transformative approach.* Part of the Advanced Sciences and Technologies for Security Applications book series (ASTSA) (pp. 279–293). Singapore: Springer.

van Hal, S. J., Foo, H., Blyth, C. C., McPhie, K., Armstrong, P., Sintchenko, V., et al. (2009). Influenza outbreak during Sydney World Youth Day 2008: The utility of laboratory testing and case definitions on mass gathering outbreak containment. *PLoS One, 4*(9).

Wallace, D., & Wallace, R. (2008). Urban systems during disasters: Factors for resilience. *Ecology and Society, 13*(1), 1–14.

White, K. E., Osterholm, M. T., Mariotti, J. A., Korlath, J. A., Lawrence, D. H., Ristinen, T. L., et al. (1986). A foodborne outbreak of Norwalk virus gastroenteritis evidence for post-recovery transmission. *American Journal of Epidemiology, 124*(1), 120–126.

World Health Organisation (WHO). (2005). Strengthening health security by implementing the International Health Regulations, section on: Global Outbreak Alert and Response Network (GOARN). Retrieved February 20, 2020, from https://www.who.int/ihr/alert_and_response/out break-network/en/.

World Health Organisation (WHO). (2014–16). Ebola Outbreak 2014–16. Retrieved February 11, 2020, from https://www.who.int/csr/disease/ebola/en/.

Xu, W. (2020). *Xi: Outbreak's lessons must be learned*, in China Daily, Top News on February 15, 2020. Retrieved February 18, 2020, from http://www.chinadaily.com.cn/a/202002/15/WS5 e46f1bfa310128217277b49.html.

Yamagata, Y., & Murayama, H. (Eds.). (2016). *Urban resilience: A transformative approach.* Part of the Advanced Sciences and Technologies for Security Applications Book Series (ASTSA). Singapore: Springer.

Yang, W. (Ed.). (2017). *Early warning for infectious disease outbreak: Theory and practices.* Elsevier: Academic Press.

Yosuf, Y., & Mustaffa, Z. (2011). Dengue outbreak prediction: A least squares support vector machines approach. *International Journal of Computer Theory and Engineering, 3*(4), 489.

Zaidi, S. S. Z., Hameed, A., Rana, M. S., Alam, M. M., Umair, M., Aamir, U. B., et al. (2018). Identification of measles virus genotype B3 associated with outbreaks in Islamabad, Pakistan, 2013–2015. *Journal of Infection and Public Health, 11*(4), 540–545.

Chapter 3
Preparedness Through Urban Resilience

*You cannot be buried in obscurity: you are exposed upon a
grand theatre to the view of the world.
If your actions are upright and benevolent, be assured they will
augment your power and happiness.*
— Cyrus the Great

3.1 Strengthening the Resilience: How Cities Should Prepare?

This book explores the topic of resilience at the city level. The focus is more on outbreak events at the city level, or how cities should prepare and react in facing the larger events of epidemic and pandemic. The latter two cover a larger scale of regions, countries, sub-regions, and global scales. For us, the study of cities and how cities can manage in such disease outbreak adversities is novel. It is, therefore, important to understand how cities can be strengthened, first in terms of their resilience strategies through preparedness, and then in terms of their city management through responsiveness (or reactions). As highlighted towards the end of the Chap. 2, there are issues that can be addressed from these two perspectives of resilience and management. Hence, the central question comes to mind that *'how cities should prepare?'*; and by such preparedness, what planning is required to strengthen the resilience in the first place? Therefore, this chapter, as the main chapter of the book, aims to highlight the perspectives of urban resilience and city management in disruptive outbreak events. In doing so, we propose practical solutions that emerge from existing literature, strategies, application, and experiences of disease outbreak and epidemic/pandemic events. The procedural approach to this important topic comes from the idea of having a comprehensive urban resilience framework, which addresses the multiplicity of city management requirements.

In order to strengthen a city's resilience, we must look into a variety of systems and intuitions that build the city's backbone of operations. A disease outbreak event may have a direct or indirect impact on each of them; hence, a plan for preparedness is essential. This can then eventually support the city's responsiveness or reactions

A. Cheshmehzangi, *The City in Need*, https://doi.org/10.1007/978-981-15-5487-2_3

to disruptive issues of the outbreak. In this chapter, we first delve into the concept of 'urban resilience' before reaching the more elaborate discussions of 'city management' matters; the former is believed to support the latter. In doing so, this chapter introduces a broader knowledge of urban resilience, and its conceptualisation, as well as practices in cities and associated studies to urbanism. These perspectives will then narrow down to three factors of resilience education, resilience characteristics, and resilience comprehensiveness. Afterward, we introduce a comprehensive urban resilience framework and progress with the introduction of its dimensions and characterisation, before further elaboration on action plan and responsiveness through the foremost mechanism of city management. These discussions will become more robust with a detailed evaluation of all associated aspects across multiple phases of the outbreak progression (as demonstrated in Chap. 2). This chapter would then conclude with viewpoints on the utilisation of urban resilience measures for city management issues, and how cities should be prepared and react in times of need.

3.2 A Broader Perspective of 'Urban Resilience'

As demonstrated already, cities face significant challenges during disease outbreak events. In such occasions, we face certain vulnerabilities that reduce the healthiness of city operations, the society, and the overall governance of the city. The resilience of the city depends on multiple factors in a different emergency, crisis, and disaster events. On each occasion, the city must react differently but with some primary and overlapping factors that need to be understood both broadly and in detail. This is also similar to earlier statement of the Rockefeller Foundation 100 Resilient Cities organization, who argue that "...*building urban resilience requires looking at a city holistically: understanding the systems that make up the city and the interdependencies and risks they may face*" (100 Resilient Cities 2018). In their report, 100 Resilient Cities (ibid) also highlight some of the main drivers of urban resilience that address the factors of infrastructure and environment precisely at the city level. These factors include:

- *Provide and Enhances Protective Natural and Man-Made Assets*

Maintain protective natural and man-made assets that reduce the physical vulnerability of city systems. This includes natural systems like wetlands, mangroves, and sand dunes or built environment like sea walls and levees.

- *Ensure Continuity of Critical Services*

Actively manage and enhance natural and man-made resources. This includes designing physical infrastructure such as roads and bridges to withstand floods so that people can evacuate, as well as ecosystem management for floor risk management. It also includes emergency response plans and contingency plans that may coordinate airports to function so that relief can be lifted in and out during a crisis.

- *Provide Reliable Communication and Mobility*

Provide a free flow of people, information, and goods. This includes information and communication networks as well as physical movement through a multimodal transport system.

In their opening statement of their study, Admiraal and Cornaro (2018b, p. 2467), highlight the importance of resilience for multiple reasons:

> The need for future cities to be resilient stems from the fact that now more than ever in history, both natural and human-made hazards are threatening cities in the forms of shocks and stresses. The ability of cities to resist or restore themselves following these events is dependent on their resilience.

This statement reflects on the importance of resilience (Hopkins 2009) in cities and city management, particularly for those that face hazards, threats, emergencies, and shocks. This requires integrated thinking (Coaffee 2013; Cheshmehzangi 2016) and a comprehensive approach (Cheshmehzangi 2020) of urban resilience in planning practice. The restoration of cities through resilience is detectable in many cases, with methods of integrating the operations, prioritising key systems, and development of a chance for the development of multi-sectoral management to inter-sectoral management. In this regard, what we see is a unique opportunity to create a healthy network between multiple domains (Ban 2012; Linkov et al. 2013; Hajer and Dassen 2014; Admiraal and Cornaro 2018a, b, 2019), through which we can reduce the adverse impacts on the social domain (in particular). Hence, the opportunity can create a new mechanism to not only maintain the primary city dynamism but also to enhance them when and where needed.

In the literature, urban resilience is also regarded as an important concept for 'sustainability' and 'sustainable development' (The World Bank 2012; Washington 2015; Knieling 2016; Tabibian and Movahed 2016; Resilience Alliance 2017; 100 Resilient Cities 2018; Admiraal and Cornaro 2018a, b; Deng and Cheshmehzangi 2018; 'Urban Resilience' webpage 2020), and particularly the environmental sustainability (van der Heijden 2014; Schewenius et al. 2014; Sanchez et al. 2018; ICLEI webpage 2020) and emerging as a component of sustainability in urban policy (Davidson et al. 2019), or associated to issues of climate change, climate change impacts, and climate change disasters (IPCC 2007; ISET et al. 2010; Roberts 2010; Moench and Tyler 2011; Cheshmehzangi and Dawodu 2018; Kershaw 2018). It is regarded as a holistic approach that recognises the systems and processes of urban metabolism (Marvin and Medd 2006; Hajer and Dassen 2014; Admiraal and Cornaro 2018a, b) and the urban being (Renner 2018); or the backbone of city capabilities and measures for the time of need. While there are many conceptualisations and policies of urban resilience (Alexander 2013; Vogel and Henstra 2015; Sanchez et al. 2018; Huck and Monstadt 2019; Nunes et al. 2019), this important topic is yet to be studied as a supplementary factor during the disruptions and adversities of outbreak events. Hence, this book partially addresses this major research gap and expands on the existing literature, which is discussed by cross-referencing to this important field of research in urban studies.

Further to what urban resilience means in the various literature, Sanchez et al. (2018, pp. 3–7) also introduce a variety of resilience conceptualisations, of which the followings are addressed in detail, which are briefly mentioned here (for more details, refer to Sanchez et al. 2018):

- 'Disaster Resilience' (also see: Manyena 2006, Coaffee and Bosher 2008, Leichenko 2011, Bosher 2014);
- 'Engineering Resilience' (also see: Klein et al. 2003, Ahern 2011, Manyena et al. 2011, Davoudi et al. 2012);
- 'Ecological Resilience' (also see: Adger 2000, Monstadt 2009, Davoudi et al. 2012, Anderies 2014, Vale 2014);
- 'Socio-ecological Resilience' (also see: Alexander 2013, Smith and Stirling 2010, Hassler and Kohler 2014, Meerow and Newell 2016);
- 'Evolutionary Resilience' (also see: Manyena et al. 2011, Davoudi, et al. 2012, Abdulkareem and Elkadi 2018, Nunes et al. 2019);
- 'Built-in Resilience' (also see: Bosher et al. 2007, Bosher 2008, Bosher and Dainty 2011, Bosher 2014); and
- 'Climate Change Resilience' (also see: Adger et al. 2011, Leichenko 2011, Asian Development Bank (ADB) 2014, Davoudi 2014).

In another example of demonstrating a multidisciplinary perspective of urban resilience, Chelleri and Olazabal (2012) put together a range of resilience conceptions, and put urban resilience as the core of all those necessities. They utilise some of the already mentioned examples of resilience, and few more, such as 'Socio-Technical Systems (STSs) Resilience', 'Individual (Psychological) Resilience', and 'Market (Economies) Resilience'. They (ibid) also refer to built-in resilience as 'Infrastructures' or 'Networks' resilience. They suggest engineering resilience as 'materials' resilience and define socio-ecological resilience as an intersection between two separated and defined resilience categories of 'Ecological' or 'Eco-systems' resilience (Alberti and Marzluff 2004; Minucci 2012; Monteiro et al. 2012), and 'Social' resilience (Waters 2012). The below diagram summaries their understanding of the multidisciplinary perspectives of urban resilience in relation to other resilience sectors (also see Chelleri and Olazabal 2012; Olazabal et al. 2012). To summarise, their analysis (ibid) includes four primary resilience of social, markets (economies), infrastructures (networks) and ecological (ecosystems) in addition to their evident intersects comprised of: individual (psychological), STSs, Socio-Ecological Systems (SESs), and engineering (materials). In their analysis, Chelleri and Olazabal (2012) evaluate resilience in multiple disciplines and then position urban resilience in the center of all those defined primary and secondary resilience studies/thinking.

There are also other less-defined categories that are highlighted by Sanchez et al. (2018, p. 7). For instance, examples of stable and unstable resilience (Angeon and Bates 2015), anticipatory and reactive resilience (Vale 2014), or what Anderies et al. (2013) consider as the 'general resilience', which "*refers to broader system-level attributes such the ability to build and increase the capacity for learning and adaptation*" (ibid, p. 4). There are also other more recent examples of "societal resilience"

(Marana et al. 2019), which is suggested as a standardisation approach to support the resilience development process. In this chapter, our focus is more towards those generalities or "general resilience" (elaborated from Anderies et al. 2013), with some overlaps with disaster resilience, and defining what they may mean for the city management considerations in the case of disruptive disease outbreak events.

Currently, as demonstrated in the review analysis by Sanchez et al. (2018), urban resilience has gained popularity in various urban-related studies, stretching from climate change studies to urban studies and urban geography (Manyena 2006; Ernstson et al. 2010; Haase et al. 2014; Boyd, et al. 2015; Meerow and Newell 2016), and often in the apparent combination with the overarching topics related to sustainability (Cheshmehzangi and Dawodu 2018). These sustainability-oriented studies also address issues of diversity and sustainability of social-ecological systems (Folke et al. 2002), (urban) sustainability governance (Sanchez et al. 2018), recovery measures (UN Habitat 2012), and associated to urban sustainability goals (Fiksel 2003; Register 2014) as well as the institutional understanding of the United Nation's sustainable development goals (SDGs) (Cheshmehzangi and Dawodu 2018; Acuti et al. 2020). In more recent years, urban resilience has gained a stronger position in policy associated studies to urban governance systems (Sanchez et al. 2018; Davidson et al. 2019), the complexity of city operations (Asian Development Bank (ADB) 2014; Tainter and Taylor 2014), climate resilience policies and governance (Davoudi 2014; Moffatt 2014; Lister 2016), etc.

As Davoudi (2014) puts it well, the topic of urban resilience emerging as the overarching field of 'resilient urbanism', which in fact is a response to a multiplicity of threats, risks/hazards, emergencies, disasters, etc. Many collections of scholarly work, cover the emerging crossovers from urbanism to urban development (Eraydin and Taşan-Kok 2013; Singh 2015; Crowe et al. 2016) or vice versa. It is also transferred from urban development to the fields of urban design (Pickett et al. 2013; Liao et al. 2016; Abdulkareen and Elkadi Abdulkareem and Elkadi 2018) and urban form (Sharifi 2019), and is integrated into the ecological, socio-economics, and planning realms (Pickett et al. 2004; Smith and Stirling 2010; Chelleri 2012). These topics are emerging fast, addressing a range of challenges, risks, and resilience (Singh 2015), such as risks associated with urban health (Singh et al. 2020), urban planning measures (Yamagata and Sharifi 2018). These are addressed from multiple perspectives of adaptation governance (Brunetta et al. 2019), security (Fekete and Fiedrich 2018), urban transformations (Westley et al. 2011; Kabisch et al. 2018), etc. Moreover, as it is highlighted by Sanchez et al. (2018, p. 10), the emergence of 'urban resilience policy' is also developing fast and is explored from multiple aspects:

> Urban resilience policy is a complex and evolving field characterised by significant challenges associated with urban governance systems, political pressures, uncertain and emergent nature of threats, speed of change and the level of complexity of long-lived networks that form cities. Added to these issues, there are a number of resilience concepts that can potentially be used to develop such policies. These various conceptualisations come with a range of critiques, the most dominant being that they have a too strong focus on, for lack of a better term, bouncing back and seeking to maintain a known way of living; that they do not align well with other urban policy goals; and that their focus is too short-term and too small-scale.

In this regard, we can verify the significance of resilience from multiple perspectives, a range of applicability, as well as from multiple approaches, multiple planning measures, and in various spatial scales (the UN-Habitat 2012; Asian Development Bank (ADB) 2014; Moffatt 2014; Resilience Alliance 2017; Sanchez et al. 2018). In addition, urban resilience surely is a very important concept for the city management scenarios of the outbreak events. It indeed responds to the crucial issues around the reduction of vulnerabilities and the enhancement of city stability. In the following sub-sections, we briefly explore the concept of resilience from three perspectives of education, its characteristics, and its comprehensiveness. It is believed these three factors are crucial to the early preparedness of the city in the case of disease outbreak events.

3.2.1 The Education of Resilience

Our approach here must be recognised and valued as an educational approach, the one that can have a positive impact on how decision making can happen, how cities can improve their resilience, and how the society can be supported in a more extensive way. The education of resilience is as important as city operations and management. Hence, we first look into what covers the education of resilience, not only at the city level but also from the broader understanding of other scales that may be relevant to the city-scale studies of resilience. More recent scholarly research already cover the important topics of integrated health education in disaster risk reduction, including information dissemination (Pascapurnama et al. 2018), public health education (Levy et al. 2017), sector-based knowledge on public health emergencies (Ung 2020), community engagement and public health ethics (Schoch-Sapana et al. 2020), and public health decision making (Ambat et al. 2019). Most studies explore these important factors from various perspectives of different scales, as their applications are very different between multiple scales. Some examples include decision making and control measures at the ward scale (Han et al. 2020), outbreaks trends in specific settings or sectors (Taoti et al. 2019; Wu et al. 2019), neighbourhood level (Reyes-Castro et al. 2017), city-level (Levin-Rector et al. 2015), provincial level (Chirambo et al. 2019), and also at the state-level through national guidelines (Grafe et al. 2018); or from other perspectives of spatio-temporal dynamics (Paripa et al. 2019) and characteristics (Reyes-Castro et al. 2017), control measures (Al-Abri et al. 2020), etc. Moreover, it is noticeable that these examples are all driven from specific outbreak experiences, and highlight some very valuable reflections of those adversities at the time of vulnerability and need. These are studied regardless of the magnitude of the impacts.

From the extent of available literature, Gomes Ribeiro and Gonçalves (2019) argue in regards to what seems to be the minimal available tools or methods for the improvement of city resilience, and particularly the resilience of urban systems (also, The World Bank 2012). Their analysis covers four basic pillars of (1) resisting, (2) recovering, (3) adapting, and (4) transforming, while addressing how urban resilience should be seen from five dimensions of: 'natural/environmental, 'economic', 'social', 'physical/infrastructure', and 'institutional' (Tabibian and Movahed 2016; Gomes Riberio and Gonçalves 2019); two more dimensions than the three traditional sustainability pillars (i.e. Environmental, Social and Economic) and more complex in terms of their positions in the education and the practice of urban resilience. Also, there are some overlaps with some of those defined domains suggested by Linkov et al. (2013, p. 10109) in their proposed 'Resilience Matrix', as discussed in Chap. 2 of this book. The example of five dimensions by Tabibian and Movahed (2016) is a comprehensive illustration of not only multiple dimensions but also multiple factors of those dimensions. The breakdown of their resilient city framework includes the following factors under the specified five dimensions (ibid):

1. **Social Dimension**
   Including multiple factors, such as 'Demographics (age, race, class, gender, occupation)', 'Social networks', 'Community values-cohesion', 'Income level', 'Faith-based organisations', 'Cultural diversity', 'Education', and 'Awareness Level'.
2. **Economic Dimension**
   Including multiple factors, such as 'Employment', 'Value of property', 'Financial stability and flexibility', 'Wealth generation', 'Municipal finance and revenues', 'Job diversity of residents', and 'Housing capital'.
3. **Natural (or Environmental) Dimension**
   Including multiple factors, such as 'Erosion rates', 'Biodiversity', 'Restoration of hydrologic flows', 'Conservation of ecologically vulnerable areas', 'Proximity of different habitats', and 'Wetlands acreage and loss'.
4. **Physical (or Infrastructure) Dimension**
   Including multiple factors, such as 'Transportation networks', 'Lifelines and critical infrastructure', 'Commercial and Manufacturing establishments', 'Water demand and conservation systems', 'Flexibility of grid', and 'Energy Monitoring'.
5. **Institutional Dimension**
   Including multiple factors, such as 'Hazard analysis and creation of hazard maps', 'Emergency services', 'Zoning and building standards', 'Emergency response plan', 'Interoperable communications', 'Continuity of operations plans', and 'Collaborative planning'.

In addition to what we see from all above five dimensions of urban resilience and their breakdown (Tabibian and Movahed 2016), there are added factors of resilient measures that need to be included in the case of emergencies, of particular outbreak events. These will be discussed in later parts of this chapter. Similarly, Huck and Monstadt (2019) suggest a critical reflection on the important topic of resilience, perhaps to be seen as a 'boundary concept', which indeed needs to be assessed and understood from a cross-boundary learning approach. In this regard, with such cross-boundary, multi-dimensional and multi-domain systematic structure, we verify the importance of resilience knowledge to a wide range of stakeholder constellations, including the governmental authorities, multiple city systems, developers, businesses, public and private sectors, and the general public. How we can reach each of these stakeholder groups is a practical challenge that may require further understanding of urban resilience practices.

The transmission of resilience knowledge to each group is possible through various means and measures, which may not be so relevant to the overall aims of this book. Nevertheless, we can argue that it is within the boundaries of resilience knowledge that we can educate those individual or clustered groups, and for each of them in a different way. Regardless of how difficult it may be in real practice, such education is essential, particularly in the event of an outbreak and its progression at the city level. It is also important to note that such challenges of education may also be context-specific as they respond to specific issues that may be related to a particular city and irrelevant to others. These include urban structures, governance structures, the existing institutions of the specific location, etc. Hence, we refrain from solving context-specific challenges by suggesting ubiquitous solutions. In this chapter, we mainly provide a framework that is adaptive and comprehensive enough for not a location, but during a particular emergency event; meaning that it responds to what urban resilience and city management may need to address during the disease outbreak event.

3.2.2 The Characteristics of Resilience

Broadly speaking, there are certain characteristics attached to the concept of resilience. Consequently, for urban resilience, as Gomes Ribeiro and Gonçalves (2019) demonstrate, there are also five relevant urban resilience characteristics, including, (1) Redundancy, (2) Robustness, (3) Adaptation (also known as 'flexibility', by Fabbricatti and Biancamano 2019), (4) Resources, and (5) Innovation. This categorisation of urban resilience characteristics share some similarities with the circular economy and resilient thinking model of Fabbricatti and Biancamano (2019) who demonstrate their model with seven characteristics, including three new characteristics of: (1) Reflectiveness (also known as 'capacity to learn', by da Silva et al. 2012), (2) Inclusiveness, and (3) Integration; and not including 'innovation'. To summarise and combine these existing studies of multiple sources, and from the extensive reports of 100 Resilient Cities (2018) and the framework of The Rockefeller Foundation and Arup (2014), the seven characteristics of urban resilience are based on the need(s) of the city's various (urban) systems. These are summarised as the followings:

(1) **Reflective**—meaning to "*use past experiences to inform future decisions*";
(2) **Resourceful**—meaning to "recognise alternative ways to use resources";
(3) **Inclusive**—meaning to "prioritise broad consultation to create a sense of share ownership in decision making";
(4) **Integrated**—meaning to "bring together a range of distinct systems and institutions";

(5) **Robustness**—meaning to have "well-conceived, constructed, and managed systems";

(6) **Redundant**—meaning to "spare capacity purposefully created to accommodate disruption";

(7) **Flexible**—meaning to have "willingness, ability to adopt alternative strategies in response to changing circumstances".

Furthermore, in their conceptual framework, da Silva et al. (2012, p. 3) create a network between three key elements of resilience, productivity, and circularity. This conceptual model is also very similar to the one proposed by da Silva et al. (2012) who developed a systems approach specifically to meet the challenges of urban climate change. This model (ibid, p. 3) creates a network of environmental factors, with social issues, risks, and vulnerabilities of the city. They create three central questions of: "*(1) How does the city work? (2) What are the direct and direct impacts?... (*of climate change for their model example), *and (3) Who is the least able to respond to shocks and stresses?*". They later link these factors to the actual purpose of urban systems, reflecting on studies of well-being conducted by Maslow (1971), Alcamo et al. (2003), Huitt (2004). Through these examples, the well-being is then demonstrated (da Silva et al. 2012, p. 6) based on five primary characteristics of:

1. Basic needs for survival (such as, biological and physiological needs)—also addressing: "Adequate livelihoods, sufficient nutritious food, access to water, sanitation and shelter, access to goods";

2. Security (such as, safety needs)—also addressing: "Personal safety, security from natural hazards and man-made hazards (terrorism, pandemics), secure resource access, order, law and stability";

3. Health (such as. healthy body and mind)—also addressing: "Feeling well, access to clean air and water, access to health care";

4. Good social relations and esteem (such as, belongingness and love needs)— also addressing: "Social cohesion, mutual respect, ability to help others, family, personal relationships, achievement, status, responsibility and reputation"; and

5. Freedom of choice and action (such as self-actualisation needs)—also addressing: "Opportunity to be able to achieve what an individual values doing and being, personal growth and fulfilment".

Another more recent example of characteristics of urban resilience is the one generated as part of the UN Habitat's 'City Resilience Profiling Tool (CRPT)' (2018), which is a general guideline to a range of adversities and events. In their framework conceptualisation, the team at the UN-Habitat (ibid, p. 21) utilise 10 critical factors in building urban resilience. These critical factors are recognised as holistic measures, in order to include a range of considerations. The first factor is for the urban resilience framework to be 'measurable', highlighting both "*tangible and intangible realities that translate into qualitative and quantitative data…*[that]*…can be analysed*". The second factor is the inclusion of 'urban systems', the ones that are defined as complex parts of the "*integrated and complex systems of systems, comprised of sectors, people and hazards…and managed through effective governance mechanisms*". The third factor is the consideration of 'inhabitants', all people who *"live, work, visit, navigate, and/or travel to the city, as we as resident or connected institutions, organisations, businesses, etc."*. The fourth factor is the importance of 'continuity', particularly including *"maintaining the protection and provision of services, flows, and structures in order to save and preserve inhabitant's lives and livelihoods"*. The fifth factor is recognised as a response to pressures and overcome 'shocks and stresses'. These are highlighted as examples of situations with risks to the city and those that may be *"sudden and slow-burning, natural or human-made, rare and regular, foreseen or not"*. The sixth factor is the important consideration of 'transforming', referring mostly to those examples of *"adopting proactive, forward-looking attitudes that turn challenges into opportunities for growth"*. These mean the methods of transformational progression(s), those that can change the situation by generating incremental and supporting transformations. The seventh factor is the recognition of 'sustainability' and its practices to include a wide range of community development factors, innovations, economic generation, and services support. This leads to the eighth factor that addresses the importance of 'access', referring to what eventually devises a range of actions, guidelines, and recommendations, those that would essentially be implementable in practice. Finally, all these eight factors feed into the two important factors of 'planning' and 'action'. Through the right planning methods, we can create effective strategies and enhance the city in order to *"tackle…*[a wide range of]*…vulnerabilities and strengthen capacities to function effectively and efficiently"*. And through actions, we are able to provide reliable and constructive assessments for the support of those strategic planning and responsiveness to the situation. Such an approach would also provide possibilities to provide inclusiveness and take into consideration a multi-sectoral approach to combat challenges. In doing so, we are able to respond to the needs of various stakeholders through their involvement. These are then represented into six distinct characteristics of urban resilience (UN-Habitat 2018), namely (1) persistent, (2) adaptable, (3) inclusive, (4) integrated, (5) reflexive, and (6) transformative—which indicate some overlaps with other studies of urban resilience charactertisation. Furthermore, this model (ibid, p. 22, 23) represents three factors of what comprises the idea of resilient (i.e. persistent, adaptable, inclusive). It also embraces three distinct processes, reflecting on how these can be achieved (i.e. integrated, reflective, and transformative). In reality, the UN-Habitat's model of characteristics of urban resilience, articulating urban resilience through describing *"WHAT comprises being resilient—by being persistent, adaptable, and inclusive–and the processes on HOW these can be achieved—through being integrated, reflexive and transformative"* (UN-Habitat 2018, p. 22, 23). Hence, it allows for a holistic thinking to resilience thinking:

Under **WHAT** aspects, these characteristics include (ibid, p. 23):

Persistent—A persistent city anticipates impacts in order to prepare itself for current and future shocks and stresses. It builds robustness by incorporating coping mechanisms to withstand disturbances and protect people and assets. It encourages redundancy in its networks by generating spare capacity and back-ups to maintain and restore basic services, ensuring reliability during and after disruption.

Adaptable—An adaptable city considers not only foreseeable risks, but also accepts current and future uncertainty. Going beyond redundancy, it diversifies its services, functions and processes by establishing alternatives. It is resourceful in its capacity to repurpose human, financial and physical capital. It pursues a flexibility that encourages it to absorb, adjust and evolve in the face of changing circumstances, dynamically responding by turning change into opportunity.

Inclusive–An inclusive city centers on people by understanding that being resilient entails protecting each person from any negative impact. Recognising that people in vulnerable situations are among the most affected by hazards, it actively strives towards social inclusion by promoting equality, equity and fulfilment of human rights. It fosters social cohesion and empowers comprehensive and meaningful participation in all governance processes in order to develop resilience.

Similarly, under **HOW** aspects, these characteristics include (ibid, p. 23):

Integrated—An integrated city appreciates that it is composed of and influenced by indivisible, interdependent and interacting systems. It combines and aligns many lenses to ensure input is holistic, coherent and mutually supportive towards a common cause. It enables a transdisciplinary collaboration that encourages open communication and facilitates strategic coordination. It supports the collective functioning of the city and guarantees far-reaching, positive and durable change.

Reflexive—A reflexive city understands that its system and surroundings are continuously changing. It is aware that past trends have shaped current urban processes yet appreciates its potential to transform through shocks and stresses over time. It is reflective, conveying the capacity to learn from knowledge, past experiences and new information. It also learns by doing and installs mechanisms to iteratively examine progress as well as systematically update and improve structures.

Transformative—A transformative city adopts a proactive approach to building resilience in order to generate positive change. It actively strives to alleviate and ultimately eradicate untenable circumstances. It fosters ingenuity and pursues forward-looking, innovative solutions that over time create a system that is no longer prone to risk. A transformative city is focused and goal-oriented towards a shared vision of the resilient city.

In the UN-Habitat's model (ibid, p. 26, 27), the implementation process of urban resilience is realised as a method of data collection, information processing, assessment, and response. Such a system is designed in a way to fit with a wide range of spatial levels, including "*multiple city scales, geographies, and types*". This then requires the specification of available data and materials for the exact conditions and the feasibility of what responses may be needed. In a way, this provides a profiling approach that needs to be integrated and holistic. It also reveals a range of aspects that are crucial to the context and the strategies needed at the (local) decision-making level (ibid, p. 26):

The definition of the assessment boundaries can be determined by the local government, based on the mandate of the local government and the relevance of the analysis in the context. The diagnosis assesses multiple geo-spatial areas and scope, as such, obtaining an understanding of decentralisation aspects is essential in our approach to clarify the administrative and financial competences of the local government. The expressions 'city', 'city area' and 'urban area' are employed throughout the tool to refer to the study area. Similarly, 'local government' refers to the government entity level that has jurisdiction over the considered study area.

The arguments here lead to the development of a holistic model, one that puts urban systems in a systematic network, and ideally in a network of networks. The implementation process then takes up the opportunity to have an iterative process that is analytical and action-driven. This requires multiple engagement activities, multiple assessment and measurements, careful monitory, and careful development of action plans in a process. This process is described in five stages (ibid, pp. 27–36), starting from 'initiation and training' that offers evaluation and collaborative opportunities for implementation of urban resilience planning. In the next stage, 'data collection and diagnosis' are considered as key aspects of *"gathering the relevant data and ensuring its traceability is an essential step in building resilience"*. This stage is then processed through data collection of four sets of (1) City ID, (2) Local Governments and Stakeholders, (3) Shocks, Stresses, and Challenges, and (4) Urban elements (ibid). Each of these sets is then provided with a set of guidelines, addressing specific elements under each set of data collection. The third stage is 'analysis', which is a major part and could be conducted in an iterative process. In the final two stages, we see 'actions for resilience' that itself may require further analysis (again through the iterative process), and lastly 'taking it further' for the actual implementation and urban resilience enhancement.

In the same context of the argument, Huck and Monstadt (2019, p. 211), suggest for *"more interaction and cross-boundary learning between respective knowledge communities"*. In this regard, we can understand the values of strategies and instruments for urban resilience (Acuti et al. 2020). These require an understanding of a co-creation process, which improves the involvement of multiple city stakeholders (Marana et al. 2019). Such an approach also enables opportunities to enhance co-benefits of multiple systems, and help to restructure co-planning methods of the city management. Moreover, this allows us to assess resilience not only from its dimensions but also in terms of its application in practice. Hence, it is recommended to have a good understanding of resilience from its comprehensive nature.

3.2.3   The Comprehensiveness of Resilience

The first toolkit developed for resilient cities (Siemens et al. 2013; The Rockefeller Foundation and Arup 2014) showcases the comprehensiveness of resilience for the first time (also see 100 Resilient Cities 2018). This was demonstrated through a multi-dimensional understanding of resilience as a concept and then was utilised as a unique practice of urban planning and city management. Through this conceptualisation, the multiplicity of urban resilience was demonstrated and then tested for a new toolkit development, addressing what it means to create resilient cities, and why:

> …[The]…events are playing out against a backdrop of global population growth and urbanization, leading to a complex knot of interrelated pressures. In emerging and established cities alike, these trends are changing the spatial pattern of risk and radically altering perceptions

of whether a city is 'safe' or 'well prepared'. Cities have a tremendous challenge to maintain social well-being and economic vitality in the face of these complex, uncertain and constantly changing risks (ibid, p. 3).

Through their explanations of urban resilience, they (ibid, pp. 5–7) then highlight three main factors behind "*creating resilient systems*" in the practice of resilience enhancement, including: (1) Robustness of new and existing infrastructure, (2) Decentralized resource supplies and distribution networks, and (3) Enhanced monitoring and controls. These three pillars are then identified as precursors to creating resilient cities that require a comprehensive action plan. These pillars cover three crucial areas of 'policies', 'governance, and 'finance', which are highlighted as:

1. Urban planning and land use policies can direct development in ways that protect people and structures from harm;
2. Governance should take a whole system approach to city management; and
3. Appropriate financing mechanisms are needed to support investments in resilient infrastructure.

Further development of this conceptualisation is shaped around multiple dimensions of urban resilience measures, of which four were highlighted to be "*people, organisations, place, and knowledge*" (The Rockefeller Foundation and Arup 2014; Arup reports in 100 Resilient Cities 2018). For people, the factors of health and well-being were taken into consideration in order to emphasise the societal values and issues that are directly and indirectly relevant to the overarching factor of public health. For organisations, the key factors are in fact cross-overs between social, economic and institutional measures, with which we can develop the operational backbone of the city and the urban systems. For place, the recommendations were mostly about the quality of infrastructure, the physicality of the built environment, critical infrastructures (European Commission 2006; Boin and McConnell 2007; De Bruijne and Van Eeten 2007; Almklov et al. 2012; Brassett and Vaughan-Williams 2015; Coaffee and Clarke 2017; Monstadt and Schmidt 2019), and multiple values of the ecosystems in the city and urban environments. And for knowledge, the considerations are mostly similar to those reflectiveness matters (da Silva et al. 2012; Fabbricatti and Biancamano 2019) and the capacity to learn from the past and take appropriate actions in return. Such an approach should be "*informative, inclusive, integrated, and iterative*" for the decision making processes of the city management (Siemens et al. 2013; The Rockefeller Foundation and Arup 2014; 100 Resilient Cities 2018). Therefore, what we create through resilience is not just a product or by-product of multi-sectoral cooperation and interaction, but also a process of making those crucial decisions at the time of need. As a result, we can argue that the comprehensiveness of resilience, and urban resilience, in particular, is a reflection of the complex city operations. Such complexity of operations can be expressed and developed through decision-making processes, policy demands (Davidson et al. 2019), and planning strategies that define the overall mechanism of city management.

To summarise, there are indeed a range of resilience values (of which educational was discussed here), dimensions, and characteristics that define the comprehensiveness of resilience. They portray an array of resilient systems, resilient measures, and resilient governance—all together can highlight the importance of resilience in urban planning and the practice of making cities more resilient. The various global examples, of which we only showed a few here, are valuable lessons for us to know exactly 'what does resilience mean for the case of emergencies?', and of particular, how can we develop it in the case of disease outbreak events? Undoubtedly, and as expressed in the previous chapters, in such events we deal with a different set of resilience issues; those that may require a procedural approach to be implemented and enhanced, and those that may stretch over multiple stages of the outbreak progression. However, the commonalities between what has been discussed in this section and what is proposed in the following sections are significant; and therefore, what we have in hand is a comprehensive approach to city management—something that is conceptualised in theory, nourished with the literature of cross-disciplines and multiple sectors, and is finally nurtured in the practice of resilience. In the following section and its subsequent sub-sections, and by utilising the knowledge gained from the existing urban resilience studies, we demonstrate a novel urban resilience framework mainly applicable for the case of disease outbreak events.

3.3 Conceptualising a "Comprehensive Urban Resilience Framework"

In facing the disease outbreak event, the city needs to be fully prepared and it should respond in the most effective way. The city faces new adversity and it should recognise its reactions are precarious. In the case of failure, the city may face a deadly situation of high mortality rates and high infected cases, costly in every sector, and harder to reconcile. The city may become fragmented, chaotic scene, and uncontrollable with prevalent disorders and dysfunctional systems. It is then that the city may gradually collapse, and it may continue to suffer for a long time. Its failure can become widespread, and may even have lager global impacts. If the city fails, its operations will fail, and then its governance will become fragile. This is no longer a case of an outbreak. The situation may turn into a disaster, with sirens in every moment of the day, emergencies in every corner, and raids in every community. There are no unbroken shops, there is no security in place, and there is no support. The destructive damages are then beyond those preliminary disruptions. The city will fail its residents, and the hardships will become unbearable through the flames of anxiety, insecurity, fear, and distress. This is how we can negatively portray a city without an urban resilience plan. This can happen to any city and there are no exceptions (Fig. 3.1).

However, in the face of such possible calamitous probabilities, the following framework is proposed to overcome the above issues (see Subsect. 3.3.2, and Fig. 3.2), and expectantly, not to let such failures happen in the situation of the disease outbreak. The period of the outbreak would certainly be disruptive and impactful (not in a positive way), and hence we require to have a holistic review and assessment of all aspects associated with urban resilience. The eventual outcomes would feed into the all-embracing action plan and strategies at the city management level. As discussed already, impacts from the event should be assessed throughout its progression, and all units/sectors should react in a continuous process of being reflective and adaptive. This requires a thought-through planning, a step-by-step evaluation, and preparedness of multiple factors. Each stage differs from the other stages, and each sector would also require certain needs or face certain challenges at different times of the outbreak progression. These factors have to be addressed holistically from both perspectives of individual sectors and the integration between them that is seemingly more viable on such occasions. The approach requires to be an indicator-based approach of urban resilience enhancement, enabling the opportunity to assess the progress and evaluate any shortcomings and challenges. Therefore, an indicator-based approach provides the opportunity for measurement, monitory, and regular (but careful and reflective) adjustments to specific parts or targets of the overall plan. Also by having an "*interrelated system of indicators*", as suggested by Fabbricatti and Biancamano (2019, p. 7), we can have a better overview of assessment and interpretation of the city complexities. Those complexities that can be regarded as measuring the city's performance and its capacity (Huang et al. 2015), similar to standalone urban sustainable development goals (USDGs), in order to make cities and human settlements "*inclusive, safe, resilient, and sustainable*" (Klopp and Petretta 2017, p. 92). By defining these measures, we can propose a set of precise targets (or target groups), specific applications for diverse actors and systems, and a practical guideline for the time of need. In doing so, the framework would help to enhance the monitory and control measures of the outbreak progress, and propose for an all-inclusive resilience enhancement. It is important to note that the central element of 'enhancement' is extremely crucial, as it responds to not only what is needed (or should be added), but also what is already in place and requires improvement or further attention. Such considerations would depend on the conditions of the context, as well as the availability of institutions, operational units, and multi-sectoral ecosystem of the place/city.

Earlier in this chapter, we explored resilience from multiple perspectives and a range of multiple aspects. Consequently, we highlighted the characteristics of both resilience and then urban resilience, and how they play their parts in both the conceptualisation of resilient systems and the practice of resilient city. Therefore, the way in which operationalisation of urban resilience (Heinzlef et al. 2019) takes place, can be developed as a territorial approach, or even through spatio-temporal, and socio-temporal methods (Hogg et al. 2016; Komugabe-Dixson et al. 2019; Jiang et al. 2020) or patterns (Ma et al. 2019; Aswi et al. 2020) of monitory and enhancement of the city. Nevertheless, Marana et al. (2019) argue that there is a general lack of operationalisation frameworks for the development of resilience in cities—something that is also addressed in this book. This approach includes the perspective of supply and demand (Sun et al. 2019), and how cities will need to continuously adapt to healthy transitions, if not major transformations. Thus, any resilience approach ought to be as holistic as possible to be able to adapt to a 'systemic approach' to a particular territory or location (ibid). This requires an integrative approach to resilience thinking, one that encompasses a range of considerations. These are expected to be arranged in a hierarchical set-up as we may change our responses or may simply alter the priorities of our operations and management. The city has its specific dynamism, and undeniably a very complex one, too. Hence, it is important to allow flexibility in such resilience planning that can support potential adjustments or amendments, when and where needed throughout the outbreak progression.

In the following two sub-sections, we first explore more about the available frameworks and what they offer and then introduce the new comprehensive framework of urban resilience in the event of an outbreak. In here, the summaries of existing studies and global examples provide invaluable observation and knowledge about this important topic—particularly that it is evident that outbreak events are exceptional cases or occasions, but they potentially can cause substantial damage to the society. Besides all these impacts, it is important to support how the city can manage during such adversities, and how the preparedness can be in place to support the overall responsiveness of the overall stakeholder constellation of the city management, public and private sectors, communities, businesses, etc. Through only a few examples that are explored here, we extract a range of commonalities and significant values that could eventually help to define what can be regarded as an explicit urban resilience framework for the outbreak events. The aim is a two-sided matter: first to reduce the negative impacts on the city, and second to reduce the vulnerabilities. By addressing these two, we are able to strengthen the city in need.

3.3.1 Available Frameworks: What do They Offer?

Current available urban resilience frameworks are mostly applicable to a variety of emergencies and disasters (The World Bank 2012; 100 Resilient Cities 2018; UN-Habitat 2018), and some are specifically designed to combat climate change impacts or natural disasters (ADB 2014). In all these global examples, outbreak and epidemic/pandemic events are either just mentioned as a category (for example in The World Bank 2012) or are not included at all. Through all examples, we see little attention is given to this global matter of biological classification, which can be recognised as frequent and disruptive adversity. Much of the work around urban resilience framework in disease outbreak events are either included as part of those overarching frameworks and tools (few examples named above) or are merely a set of guidelines and recommendations that may not be necessarily specific enough to address the resilience of multiple systems and sectors in the city. They usually appear as general guidelines or national strategies, which are somewhat different from what a city-level framework normally includes. Hence, there are certain limitations in the practicality of those frameworks, specifically for the outbreak events. In all cases, however, there is consideration of multiple aspects that show the capacity to have an integrated approach to urban resilience (Chelleri and Olazabal 2012; Olazabal et al. 2012; UN Habitat 2018; Zheng et al. 2018; Bush and Doyon 2019; Davidson et al. 2019; Sun et al. 2019; Cheshmehzangi 2020).

In their all-inclusive city resilience framework (CRF), the Rockefeller Foundation and Arup (2014) developed an invaluable set of essential urban systems that highlight the complexity of cities from various drivers and necessities to the city. Their developed CRF then includes four dimensions of *"(1) Health and Well-being, (2) Economy and Society, (3) Infrastructure and Environment; and (4) Leadership and Strategy"*, with each dimension comprised of *"three drivers, which reflect the actions cities can take to improve their resilience"* (ibid; also 100 Resilient Cities 2018). These drivers include:

1 Under **'Health and Well-being'**: "Everyone living and working in the city has access to what they need to survive and thrive", including:

    1.1 **Meet basic needs** (which are suggested as methods to: *"particularly in times of crisis, ensure that people have access the basic resources necessary to survive—food, water and sanitation, energy, and shelter"*).

    1.2 **Support livelihoods and employment** (which are suggested as mechanisms to: *"assist individuals to access diverse livelihood and employment opportunities, including access to business investment and social welfare. This includes skills and training, fair labour policy, and development and innovation"*).

    1.3 **Ensure public health services** (which are suggested as plans to: *"provide access to effective public healthcare and emergency services to safeguard physical and mental health. This includes medical practitioners and plans, as well as clinics and ambulances"*).

2. Under **'Economy and Society'**: "The social & financial systems that enable urban populations to live peacefully, and act collectively", including:

    2.1 **Promote cohesive and engaged communities** (which are suggested as methods to: *"create a sense of collective identity and mutual support. This includes building a sense of local identity, social networks, and safe space; promoting features of an inclusive local cultural heritage; and encouraging cultural diversity while promoting tolerance and a willingness to accept other cultures"*).

    2.2 **Ensure social stability, security, and justice** (which are suggested as mechanisms to: *"ensure a comprehensive and inclusive approach to law enforcement and justice that fosters a stable, secure, and just society. This includes fair and transparent policing and deterrents to crime—specifically in times of crisis, as well as enforcement of laws such as codes and regulations"*).

    2.3 **Foster economic prosperity** (which are suggested as plans to: *"ensure the availability of funding and a vibrant economy as a result of diverse revenue streams, the ability to attract business investment, and contingency plans. This involves good governance, integration with the regional and global economy and measures to attract investment"*).

3. Under '**Infrastructure and Environment**': "*The man-made and natural systems that provide critical services, protect, and connect urban assets enabling the flow of goods, services, and knowledge*", including:

    3.1 **Provide and enhance protective natural and man-made assets** (which are suggested as methods to: "*maintain protective natural and man-made assets that reduce the physical vulnerability of city systems. This includes natural systems like wetlands, mangroves and sand dunes or built infrastructure like sea walls or levees*").

    3.2 **Ensure continuity of critical services** (which are suggested as mechanisms to: "*actively manage and enhance natural and man-made resources. This includes designing physical infrastructure such as roads and bridges to withstand floods so that people can evacuate, as well as ecosystem management for flood risk management. It also includes emergency response plans and contingency plans that may coordinate airports to function so that relief can be lifted in and out during a crisis*").

    3.3 **Provide reliable communication and mobility** (which are suggested as plans to: "*provide a free flow of people, information, and goods. This includes information and communication networks as well as physical movement through a multimodal transport system*").

4. Under '**Leadership and Strategy**': "*The processes that promote effective leadership, inclusive decision-making, empowered stakeholders, and integrated planning*", including:

    4.1 **Promote leadership and effective management** (which are suggested as methods to: "*encourage capable leadership and effective urban management within government and civil society, particularly during an emergency. This involves strong leadership, cross-sector communication, and evidenced-based decision-making*").

    4.2 **Empower a broad range of stakeholders** (which are suggested as mechanisms to: "*ensure everybody is well informed, capable, and involved in their city. This includes access to information and education, communication between the government and public, knowledge transfer, and timely and appropriate monitoring*").

    4.3 **Foster long-term and integrated** *planning* (Which are suggested as plans to: "*align sectoral plans and individual projects with the city's vision to be coordinated and appropriate to address the city's needs. This includes city strategies and plans*").

The examples of urban resilience frameworks, developed globally by major organisations, or nationally as part of national strategies, initiatives or guidelines, and in academia through conceptualisation and theory. They represent a wide range of resilience thinking and approaches that address adversities of various types. However, by decoding them carefully, we can see common messages that suggest a holistic approach that needs to be: inclusive, integrated, reflective, and systematic. Therefore, in the first steps of developing an explicit framework, we have to take into consideration four key aspects, including: (1) the playfulness of synergies between multiple sectors and the effectiveness of integration between them, (2) the comprehensiveness of the overall structure that includes an adequate range of urban systems and measures, (3) the positions and values of stakeholders and multiple means of engagement and empowerment, and (4) the system-based mechanism for institutions that suggest a persistent network of analytical information, infrastructures, materials, and knowledge. In the following sub-section, we put together the above assessed knowledge into a major proposition, which is the development of a novel urban resilience framework against the outbreak events.

3.3.2 The New Framework: What Can We Offer?

Frameworks are known to be effective instruments, which can demonstrate a comprehensive plan in urbanism, city management, and governance. On many occasions, it is the lack of planning and framework that reduces the effectiveness of urban resilience measures (in practice). Even if those measures are in place, they may not be easily implementable without a central mechanism or structure. Hence, by having a comprehensive assessment of multiple factors, a framework seems to be a vital instrument to enhance the city's resilience, boost its preparedness, and implement its resilient strategies. A framework development should occur in a procedural approach (Fig. 3.1), understanding what needs to be assessed at first. Those factors may be essentialities of the context, or externalities that would nurture the idea of 'resilient thinking'. It is then through a reflective approach that indicators and dimensions are developed, before we can suggest how they can be integrated by a breakdown of target groups. This procedural approach summaries the urban resilience framework that should capture the realities and respond to them effectively and efficiently.

In the first step, we have to verify the main structure of the framework, which is generated from an indicator-based system. In this structure, we define major groups of indicators and categorise them based on their relational position to urban systems. In this regard, the urban resilience indicators are defined into two distinct (indicator) categories of '*management indicators*', and '*provision indicators*'. These are selected through an analysis of the city's complex systems and arrangements; through which, we can assess what can be regarded as a matter of management, and what can be identified as a provision matter. This categorisation enables us to distinguish what needs to be managed from multiple levels and by multiple actors, and what should be provided and by whom. The management indicators are central to city management structures, of which we can include a range of relevant commissions, organisations, bureaus, and municipal or district-level authorities (depending on the scale of the city). On the other hand, the provision indicators are essentialities of cities that include a range of public and private sectors, businesses, and a wider range of stakeholders in the city. The management indicators are meant to help the city by reducing the negative impacts of the outbreak event and its progression,

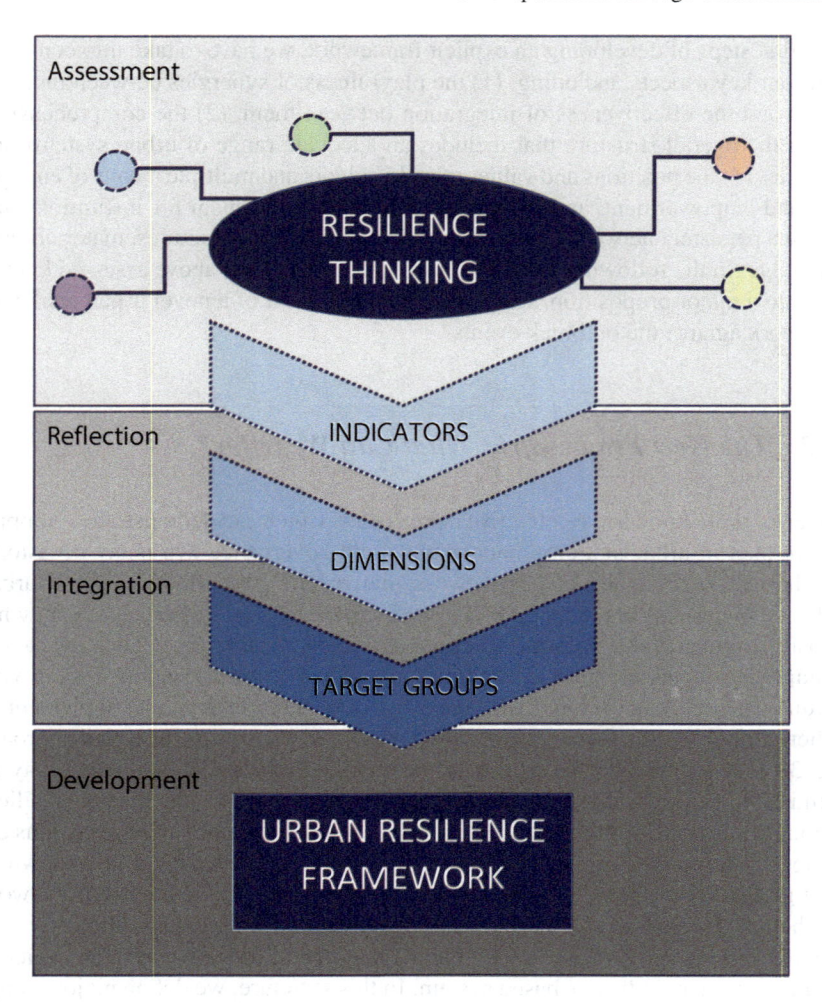

Fig. 3.1 The procedural approach to urban resilience framework development, from assessment stage and including externalities to resilience thinking of the specific situation and context, to stages of reflection, integration, and development. These include the steps of identifying and categorising indicators, development of dimensions, and allocating relevant target groups for each of those defined dimensions of the urban resilience framework. *Source* The Author's

while the provision indicators are meant to enhance the city's capacity to reduce the vulnerabilities of multiple sectors and groups (in the city).

In the second step, and expanded from the initial categorisation of indicators, we define the main dimensions of each category. In this regard, we identify two critical dimensions for each of the indicator categories. For management indicators, the two dimensions are regarded as 'operational' and 'institutional' dimensions, both representing the crucial factors of the city management. For provision indicators, the two dimensions are identified as 'services' and 'supplies' dimensions, both reflecting on the needs and necessities of the society and urban system operations. In doing so, we distinguish in a reflective manner how each of these dimensions then comes with specific needs, pressures, and vulnerabilities. For operational and institutional dimensions, we identify certain factors that feed back to the overall management matters. Subsequently, for services and supplies dimensions, the same applies that reflect on the provision factors and those essentialities of the city. The identification of urban resilience dimensions is important in terms of multiple aspects, which address:

(1) For 'operational dimension'–'How cities operate?' And 'whom is in charge of what operation?';
(2) For 'institutional dimension'–'How cities are managed?' And 'which managerial group/team is in charge of those individual management mandates?';
(3) For 'services dimension'—'How services are provided?' And 'Through which means do we provide those services in the best possible way?';
(4) For 'supplies dimension'—'what supplies and demands are needed?' And 'How does the city maintain the provision of all primary supplies to all groups?'.

In this process of dimension selection, we have to match questions of 'what', 'how' and 'whom' to their right sectors and in most cases to multiple sectors in each dimension. Hence, integration is essential, which allows us to have a holistic plan that encompasses a wide range of operations, institutions, services, and supplies.

In the third step, we allocate specific target groups for each dimension. In this arrangement, each dimension consists of two specific target groups, and with one main target group shared between two dimensions of a singular indicator category; i.e. a total of three target groups per dimension, or a total of five target groups in each indicator pillar. In reality, the shared approach in one key target group provides a mechanism for the synergies between two dimensions of each indicator category. The breakdown is in the following. For the operational dimension, the two specific target groups are 'asset management' and 'services management', both responding to the most important groups under the operational needs of the city in the outbreak. For the institutional dimension, the two specific target groups are 'media management' and 'economic management', both responding to more socio-economic aspects of management and from the broader understanding of institution enhancement. The shared target group between the two dimensions is the overarching 'healthcare management', which is a crucial target group embedded in both operations and institutions of the city, and particularly in the outbreak event. In addition, in the indicator category of provision, we have a similar layout of target groups. For the services dimension, the two specific target groups are 'transportation services', and 'safety and security services', reflecting on primary services of the city operations from the perspective of the services. For the supplies dimension, the two specific target groups are 'food supply' and 'amenities supply', carefully including those primary needs of the society and multiple sectors. The shared target group between the two dimensions is the predominant target group of 'social services', which itself is comprised of multiple factors and sub-target groups. The social services target group includes the often neglected factors in the event of an outbreak, hence it is important to include it as a shared target group in between the two dimensions of services and supplies.

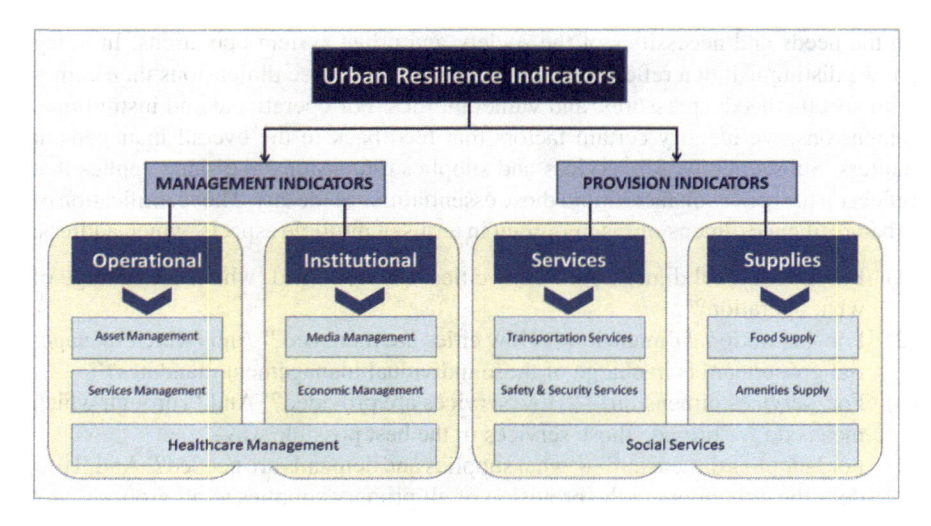

Fig. 3.2 The breakdown of the proposed urban resilience framework for the outbreak events. *Source* The Author's, adapted from Cheshmehzangi 2020

The above breakdown of indicator categories, dimensions, and target groups is provided through a comprehensive urban resilience framework, specifically developed for the context of the city (Figs. 3.2 and 3.3). It is also applicable to smaller communities or cities of various sizes, yet it should be carefully adapted and reflected on the actual context and conditions of the outbreak event (and its progression). The breakdown of this proposed urban resilience framework is demonstrated in Figs. 3.2 and 3.3. The following two Sects. 3.4 and 3.5 will delve into all factors associated with both dimensions of management and provision.

3.4 Understanding the 'Management' Indicator Category

In this section, we elaborate on all target groups under the main indicator category of 'management' and its two dimensions of operational and institutional. Regardless of the type of outbreak, and even from the examples of non-human disease outbreaks, we deal with certain issues of large-scale disturbances and the implications for management practices (de Groot et al. 2018). Hence, the actual momentum is recognised as a disruptive period. The situation also encompasses a range of management issues that require urgent attention in order to sustain the performances of various systems in place. For human-related disease outbreaks, there are examples of unknown and sudden outbreaks (Schuster and Newland 2015; Ansumana et al. 2017; Lowe et al. 2019; Shears and Garavan 2020) as well those that are more common or seasonal (Firestone et al. 2012; Pires Maciel et al. 2014; Ohuabunwo et al. 2016; Godefroy et al. 2018; Rizkalla et al. 2020) but often occur in a sudden instance, too. The situation

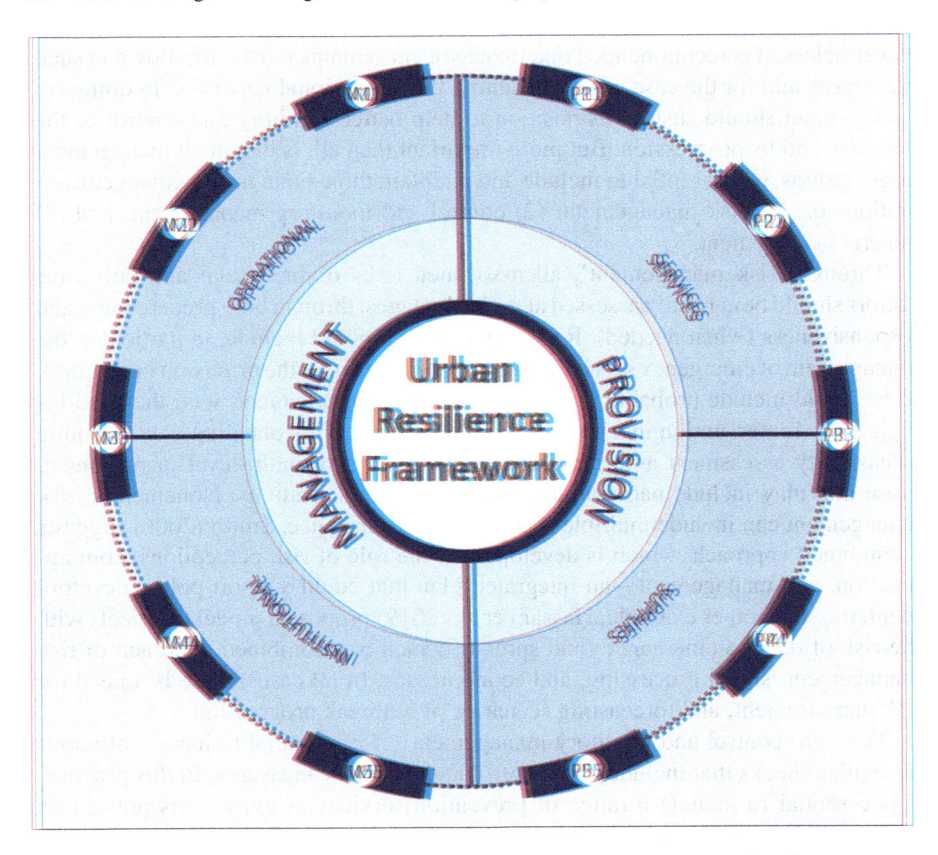

Fig. 3.3 Simplified version of the framework comprised of two indicator categories of management and provision, four dimensions, and a total of 10 target groups. Legend for target groups in the Management (**M**) Category include: **M1**: Asset Management, **M2**: Services Management, **M3**: Healthcare Management, **M4**: Media Management, and **M5**: Economic Management—Legend for target groups in the Provision (**P**) Category include **P1**: Transportation Services, **P2**: Safety and Security Services, **P3**: Social Services, **P4**: Food Supply, and **P5**: Amenities Supply. *Source* The Author's Own

escalates further once the outbreak expands further to become an epidemic or even a pandemic situation. Some of these cases are novel in terms of the disease characteristics (Wang et al. 2012) and transmissions; meaning that they require a set of watchful control and management measures. The occurrence of such events, and the upsurge in their frequency and intensity, are alarming factors for the overall management of the situations as well as the management of the systems. Undoubtedly, there are certain conflict and control issues (Shears and Garavan 2020) that reflect on issues of community involvement, response programme development, strategy enhancement, integration of plan to existing health structure, and enhancement of response effectiveness. These aspects require management not only at the top, or from a top-down approach, but also at multiple sectors or systems and sometimes at the bottom levels.

Nevertheless, it is recommended that management remains mostly top-down in such incidences and for the case of strengthening the institutional networks. In doing so, management should sustain its position to help better monitory and control of the outbreak and its progression. But more important than all, is that in all management target groups, it is essential to include and maintain three main management considerations of: (1) risk management, (2) control and monitory management, and (3) general management.

Through 'risk management', all associated risks of individual and collective sectors should be carefully assessed at multiple stages, through both preparedness and responsiveness (when needed). Risk management should include, in particular, the management of emergency services (which also come under the provision indicators). This should include probable plans for evacuation management, with the need for higher-level intervention in a possible escalated event. This planning would require a feasibility assessment as it may require the support of multi-level management, those that may include national and international organisations. Nonetheless, risk management can include multiple aspects, too. For instance, Smith (2006) suggests a combined approach, which is developed on the role of risk perceptions, communication, and management—an integrated plan that could support policy development, too. In another example, Hassani et al. (2019) propose a model that deals with the risk of disease emergence (and spread), which is a combined approach of risk management, signal processing, and econometrics. In all cases, there is a need for risk measurement, and forecasting scenarios of outbreak progression.

Through 'control and monitory management', it is essential to have continuous or regular checks that include both control and monitory measures. In this practice, it is essential to include a range of prevention services or even crisis prevention measures (Choi and Kim 2016) that can fully cover the management of the outbreak, which should help to avoid reaching a crisis stage. As part of prevention services, the foremost example is having structured isolation management in place, which is more practical at smaller scales of various range (Anderson 2009; Risa et al. 2009; Doménech-Sánchez et al. 2020) even though it can get prolonged and costly (Danial et al. 2016). At the city level, it is important to have a systematic investigation planning in place (Korte et al. 2016), which can increase the efficiency of monitory and detection methods. In doing so, we can promote case management (Ohuabunwo et al. 2016) from various viewpoints of on-site mobilisation, training development, management of care-providers, and rapid containment planning. Moreover, in order to promote control and monitory management, it is also vital to utilise the integrated digital platforms (from the experience of COVID-19) and have agent-based models in order to detect multiple important outbreak patterns, such as (case) mobility, growing symptoms, transmissions, etc. Finally, in all these examples, it is crucial to sustaining control and monitory measures across all management sectors of urban systems. Some of these prevention measures/services, should develop as isolation management plans, and if required, with urgent need of collective measures and management practice across multiple sectors or systems of the city. This means isolation management should be embedded in the overall management plan.

Lastly, through 'general management', it is essential to sustain the generalities of management demand, such as those supply and demand factors that can feed into policy adaption and policy implementation. Such an approach to policies should include the development of general public guidelines, general public education, enhanced control, as well as the development of restrictions and further inspection (if needed). The general management should act as a major mechanism in between multiple objectives and multiple criteria of urban systems, to ensure their management is monitored, supported, and updated (or adjusted) in the times of need. Here, the primary functions of general management are beyond the usual finance, marketing, and operations. In the outbreak, the general management ought to maintain a range of activities that can promote the involvement of a larger group of stakeholders who can feed into the critical factors of planning, strategy development, conflict management, and decision making.

In the following five Subsects. 3.4.1–3.4.5, we explore all five individual management target groups, across two dimensions of 'operational' and 'institutional'. As expressed previously, while these target groups are recognised as individual systems or factors of the overall (city) management, they should also be seen collectively, and if not, at least in integrative thinking. This is certainly realised as a crucial factor for the enhancement of urban resilience from the perspective of multiple management.

3.4.1 Operational Dimension—M1: Asset Management

Asset management is all-encompassing management that includes a range of products, values, investments, and the overall profile of one's resources. While it is a terminology mostly used in the finance sector, it is highly important for the operational dimension of urban resilience, particularly that it provides a wide-range understanding of urban resources. More broadly, asset management can include asset vulnerability framework (Moser 1998), asset allocation and performance (Sharpe 1992), and measures that address issues of maintenance, replacement, and reliability (Jardine and Tsang 1973). The latter is the crossover between two crucial aspects of maintenance management and physical asset management, from the understanding of productivity and maintenance. In urban resilience, it is important to have an overview of what is available? In what capacity are they available? And how they can be maintained as resources for better resilience? Hence, asset management can play a major part in how we identify necessary resources, and then manage where and how we should allocate them in the practice of preparedness and the later responsiveness.

In her presentation on issues of asset management in cities, Mian (2018) profoundly connected the dots between urban resilience development, city-scale governance, and asset management (from the general resilience perspective). She addressed how asset management cannot be dealt with in the isolation of individual sectors, but should be seen holistically across multiple sectors of urban resilience. Hence, there is a need for us to identify how we assess our assets and how we then manage them in the case of city preparedness through urban resilience enhancement. The asset management decisions should be made across multiple sectors and in line with the critical infrastructures' resilience plan. Similarly, other examples link asset management to measures of risk management and risk reduction that enable the possibility for urban resilience enhancement. For instance, AECOM (2020) explores a range of representative services, comprised of critical infrastructure protection, physical security and hardening, climate change adaptation, cybersecurity, disaster risk reduction, hazard mitigation planning, and disaster recovery. In all cases, they (ibid) highlight the position of asset and asset management in the practice of urban resilience, by addressing: *"investment in resilience today means protection and cost avoidance in the future. We help clients and communities around the world build resilience through an extensive understanding of assets, risks, and vulnerabilities"*. This is similar to earlier reports by Mitchell and Harris (2012), who described resilience as a risk management approach. They included asset management as part of risk management across multiple policy areas, suggesting how they can support the plan for resilience enhancement. This is a reflection on asset resilience and how it helps to reduce the vulnerabilities through prompt management (MMI Engineering webpage 2020); i.e. asset as a service to help to manage the risk. In their narrative, assets are valuable for resilience and the resilience of assets are valuable for good operations (ibid). Other examples are those that relate to specific sectors, such as building community resilience (Municipal Affairs of Alberta 2015), transportation asset management scenarios (US Department of Transportation 2013), or understanding infrastructure resilience from the perspective of asset management (New South Wales Government webpage 2020). In the latter, the argument is that *"infrastructure resilience is focused on the resilience planned for, designed and built into assets, networks and systems"*, namely those assets that deliver *"positive service or amenity to their communities e.g. transport infrastructure, water and wastewater, stormwater drainage, waste facilities, dams, community buildings,* etc." (ibid). Therefore, we can argue asset management plays a major part in general city operations, particularly in the time of outbreak events.

To summarise what needs to be taken into consideration for asset management, we argue in terms of the overall operationalisation perspective (Marana et al. 2019), which includes a variety of systems, resources, and capacities at the city level. Hence, we should include factors that include human resource management, particularly for those weaker healthcare systems or those who lack enough human power to handle the outbreak event. The boosting support from human resources will be a major asset for the overall operational factors, and as such, it is evident that early assessment of assets would help to overcome the later problems. One of the other aspects significantly crucial in the event of an outbreak is having a structured plan to implement a well-informed asset resilience index (Argyroudis et al. 2020), which should be in place to address any damages or deficiencies to the overall asset management. This system-based approach is very much relevant to those specific critical infrastructures, and those that can help to support the operations of the city. In addition, having an adaptive plan in asset management would help to overcome one of the main resilience shortages, which is to ensure having the capacity to adapt or to even transform the urban systems (Cariolet et al. 2019). This comprehends a greater understanding of a 'reflective approach' that suggests the systematic property of resilience (ibid), which is particularly relevant to the city operational factors. Moreover, it is important to put together a combined approach of resilience-based infrastructure planning and asset management, as proposed by Rasoulkhani et al. (2019). Again, this approach responds to those supply and demand mechanisms and the overall operational needs of the city. Lastly, and more importantly, we can argue that much of the asset management comes down to resource management (Berg and Nycander 1997). Hence, it is crucial to have a holistic plan of resource-asset management that is multi-sectoral, multi-objective, and multi-spatial.

3.4.2 Operational Dimension—M2: Services Management

As critical as it sounds, services management is commonly recognised as the backbone of city operations. Services themselves are identified as one of the two dimensions of provision indicators. But for the specific target group of 'services management', we have to understand how services should be managed, and how they play their part from the operational perspective. The city normally operates through a range of daily operational services, some can be in full operations and some are partial, seasonal, and temporary. Their management plays a major part in how they run, as well as how they may impact the societal need. Hence, the management aspect should consider the basic services (Etinay et al. 2018), and if not all. In the case of emergency, the management may exclude secondary services, usually the ones that run in parallel to the basic services as they either support the secondary activities or merely maintain the sustainability of the basic operational activities. Therefore, the basic services are defined as the most accessible ones (McPhearson et al. 2014) and are usually the ones essential for the daily needs of society. In most cases, infrastructure and services come together (The World Bank 2012; Ogie et al. 2017; Chaturverdi et al. 2019), suggesting how they jointly address the primary needs of city operation.

In urban resilience, services management refers to the enhancement of societal resilience (Marana et al. 2019). This needs to develop through a co-creation process, holistically planned to involve a larger body of stakeholders (ibid). In return, the co-benefits of each service management criteria should ideally reflect on the development of an integrated management system. At first, we may require to take on a board a multi-service differentiation plan (Wang et al. 2020) before the middle stage of service management co-creation through integration, and the later stage of network development for multi-services management. Hence, this needs to develop into a holistic services management system (SMS), which is defined as an inclusive management system bringing together a range of organisation management aspects, such as planning, policies, strategies, objectives, documentation, and processes. Developed first in manufacturing and then vastly utilised in the information technology systems, services management (for cities), can be an integration between operational aspects of production and activities with information-based processes. Hence, there is an evident advantage of utilising this particular system of multi-management and integrate it with a digital platform for operational monitory purposes.

In the case of an outbreak, services management can be a flexible platform for control and monitory of operations in a multi-aspect approach. There are also some minor overlaps with the earlier asset management target plan, which refer to services as resources for operation. In smart city examples, we can see such a mechanism as integrated management of public services (Smart Water Magazine 2019), which is aimed to optimise the operations:

> ...the efficiency in terms of managing new and existing infrastructures has to reach levels previously unseen. This requires the integration of all infrastructure, both public and private, with regard to operation and maintenance. The purpose is to have a sensible impact and improve the quality of life of citizens.

However, we must conduct a feasibility assessment to certify the effectiveness of such an integrated management system model, or aim to enhance the capacity of multi-management approach when needed. This should be seen beyond the typical provision of quality services, as it ought to undertake the role of management beyond just one sector or one need. Similar to those water services management examples (highlighted in Katko et al. 2013), a system approach of multi-services management would help to strengthen the governance of public services. This approach indicates a move away from risk management to resilience management of services, one that can boost the *"deployment of crucial services"* (Linkov et al. 2014) in the time of emergency. There are some similarities of such approach with those in ecosystem services and social-ecological systems (Lundberg and Moberg 2003; Briggs et al. 2012, 2015), which indicate the benefits of multi-management approaches in the resilience paradigm, particularly in the sub-field of resilience planning. To summarise, in the case of an outbreak, services management should enhance the operational robustness, particularly in public services. The approach, as expressed through other resilience examples, should be integrated into a multi-sectoral and multi-management mechanism. Through a co-creation approach, the mechanism should develop a dynamic network-based system for resources, basic services, critical infrastructures, productions, and primary city operations.

3.4.3 Operational/Institutional Dimensions—M3: Healthcare Management

A major management aspect of outbreak events is healthcare management, which is jointly shared between operational and institutional dimensions of management. In reality, the healthcare system is a major player in outbreak events, particularly that it requires to provide a variety of aspects, including health care services, resources, physical support (infrastructure), facilities, emergency services, isolation support, health care logistic response (Cutter et al. 2010), etc. In many cases, healthcare units become obvious hotspots of disease outbreak, and their workforces and units are usually subject to high(er) risk of disease transmission (Ghebrehewet et al. 2016; Hiller et al. 2019; Orsi et al. 2019; Taori et al. 2019). The healthcare system generally receives one of the early sudden shockwaves of the outbreak. In return, they face a higher probability of becoming a vulnerable system at the early stage(s) of disease outbreak progression. If not managed well, the impacts on health systems can cause other issues that may have a wider societal impact (Johnston et al. 2007; Elston et al. 2016). This may even cause further psychological and emotional issues that may increase the society's level of anxiety; also often leading to extreme examples of health conditions (Atkinson et al. 2009). But despite the difficulties it may face, the healthcare system plays a vital part in the delivery of health services to the general public. In the case of an outbreak event, it is crucial to monitor the quantity and quality of healthcare units and centers; hence, careful management is essential. Kruk et al. (2015) also suggest how fragile the healthcare system can become during disease epidemic situations, highlighting the ways we should enhance the resilience of the health systems to ensure their preparedness and continuity in operation and services. In reality, the management of the healthcare system is invaluable as part of the primary city preparedness measures.

In urban resilience enhancement, we ought to comprehend healthcare management as both an operational matter and an institutional factor. Healthcare is an operational target group based on its position in dealing with relevant training, treatment, control, and even containment procedures. It is also an institutional target group (i.e. identified as a healthcare system), which is defined as a formal structure comprised of multiple aspects. These aspects include certain associated organisations, workforce units, people and actions, principally intended to maintain, restore, or promote health. It should not be mistaken with health infrastructures and/or facilities (e.g. hospitals, clinics, etc.) as healthcare is broadly described as a system more than just the facilities it should normally offer. Also, there is a common misinterpretation between the two primary terminologies of 'healthcare system', and 'public health', which in reality, are linked together in resilience thinking. However, public health is closely linked with factors of governance, regulation, and support. It is also defined as a branch of medicine, which includes disease prevention (Hall et al. 2011) or crisis prevention (Quah and Lee 2004) as one of its mainstreams. The other main difference between the two terminologies is the focus areas that define them more clearly. While public health focuses more on disease prevention, it mostly focuses on safety measures and health improving mechanisms. It is also commonly used as a platform to detect health issues and provide necessary responses at the time of need. On the other hand, the healthcare system or health system is the actual capacity of operations. At the same time, it is a major institution within the array of public services and urban systems. Hence, during the outbreak, its management requires a two-sided approach, both as part of the city operations and also as the institution that supports the delivery of health care services. In their studies, Lurie and Fermont (2009) suggest building bridges between health care and public health, including also the associated medical care systems (Woolf 2013) that are vital to cases such as the outbreak event. In their recommendations, Lurie and Fermont (2009) highlight the importance of synergies between health care and public health, including the possibility to integrate the two in practice. As such, they argue in favour of the *"effectiveness of shared efforts between health professionals and other stakeholders, including community-based organisations (COBs) and health plans"* (ibid).

Resilience, as an integrative construct (Zautra et al. 2010), should ultimately sustain the health and well-being of the society in the case of an outbreak. This should help to further develop the capacity of public health and its effectiveness in practice, including specific programmes and interventions that are specifically designed for the promotion of health and societal well-being (ibid). If healthcare management is not sustained promptly, it can eventually fail the healthcare system and its services. This will lead to a shortage of workforces, shortage of medical support, shortage of treatment support, and many health disadvantages that can intensify the adversities of the outbreak event. The immediate impacts can be increased in the number of infected cases and mortality rates. As a result, healthcare management would need to sustain all healthcare operations. In return, the healthcare system should persistently operate as a fundamental institution in support of public health and medical demand (particularly in the case of emergency). It should also continue with optimal quality of medical care, providing support to those in need for various health conditions, and not only for the specific disease of the outbreak. The involvement and integration of both private and public sectors would help to increase the health system capacity to ensure all healthcare services are maintained, offered, and managed in the most optimal way. In doing so, we require to aim at safeguarding the society from any risks associated with the outbreak event and provide healthy management of the resilient healthcare services.

3.4.4 Institutional Dimension—M4: Media Management

In our age of media control (Rosa and Rosa 2011), we have to understand the two-sided control of media in the case of the outbreak—i.e. how media can control? And how it can be controlled? Such control comes from the idea of media management that signifies the important role of media as an institution, which ought to be influencing and informative. Strongly tangled with public policy and education, media is a playful part of an outbreak, from the dissemination of correct and false news, to strengthen the public trust, reassurances, support, education, etc. The way how media plays its part during the outbreak is very crucial for the maintenance of a societal mindset, educating the general public, and informing them of the guidelines, policies, requirements, and updates of the situation. The consequences of lack of trust in governance may cause adverse impacts on a wide range of areas of policy effectiveness, economic policy, the economy, the economic crisis, compliance(s), accountability, regulations, education, and social capital. Hence, it is important to maintain healthy public diplomacy (extracted from multiple sessions by Algan 2013, Burns 2013, Coyle 2013—all from Organisation for Economic Co-operation and Development (OECD) workshop on "joint learning for an OECD Trust strategy" 2013), which is trustworthy in every step and is informative to a larger group of the community.

While the sensitivity of the information should be taken into full consideration, it is important to maintain the right level of transparency and community engagement through the right sources of media. The rightfulness of this particular management is vital to avoid any further uncertainties, confusion, fear, and anxiety.

Trust in governance is essential from various perspectives, and should be maintained through a robust media management plan. As highlighted by the OECD (2019), there are six areas that governments can increase the public trust, including (1) reliability, (2) responsiveness, (3) openness, (4) better regulation, (5) integrity and fairness, and (6) inclusive policy-making. In each area, there are certain aspects that should be identified and practiced. These are key aspects that can be utilised later on specifically for the practice of media management, and help to increase its effectiveness:

– **Reliability**—*governments have an obligation to minimise uncertainty in the economic, social, and political environment*;
– **Responsiveness**—*trust in government can depend on citizen's experiences when receiving public services—a crucial factor of trust in government*;
– **Openness**—*open government policies that concentrate on citizen engagement and access to information can increase public trust*;
– **Better Regulation**—*proper regulation is important for justice, fairness, and the rule of law as well as in delivering public services*;
– **Integrity and Fairness**—*integrity is crucial determinant of trust and is essential if governments want to be recognised as clean, fair, and open.*
– **Inclusive Policy Making**—*understanding how policies are designed can strengthen institutions and promote trust between government and citizens* (ibid).

Furthermore, what we can extract from this trust-building strategy are multiple factors of inclusiveness, openness, and engaging attributes that can help to strengthen the media and its management. The official sources should remain valid and trustworthy so that the general public can have consistent access to the right information and updates, and have a clear understanding of the outbreak progression, its impacts and the guidelines to control and monitory measures. As an institution, media management should be effective to address these factors from the inception of the outbreak event. Failures to do so would shift the general public mindset to different directions. It could also create added confusion and anxiety, and could intensify the mistrust with the government. Hence, it is vital to avoid the inappropriateness of empowering the unofficial media, as they may have a higher chance of creating and disseminating false news and information. Thus, the role of media management should be effective enough to avoid unnecessary distributions caused by the public or even professional assumptions. Media management should rather focus on the provision and management of rightful sources of information that people can trust and follow meticulously. Adhering to such means of media management would help the society to play a part through health engagement, too.

To summarise, media management should propose for certain media arrangements during an outbreak, those than can provide access to right resources of information, knowledge of urgent matters, incident response arrangements, information on possible restrictions, and relevant considerations for outbreak progression (Biose-curity Incident National Communication Network 2019). This dialogue of public communication would normally be widespread in the mass media (Watson 1993), which in reality is consisted of multiple media means, including: digital media (of various examples), social media, unofficial media (e.g. forums, blogs), and official media (e.g. news and official dissemination). Mass media also operates at different scales of community-level, local level (district or city level), provincial/regional, national, and international. The media may also be framed and generated by different groups, such as local or federal officials, academics, medical experts, public health authorities, etc. (Kott and Limaye 2016). In an outbreak event, all these sources play a part in the dissemination of information. While social media is appraised as a valuable tool from certain viewpoints, such as for supporting public health practices (Charles-Smith et al. 2015), it is important to manage its validity and effectiveness through the practice of media management. Hence, there needs to be an integration between a particular professional sector and the social media, to ensure valid knowl-edge is transformed into the wider community. A similar example of social media use during Ebola outbreak was assessed by Hossain et al. (2016, p. 2136), suggesting to have a formalized channel of communication as a necessity and to have public health responses as the priority:

> The West African 2014 Ebola outbreak has highlighted the need for a better information network. Hybrid information networks, an integration of both hierarchical and formalized command control-driven and community-based, or ad hoc emerging networks, could assist in improving public health responses. By filling the missing gaps with social media use, the public health response could be more proactive rather than reactive in responding to such an outbreak of global concern.

Furthermore, as Tumpey et al. (2018) put it well, the evolving outbreak also comes with evolving communication. Hence, careful attention should be given to those widespread news/information broadcasting networks such as the digital media envi-ronment and those means of constant news dissemination (ibid; also see: Reynolds and Seeger 2014; World Health Organisation (WHO) 2018). Therefore, it is impor-tant to involve the official public health authorities (Collinson et al. 2015) in the whole process: "*because the ways in which receipt of news is evolving, the ways in which public authorities communicate with the media and public needs to adapt in similar ways*" (Tumpey et al. 2018, online source). Such involvement should help to enhance the dissemination of important public health measurements (Collinson et al. 2015), and avoid any disparities between multiple official sources. The media management should be effective enough to identify and avoid any unforeseen conflicts between multiple sectors or professional bodies, particularly through the public health system and their respective authorities. Any differences could impact risk perception, atti-tude, and subsequent behaviours (Kott and Limaye 2016), which can be risky itself and can create unwanted tensions. Any misinterpretation could cause pressure on the authorities, hence it is important the communication is legible and inclusive to all. In this regard, the public perception of risk (Sandman 1989) and risk informa-tion (Kott and Limaye 2016) should be accurate, and communications should be conducted through the right channels. This should be processed through the development of a healthy communication planning (Covello 2002; Reynolds and Quinn Crouse 2008; Mitchell et al. 2016; WHO 2008; CDC 2018). In line with such planning and arrangements, Singleton et al. (2000, p. 267) highlight a range of media management suggestions, including ten key recommendations that reflect on their experience of an outbreak:

> Four of these…[recommendations]…are around managing the media, including using a proactive press release, providing detailed briefings, using a single spokesperson and coordination of the response by a press officer experienced in media management. Another four describe how to deliver an appropriate on-site response, often requested during community intervention programmes. The two final recommendations relate to ensuring good communication and supporting staff during what is an intensely stressful period.

The above should reflect on the details of communication planning. Therefore, we can argue that the reliability of information sources is extremely important for multiple stakeholders, multiple uses, and multiple levels. As it only takes one single ill-informed information to cause a chaotic situation with some added disruptions. In this regard, media management as a major institution, should be a constant supporting means during the outbreak progression. Finally, ignoring is not the right tool of communication, and media management should be taken seriously from inception. Media management should be done for the enhancement of reassurance, provision of guidelines and support, and information sharing with the right knowledge and updates to multiple stakeholders. While the information dissemination is important, its management is even more so.

3.4.5 Institutional Dimension—M5: Economic Management

The correlation between epidemics and economics is widely studied (Bloom and Mahal 1997; Bloom et al. 2004; Bloom and Canning 2006; Beutels et al. 2009; CDC 2016; Fan et al. 2017). Much of this correlation is studied in the field of 'Economic epidemiology', an intersection of epidemiology and economics. Yet, the overall economic assessment needs to be studied from the management perspective. Branched out of management sciences, economic management reflects on the functionality of economic operations, such as the management of resources, finances, income, expenditures, businesses, various enterprises, etc. Besides severe impacts on the society, mortality rates, and the health system, there are long term economic impacts from outbreak events; those that put production on halt, and impair the city's operations at most. Regarded as 'economic burden of disease', the World Health Organisation (WHO) (2009) identifies a range of measures that are oriented around productions (market and non-market), controls, and coping strategies through incorporation between multiple systems. Their report (ibid) focuses on disease and injuries at the micro level, while it can be understood and reflect on a higher level of a disease outbreak (Huber et al. 2018), and how it may have economic consequences for the overall city management. This is related to both short-term and long-term impacts on cities and communities.

The economic costs of outbreak events are generally significant (Thomson et al. 2003), and the economic burden often comes in combination with 'social burden' or 'social costs' (Huber et al. 2018; Qiu et al. 2018). Hence, the impacts can also be identified as socio-economic impacts (United Nations Economic Commission for Africa 2014). Similar to animal outbreak events, the *"general economic and financial instability"* in a particular location is the result of weaker institutional structure (The World Organisation for Animal Health 2007, p. 112); hence, they become a bigger burden in the outbreak management during different terms. Similarly, in their report on 'The Economic Impact of the 2014 Ebola Epidemic', The World Bank Group (2014) assessed the economic impacts of the outbreak at short and medium terms, estimating the various channels of impacts on the society. In their report, they suggest two distinct channels of impacts, direct and indirect (ibid, p. 7):

> The impact of the Ebola epidemic on economic well-being operates through two distinct channels. First are the direct and indirect effects of the sickness and mortality themselves, which consume health care resources and subtract people either temporarily or permanently from the labor force. Second are the behavioral effects resulting from the fear of contagion, which in turn leads to a fear of association with others and reduces labor force participation, closes places of employment, disrupts transportation, motivates some governments to close land borders and restrict entry of citizens from afflicted countries, and motivates private decision-makers to disrupt trade, travel, and commerce by cancelling scheduled commercial flights and reduction in shipping and cargo service.

In this regard, economic management needs to be considered at multiple levels, and it is beyond the boundaries of the city. It often takes into consideration certain national strategies, regional planning, and in more severe cases the attention of the global level. In the latter level, global trade may become disrupted (Amadeo 2020) and affected regions need to implement the emergency economic plan. While we do have an international mechanism of Global Health Security Agenda (GHSA) to improve health emergency preparedness globally (GHSA 2017), its operations do not necessarily support the economic losses of specific locales. It is unfortunate that in some cases of an outbreak, the international cooperation (usually) responds with a delay; and the outbreak only becomes a matter of emergency once the economic impacts are felt at a larger scale (through development at epidemic and pandemic events). Throughout such a process, the city still needs to cope with its economic losses, financial burdens, social losses, and production reduction; and these require widespread economic management. In addition to these factors, the city needs to deal with external economic considerations (Awalime et al. 2017) and manage the economy at micro and macro levels (Luo 2013), if not anything further. Some suggestions on economic management are associated to the implementation of a robust risk management framework (Santos et al. 2013; Cheshmehzangi 2020), which can essentially inspect *"the tradeoffs between minimising sectoral inoperability and minimising economic loss"* (Prager et al. 2017, p. 7).

To summarise, economic management should carefully consider issues of: reduction in trade and transportation, reduced tourism, reduced mobility, decreased agricultural production, decreased production activities, fewer investors, business temporary closures, market decline, high fiscal impact, higher unemployment, and redundancies (some extracted from lessons by Mercy Corps 2019). The city management ought to assess various costing and demands of multiple sectors (Luyten and Beautels 2009), particularly those that are directly affected by the outbreak. A relevant cost analysis would help to decide on certain adjustments or priorities that can help the city's economic stability. From the institutional perspective, economic management should be conscious of slowing down the economy, assess the risks, and evaluate the long term economic impacts. There is an urgent need for emergency economic plans (i.e. through a multi-sectoral framework) to have an immediate action plan, short term contingency plan, and long term recovery plan. A multi-sectoral approach is essential (Smith et al. 2019) to evaluate co-benefits between various urban systems and have effective risk management beyond medical responses (World Economic Forum 2019, p. 13). Finally, economic management response strategies should be inclusive and should create an ecosystem of multiple businesses, sectors, and organisations.

3.5 Understanding the 'Provision' Indicator Category

In this section, we elaborate on all target groups under the main indicator category of 'provision' and its two dimensions of services and supplies. Apart from the multi-management requirements of urban resilience enhancement, the city needs to sustain the provision of certain services and supplies. In doing so, we respond to a situation of supply and demand, while dealing with potential shortages across various sectors. Thus, it is inevitable that the outbreak event results in the decline of services and a shortage of supplies. For services, there is a significant decline in regular operations of multiple units/sectors and their provisions, as we see mostly in weakening health care services and primary social services. We anticipate potential fragmentation in regular operations that impacts not only the growth but also the stability of the society. While emergency services are on full alert, the operation of other services may be significantly affected. Some services may have to reallocate their resources (including human resources), change their operations (i.e. towards reduction, interim halt, and provisional closure), and may be required to shift their capacity of service delivery in the time of need. Correspondingly, the most perceptible and common impacts on the provision of supplies are on medical supplies of various kinds (both related and unrelated to disease), anti-infective products, food supplies, and associated amenities that may cause major disruptions in the society. In the case of conflict zones/areas (and also those experiencing wars, sanctions, etc.), the impacts are generally more significant. For instance, the case of Venezuela (Lodeiro-Colatosti et al. 2018, p. 1343) highlights:

scarcity of basic resources largely affecting the public health infrastructure, resulting in long-term shortages of essential medicines and medical supplies, including vaccines for universal immunisation programs and the immunisation of specific risk groups against specific diseases.

Hence, the already affected areas due to other structural factors such as conflicts, war, sanctions, economic crisis, or societal/political unrest, may face more severe impacts of the outbreak on the provision of their services and supplies. Therefore, it is harder to maintain the security of those communities as the impacts are expected to cause more adversities; and vulnerabilities are expected to be much higher. This factor also puts pressure on the effectiveness of the most pragmatic solutions or strategies. Thus, we can argue that the general impacts on those communities (as well as other cases of vulnerable cities or communities) become more perceptible in the control of the outbreak. Such vulnerability also triggers the event to have a more malicious progression, at a larger scale and a faster pace. In addition, the eventual impacts show a rapid sign of development on the supply chain, business enterprises, market, and their respective productions. Hence, we have to utilise both quantitative (Morgan 2019) and qualitative assessments of our services and supplies. The delivery of services and supplies, even in the most disruptive conditions, is meant to create a healthy continuity of daily activities. The patterns of such provisions will certainly be different from customary circumstances. More importantly, all provision target groups should include three fundamental considerations of: (1) risk assessment, (2) adaptive maintenance, and (3) prioritisation plan.

Through 'risk assessment', we have to identify all risks and provide an immediate action plan. In general, the risk assessment needs to be quantitative and evidence-based (Athar et al. 2005; Guillier et al. 2013). Hence, a qualitative risk assessment (QRA), as proposed by Guillier et al (2013), can enhance the understanding of the outbreak and its risks on services and supplies. Ultimately, this approach sheds light on possible deficiencies in provision target groups. In their disaster resilience assessment, UN's International Strategy for Disaster Reduction, UN/ISDR (2014, 2017) put cities as primary audience and suggest several risk assessment criteria in six distinct categories of:

- Research–including evidence-based compilation and communication of threats, and needed responses;
- Organisation–including policy, planning, coordination, and financing;
- Infrastructure–including critical and social infrastructure and systems, and appropriate development;
- Response capability–including information provision, and enhancing capacity;
- Environment- including maintaining, and enhancing ecosystem services; and
- Recovery–including triage, support services, and scenario planning.

We can argue that such a criteria-based approach can enhance the possibility of a thorough assessment. Also, through its prediction modelling approach, QRA could develop possible scenarios across the provision target groups. Existing research suggests risk assessment for the applicability of evidence-based practice in public health (Forland et al. 2012), which can improve the multi-sectoral assessment and their status, and can provide a reporting procedure to ensure better availability of necessary data (Cassady 2006). In doing so, we may be able to have a better assessment in place in order to reduce potential burdens, as these are expected to be widespread and disruptive to all primary services and supplies. If needed, certain adjustments should be proposed to reduce vulnerabilities.

Through 'adaptive maintenance', it is essential to increase the resilience of services and supply provisions. By responding directly to the changing environment and conditions, this factor includes both the modification plan and implementation plan for changes. In their study, Wilson et al. (2013) describe adaptive maintenance as 'resilience', which should not be confused with the transformative capacity of specific systems. In this regard, we may be able to provide a chance to "*deliberately transform systems and society*" (ibid, p. 1) or deploy an array of supporting measures when it is required. The practicalities of doing so are mostly associated to monitory of changing situations, responding to each stage of progression, the flexibility of provisions in multiple scenarios, and modification of resources in between various sectors/units in our urban systems. Hence, in order to enhance the city's resilience, adaptive maintenance should create a systematic model of multiple sectors in a hierarchical network and ensure adaptability is prudently measured across each of them. This approach would enable us to maintain the flexible operation of necessary urban systems in the time of need.

Lastly, through 'prioritisation plan', it is essential to identify what are the priorities in services and supplies? Where are potential deficiencies and how to address them? And when do we strengthen certain provisions? To answer these, and following from the utilisation of adaptive maintenance, a prioritisation plan would help to assess the needs of the urban systems. In a responsive approach, their needs should be carefully considered without negative impacts on other systems. The prioritisation plan can include a variety of factors, such as temporary arrangements, different spatial use, services support, economic support, supply increase, etc. In doing so, we can provide a holistic plan that encompasses multiple urban systems, and help to boost public investment programmes (if applicable), infrastructural support, financing (Cities Development Initiative for Asia (CDIA) 2010), as well as facilitating necessary strategy programmes to certain priorities, such as health care services, emergency medical services, etc. In the outbreak, the prioritisation plan ought to be reflective of the realities, and help to reduce vulnerabilities of those priority areas. This essentially helps decision-makers and enhances the aptitude of response strategy programmes.

In the following five Subsects. 3.5.1–3.5.5, we explore all five individual provision target groups, across two dimensions of 'services' and 'supplies'. Similar to management target groups, we urge to consider the provision target groups both individually and collectively in integrative thinking. This is considered to be effective for the reduction of vulnerabilities and enhancement of provisions at the city level.

3.5.1 Services Dimension—P1: Transportation Services

In cities, transportation services include a large body of transportation networks, and a wide range of public transportation systems (e.g. buses, underground/metro, monorail, trams, taxis, boats, etc.), intra-city and inter-city transportation systems (e.g. airports, highways, motorways, ports, etc.), shared transportation systems (e.g. shared cars, shared bikes, shared taxis, etc.), private transportation (not limited to private cars), emergency systems (like ambulance, fire services, police, etc.), and logistic system. Such services are also linked to transportation infrastructure (or physical infrastructure), such as roads, air transport, warehousing, water bodies, railways, etc. Hence, it is clear that transportation services operate for multiple purposes, for multiple groups of stakeholders/people, at multiple periods, and for multiple occasions. In the early phases of the recent COVID-19 outbreak in Chinese cities, Xu Yanhua, an official from the Ministry of Transport (MoT) had to intervene with an official announcement in support of the outbreak control progress. His statement was given as a reassurance to the general public, highlighting, in particular, the operational plans of transportation services: "*all-out efforts are being made to keep the transportation network and the green channels operational while curbing the spread of the virus through traffic control*" (The State Council 2020). There is a two-fold from this statement, one that suggests the approved operations, and one that proposes extended control through traffic control measures.

In addition, transportation services cover a large body of logistical support, those that are beyond internal productions and consumptions of the city and its residents. From emergency medical operations to food supply delivery, transportation services are comprised of many necessary factors that should be addressed during outbreak events. Many of our daily city operations are based on transportation services, and without them, the city may not be able to maintain the needs of its society, organisations, and businesses. On one hand, they are essential to the city, and on the other hand, they play a major role in speeding the spread of disease (Lowe et al. 2014). The latter is evident in many cases as the transport process is recognised as a source of transmission, particularly "*if adequate hygiene measures are not implemented*" (ibid, p. 872). It is also proven that transportation is the first cause of disease spread and scaled up situations (i.e. of disease transmission); leading to eventual change of the event status, i.e. from outbreak to the epidemic, and from epidemic to potential pandemic. Hence, the ins and outs of city areas, city districts, and communities can easily boost the outbreak transmission across the city and beyond its boundaries.

In most outbreak cases, there are limited measures that can be effectively implemented in time. Controlling and limiting transportation system services are difficult tasks for any city that should be driven by top-down decisions. The transportation system is extremely dynamic that we cannot predict how fast and in what direction the spread can take place. Similarly, as suggested by Rodrigue et al. (2020, Chap. 11), "*transportation systems due to their speed and ubiquity act as a vector in the diffusion of pandemics*", but they also need to continue with freight distributions. This means there is a likelihood of a great risk "*resides in the geographical and functional structure of supply chains because the continuity of freight distribution could be compromised*" (ibid). Some of these factors are associated with the increasingly global economy and its pros and cons, particularly that many aspects of our contemporary life are now dependant on global supply chain networks. As Rodrigue et al. (ibid) suggest, "*even the slightest disruption in the availability of parts, finished goods, workers, electricity, water, and petroleum could bring many aspects of contemporary life to a halt*".

Moreover, the spread of disease through transportation is somewhat inevitable, as relevant control measures are often only applied at the later stages of outbreak progression. Therefore, our preparedness should rely on how the city can maintain the optimal operation of transportation services. An example of this is the utilisation of redundant transportation services to help with other provisions, such as the delivery of food and medical supply (into the city). While air transport becomes more restrict or reduce in operations, it is suggested to use those services for the provision of primary supplies. In more extreme cases, they can operate in a one-way system to ensure the city is not left without its primary supplies. Transportation, as a system is essential for multi-level operations, and as a service requires to support the demand of the society. Nevertheless, there are some challenges in regards to transportation services, such as, control measures for populated public transportation networks, density control plan, healthy ventilation systems of often enclosed spaces for public transports and their stations, provision of emergency transportation services, provision of safe transportation of infected and deceased people, closure of specific transportation networks, and reduction in general mobility across the city. Moreover, transportation services of all kinds may reduce in operation, meaning that adaptive measures should be taken into consideration to ensure certain transportation services are prioritised, some are maintained, and some are carefully controlled.

3.5.2 Services Dimension—P2: Safety and Security Services

A crucial part of the services dimension is related to both the provision and operation of safety and security services. It is important to note that provision alone is not effective, similar to what we can identify as the non-operational institutional dimension of the city (Cheshmehzangi and Dawodu 2018). The provision and operation of such services respond to the role of co-benefits resiliency actions (Siemens et al. 2013), which are obligatory for decision-making procedures. The safety and security services are widespread, encompassing a range of imperative systems beyond just policy and security forces. Many safety services come under this particular target group, specifically focused on crucial factors of food safety, occupational safety, medical services, distribution system safety, logistics safety, emergency services, etc. Many security services include primary and secondary units/forces, such as security forces, police, fire services (fire detection, fire engine, etc.), bomb detection units, special forces dealing with various threats and emergencies, border offices (including immigration control), community security, communal guards, and military forces/army. Altogether, these factors shape the overarching factors of societal safety and human security, which are exceptionally important during outbreak events. By linking human security and public health, and by quoting John Donne's inspiring adage "no man is an island", Curley and Thomas (2004, p. 30) argue that *"the human security concept offers one possible model that can equitably meet the analytic needs of the international community without devaluing or distorting its myriad elements"*. In this regard, it is important to maintain the safety and security of society, as it determines the overall safety and resilience of the city.

Most studies associated with safety services of the outbreak are linked with food safety and how they should be maintained in healthy conditions (Jung et al. 2014), particularly in agricultural businesses and food processing firms (Jin and Kim 2008). This also requires to address any possible threats to food and nutrition security (Chen et al. 2020). However, it is evident safety services are more than just food safety. Many safety measures should be in place through structured services (i.e. as per the above descriptions), and they should be up and running to maintain control and monitory of the city and its communities. Such services should directly respond to various societal needs, and deal with emergencies in the best possible way. Moreover, due to potential conflicts in communities as a result of increased anxiety and uncertainties, the safety and security of cities should be continuously sustained throughout the outbreak. Communities become more vulnerable under the shadow of fear and anxiety, and those in less affluent communities may suffer the most. In such occasions, many opportunists may consider taking advantage of the city's vulnerabilities; hence, careful attention is needed from those respective services to ensure reflecting on the society's gradual and sudden changes, and those that may potentially cause larger-scale disturbances. Also, another role of safety and security services is to monitor the compliance of the general public to regulations and rules that are in place. These should be conducted through regular and careful monitory procedures throughout specific phases of the outbreak progression.

More importantly, it is vital to reduce the chances of chaos and monitor any unexpected situations. The end result of such chaotic situations would enhance the vulnerabilities, some that can potentially develop as societal disparities, unrest, looting, and potential societal downfall. Hence, many safety and security issues are dependent on society's compliance and their persistent support. Without such compliance, the outbreak can increase the chance of disorder, which ultimately increases the intensity of disruptions and vulnerabilities. Such a situation may turn some of the critical urban systems into dysfunctional units/services. Hence, precautionary measures are highly crucial. Moreover, the safety and security services, should reinforce the guidelines to ensure they are appropriately implemented through the right channels. In doing so, the effectiveness of those enforcements is essential. Such measures are assessed by Condon and Sinha (2010, p. 50), as voluntary and mandatory public health measures:

> …there was not a significant difference in compliance with mandatory and voluntary public health measures where the effect of the mandatory measures was diminished by insufficiently severe penalties, the lack of market forces to create compliance incentives and sufficient

political influence to diminish enforcement. Voluntary compliance was diminished by lack of trust in the government.

Hence, healthy community engagement is important to gain and cultivate community support; one that helps to nurture the micro-level safety of the communities in the city, if not the city-wide scale. In outbreak events, the priority is usually given to emergency services, and it remains so to ensure providing necessary support to specific emergency medical services. However, we still require to preserve the healthy operations of other services; particularly those that uphold the safety and security of our communities. Thus, a holistic understanding of multiple safety and security services would help to reduce their vulnerabilities and make necessary arrangements or adjustments in the time of need. Finally, it is important to strengthen the resilience of prioritised units and services through a hierarchical structure, which considers both the adaptive maintenance strategy and effective prioritisation plan. This approach should be practiced across multiple urban systems or sectors, to ensure all vulnerabilities are carefully assessed and addressed.

3.5.3 Services/Supplies Dimension—P3: Social Services

A major provision aspect in outbreak events is social services, which is jointly shared between services and supplies dimensions of provision indicator category. In reality, social services are often neglected in early stages of outbreak events, but become significant players of the later recovery and post-recovery phases (more details provided in Chap. 4). Social services are widespread and include many important aspects of our cities and communities. For the case of an outbreak, a selective number of social services are very crucial and require supporting plans/strategies. Hence, the need to prioritise primary social services is essential for more effective outbreak control. In fact, social services deal with a wide range of services primarily accountable for providing, managing and evaluating social care and support services. They also include a range of public health services, health care, and safety services. In the latter category, only a few social services overlap with what has been already covered in safety and security services, such as police and fire services. Other relevant social services (and some applicable to outbreak events) are the provision of benefits and facilities that include education, food subsidies, health care, job training, subsidised housing, adoption, community management, policy research, and lobbying. For the case of an outbreak event, the priority social services are to ensure the right education is provided through appropriate means. This is a twofold matter, including the provision of relevant community training opportunities and supporting the city management in the development of education for medical services, emergency units, and health care systems (such as precautions.) This should effectively influence the density of awareness through the development of social-physical networks (Yi et al. 2019), as well as through the right education.

For food subsidies, relevant social services may not be able to cope with their regular subsidies as the adversities become widespread and embrace a much larger group of people. As a substitute, those relevant bodies can provide support to the city management units, to ensure the following three provisions are maintained: (1) price monitory of primary products, inclusive of food, sanitation, and anti-infection products, medical products (inclusive of protection products, such as masks and gloves), medicine, etc.; (2) ensuring vulnerable groups have accessibility to such provisions; and (3) securing external support for emergency needs and supplies. The other essential factors under social services that should be fully maintained are community management, through which relevant social services can empower the community leaders or community management units (if any); or otherwise, can provide the necessary support to disadvantaged groups and vulnerable communities.

Above all, social services must play significant roles in the provision of health care, from both perspectives of services and supplies. This should reflect on issues of social protection and public health, a combination of effective measures that can be inclusive and supportive at the larger scale of the city. This provision could help to provide necessary precautions, specifically to those groups usually recognised with a higher risk of infection, such as health care workers (Cohen 2018), elderly, and those groups that may have no access or limited access to health care systems. In specific cases, social services should address issues of social determinants of health, which in general are the "*complex, integrated and overlapping social structures and economic systems that are responsible for most health inequities*" (Hepburn 2017). Thus, social services as an activity designed to promote social wellbeing can provide health equity (Richardson et al. 2017) and the crucial health care provision, which are needed across all communities. Finally, it is important to note that not all social services can cope with the adversities of an outbreak. Some may need to temporarily stop their operations, and some may have to deal with a set of longer-term outbreak impacts. Therefore, it is recommended to seek external support (i.e. at the national and/or the international levels) in early stages. In doing so, we should make certain that the city's primary social services, particularly those that can help to overcome the societal vulnerabilities, are resourcefully maintained, strategically strengthened, and properly managed. As a major provision target group, social services are expected to provide multi-objective support to the society and the city's resilience.

3.5.4 Supplies Dimension—P4: Food Supply

Apart from food safety, which is already covered under the previous target group, one of the main target groups under the supplies dimension is food supply. Undoubtedly, food is a primary consumable product, which can become a scarce product at certain stages of an outbreak event. The considerably large demand for food in cities is a major challenge for regular provisions of the city. This is particularly evident in the case of lockdown situations of communities and cities. In general, the main disruptions associated with food supply are associated with its 'production', 'distribution', and 'provision'. To date, there is little scholarly research regarding food supply during outbreak events, and the majority of existing literature only focus on food safety and not from the perspective of how it is supplied or distributed in affected areas.

For food production, we see a major decline in or around cities, particularly in peri-urban areas and regional food production hubs. Due to temporary closure or non-operational status of many agricultural businesses, as highlighted by Jin and Kim (2008), we see tangible reductions in food stocks, variety of food, and freshness of essential food or commonly known as 'fresh produce' (such as fruits, vegetables, (some types of) meat, etc.). Another factor associated with this conspicuous shortage is the common reaction of the society to uncertainties of the outbreak progression. This is caused by growing anxiety and the fear that the outbreak may last longer than initially anticipated. In most cases, people rush to shops and supermarkets and store as much food as they think it may be required. With limited impacts on fresh produce, we see a larger impact on other food products. This somewhat expected behaviour causes immediate food shortages. Hence, this inevitable reaction changes the equilibrium of supply and demand. On certain occasions, even if food production is only slightly reduced, we see a complete imbalance in what has been purchased and stored and what is remained on supermarket shelves. In this regard, the city may face sudden food shortages that cannot be recovered by partial operations of transportation systems. As we still lack wide-spread urban farming in our city environments, greater pressure would remain on food distribution; i.e. how do food supplies get to the city? How can they be distributed in time? And how regular can they be provided and distributed? Hence, ensuring smooth logistical operations are vital for the distribution of food supply, particularly of the regional agricultural and food supply chains (Chen et al. 2020).

In the recent COVID-19 pandemic outbreak in China (which was just epidemic for almost the first two months), the national-level Ministry of Agricultural and Rural Affairs (of the Chinese central government) released an emergency notice to all relevant and responsible departments to sustain order in markets, provision of supplies, and their delivery (ibid). In this specific event, the test for e-commerce, particularly for food and fresh produce delivery, was a success; highlighting in particular how food distribution can happen with minimised personal contact:

> …as lockdown measures have led to a huge spike in demand for home delivery of fresh groceries, e-commerce companies have announced an in-app feature for contactless delivery, allowing the courier to leave an order in a convenient spot for the customer to pick up, without having to interact. Making use of these delivery platforms could address many logistical challenges for obtaining food, while minimising the potential risk of infection from visiting crowded markets to buy groceries (ibid, 2020, online source).

The other suggestions by Chen et al. (2020) about issues related to the provision of food supply are concentrated on "close monitory of food prices and market supervision", "enabling policies for spring planting and increase support for production entities", "smooth flow of trade and make full use of international markets as a vital tool to secure food supply and demand", "protection of vulnerable groups and provision of employment services to (migrant) workers", and "further regulatory measures on wild food markets to curb the source of the disease". Under provision matters, agricultural enterprises are often hit by the complexity of outbreak issues; hence, there is an urgent need for financial support to reduce potential risks of delays and production closures. Chen et al. (2020) even suggest for possible temporary subsidies that could reduce burdens on those specific primary productions, such as food production. At the epidemic level, the increase in the import of food supply can help to increase supply stocks. However, this should be done with extra care to avoid any unexpected spread of disease between different countries or regions. In particular, occasions when food supply may be the carrier of the disease, more measures need to be implemented to ensure scanning and full monitory of food import and export throughout the outbreak.

Furthermore, it is evident that the shortage of food supply may turn into major socio-economic disruptions. The possible ripple effects on food prices and markets, as highlighted by Martin (2020), is a major concern to issues of supply and demand in food production and its provision. With restricted logistical operations, food supply and associated businesses face significant losses; hence, extra support from multi-levels of the government is essential. Similarly, Jung et al. (2014) argue the impacts of the food production chain from various aspects, as large as agricultural practices to widespread processing of certain food supplies. From their perspectives, there is scope for the development of science-based best practices to ensure food safety is maintained (ibid) before we can sustain the food and nutrition security of society. To summarise here, and in order to address issues of food supply, we have to consider the multiplicity of measures, practices, and provisions. Food supply remains a priority that requires careful attention, including both production processes and consumption demands.

3.5.5 Supplies Dimension—P5: Amenities Supply

In addition to the primary target group of food supply, other supplies require attention and support during outbreak events. In this target group, we classify these other supplies into an overarching category of amenities supply, which in reality are in forms of supplies, facilities, and services; hence, altogether defined as amenities. These amenities are categorised under four groups of 'basic amenities', 'secondary amenities', 'mobile amenities', and 'virtual amenities'. Through these amenities (including also public amenities), we have to sustain the provision of certain facilities, services, and supplies that are crucial to the well-being (Dyakova et al. 2017) of our communities and cities. From the provision perspective, we have to prioritise the societal needs and conditional essentialities of the outbreak. In particular, basic amenities tend to be more effective in maintaining socio-economic values of society. Secondary amenities can be regarded as a non-priority group, while the other two groups of mobile and virtual amenities can be provided with a certain level of flexibility and through what Alterman (1988) refers to as 'adaptive planning'.

Basic amenities are commonly recognised to be essential for the provision of societal well-being and quality of life (QoL). While the latter may be significantly affected during the outbreak, it is essential to sustain the former (i.e. societal well-being) through certain socio-economic benefits, facilities, and provisions. Some crucial examples of basic amenities are roads (particularly with access to highways and motorways), running water, energy (electricity and natural gas), telephone services (or connection networks/telecommunications), internet, as well as main public amenities, such as hospitals, clinics, medical facilities, pharmacies, sanitation units, waste collection services, and waste management units, general and specialised shops and markets, goods collection points, maintenance units, long-term care facilities (such as nursing homes and assisted living), and provision of clean environments. In this category, such amenities provide necessary values of societal well-being, those that are essential for smooth or partial operations of urban systems, multiple sectors, and communities. The city and its residents are dependent on such basic amenities, and such provisions are fundamental, with significant impacts both at the micro or individual level (such as individuals, households, and specific groups/sectors), and collective level of the city and its communities. There are also specific active facilities during an outbreak (from Interior Health website, online source 2020) that only operate in addition to those basic amenities, and are more specific to the disease treatment (including the provision of antibodies if applicable) or facilities that are temporarily needed during the event of an outbreak (or in certain emergencies).

In addition, the secondary amenities are the provision of non-priority facilities, mostly comprehended as general public facilities, such as public places or communal areas, health centers/clubs, sports centers, restaurants and food courts, street food areas, community clubs, public toilets (or facilities of such kind), public baths, temples and religious centers, (urban) parks, leisure areas, libraries, cinemas, common public buildings (such as museums, galleries, malls, outlets, etc.), hairdressers and barbers, banks, post offices, recycling centers/units, connections to major transportation hubs (such as airports and ferry terminals), local bus and railways stations, training units, educational areas (inclusive of nurseries, schools, colleges, institutes, research centers, and universities), and other public areas for recreation. Certain businesses may also offer some amenities in this category, such as marketing, finance, etc. In this category, we deal with certain non-priority amenities, hence their provision during the outbreak is not advisory, or should be limited in main phases of the outbreak progression. In most cases, it is important to limit such amenities and their operations, particularly the ones that encourage populated use or gathering, multiple functionalities of the environment, closed environment activities, and non-essential public uses. Apart from these two categories, we also have mobile and virtual amenities that include certain facilities and provisions. For mobile amenities, they mainly include mobile facilities such as fuel services, oil change units, recycling facilities, food trucks, auto washing and detail service, retails sales of specific products, mobile repair trucks/units, mobile retail of any kind, and mobile health care services. For virtual amenities, we can refer to any of the above facilities that can be provided online or virtually, including in particular: retail (both buying and selling), selective educational means, training, and entertainment.

To summarise, we refer to amenities as both priorities and non-priorities in the case of outbreak. As defined in their respective categories, certain amenities are more crucial for the societal well-being. It is also essential to effectively respond to context-specific matters, including–but not limited to–climatic conditions, availability of amenities, cultural and social factors of the society, behavioural attributes, institutional arrangements, and socio-economic structure of the city. To maintain the operations of basic amenities, as the least provision, we ought to ensure having an effective prioritisation plan in place or else the impacts would be widespread and may cause additional adversities to the society. As a matter of fact, these may potentially increase the vulnerabilities and reduce the values of essential socio-economic benefits to the societal well-being. Finally, to avoid any further adversities, the city should prepare comprehensively through strengthening resilience and an effective action plan.

3.6 Preparing an Action Plan and Supporting City Management Decision Making

Through the conceptualisation of the urban resilience framework and its details, we verify tailor-made measures and practices for city preparedness during outbreak events. This conceptualisation is a novel approach in scholarly research, which benefits the process of preparing an action plan. Through the complexity of disease outbreaks that suggest their occurrence frequency, multiple types of infection, the intensity of the event, and various transmission mode, we can verify the need to understand such events from case by case and locale to locale; but more importantly, from the overall approach of 'preparedness'. Disease outbreaks are usually infectious and vary in their nature of transmissions, such as animal-to-person, person-to-person, from the environment, or other sources. They can create critical conditions, put cities and communities under disruptive threats, and impose high risk on our socio-economic values and humanity. From their findings in comparing disease outbreak issues against natural disasters, Alwidyan et al. (2020) highlight the widespread lack of knowledge in outbreak events. Hence, we can argue that the combined effects of uncertainties (as highlighted before) and unawareness of how we should deal with the situation, make the society more vulnerable. It is worrying that in most cases, most people have very little knowledge of the situation of disease outbreak, pay very little attention to precautionary circumstances, and delay their responses in the face of mounting issues and adversities. We can argue that in all cases the society is not expected to be or cannot be fully prepared, but the city can. Thus, the city ought to prepare itself through comprehensive planning—or in other words, with a robust action plan.

Our so far analysis indicates the obvious need for preparedness in multiple ways (e.g. with multiple sectors, dimensions, characteristics, actors, systems, and target groups) and by strengthening resilience in the practice of disease control and containment (summarised in Fig. 3.4). In order to do so, we need to develop an action plan, which is comprehensive (Cheshmehzangi 2020), quick (Ung 2020), and effective. In reality, having effective responsiveness without an action plan is absurd, and having an action plan without preparedness and knowledge of the situation is even more so. Thus, the primary role of an urban resilience framework is to ensure it has the capacity to prepare the city either before the actual outbreak or with immediate effect at the beginning of the outbreak. This preparedness helps the city to develop an immediate action plan, which is reflective enough to the conditions of the outbreak. More specifically, we should aim to break prevailing boundaries between singular operations of the city, as they no longer can sustain their customaries or they may eventually fail to operate on their own. In this approach of integrated thinking, we could be in a much stronger position to support decision-making procedures at the city management level. Whatever procedure this may be, it should be all-inclusive and integrative in order to enrich the approach to embeddedness thinking (Thomson-Dyck et al. 2016) and comprehensive consideration of multiple sectors, multiple systems, and multiple stakeholders; and without having anyone or anything left behind.

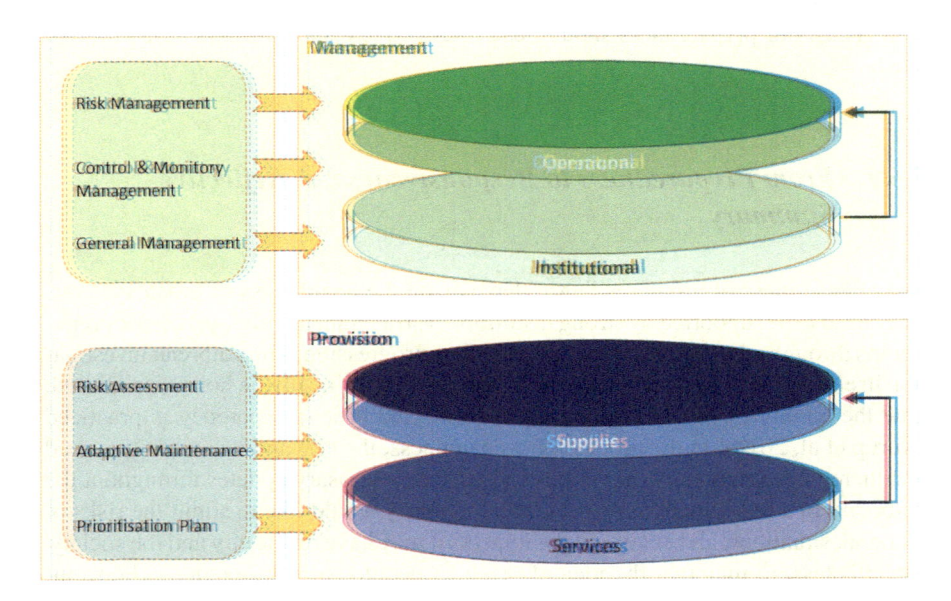

Fig. 3.4 The summary of four key dimensions of urban resilience and their relationships in their two indicator groups of management (above), and provision (below). The summary also includes specific essential factors for each indicator group, namely three factors of 'risk management', 'control & monitory management, and 'general management' for management indicator group; and three factors of 'risk assessment', 'adaptive maintenance', and 'prioritisation plan' for provision indicator group. *Source* The Author's Own

Our discussions in Sects. 3.4 and 3.5 of this chapter are narrated from a comprehensive perspective, ensuring that all city-oriented aspects of management and provision are included and addressed. By proposing this comprehensive framework, we suggest a people-centric approach, addressing: how it is implemented as an instrument (or tool) to guide the city's preparedness plan, and how it can instantaneously strengthen the city's resilience during disruptive disease outbreaks. Unquestionably, we identify the important position of people. We also highlight the major need for comprehensive control measures and communication, including a healthy process of community engagement, which is *"increasingly recognised as a key component of outbreak response, allowing responders to engage the affected population and to alter behaviours which may propagate an outbreak"* (Houlihan and Whitworth 2019, p. 142). In this regard, cities and city authorities should be adaptive, reflective, and more importantly, they should be prepared and resilient. They should prepare to: (1) make effective adjustments, (2) enhance the institutional arrangements, (3) provide the necessary support, (4) reduce uncertainties and vulnerabilities of the society, and even (5) make new policies—even if all these have to be temporary. Just because relevant policies are absent from the national agenda or local plans, it does not mean new policies or adjustments cannot be made during emergency/crisis situations, such as in outbreak. If we neglect to make necessary adjustments, it indicates the lack of reflectiveness, and hence, we may neglect the actual meaning of having an action plan. Thus, our preparedness should lead to effective responsiveness at the time of need.

3.6.1 From Preparedness to Responsiveness: A Reflective Summary

The effectiveness of urban resilience depends on the healthiness of urban systems. They need to be supported by strong institutional arrangements, and systematic preparations through adaptive planning and reflective progression. The outbreak investigation itself should be conducted promptly, reflecting specifically on how it is reported, how the quality of procedures are maintained, and how transparency is practiced (Kurup et al. 2019). In the process of preparedness, the city and its authorities need to adhere to precautionary guidelines and provide necessary updates throughout the outbreak. In a real situation, no official or authority would like to admit the risks of outbreak situations, all because of the impacts it may have on the city and the society. Nevertheless, to maintain the idea of bringing people back in practice (Thomson-Dyck et al. 2016), we have to ensure any preparedness planning is a response to immediate socio-economic contexts. Therefore, in parallel to outbreak science (Houlihan and Whitworth 2019), we have to understand the complexity of city management and how it reflects on the societal well-being and the enhancement of urban systems. This requires resilience thinking, through which we can successfully progress from preparedness to responsiveness.

On the other hand, many technological advances and instruments can be utilised to strengthen the city's resilience. There are new technologies for outbreak investigation (Srikantiah et al. 2005; Chester et al. 2011), data-based notifications (Kahn and Kinsolving 2010), new vaccine technologies (Rauch et al. 2018), enhanced information and communication technologies (ICTs) (Tom-Aba et al. 2015), informatics for detection (Zeng et al. 2005), various spatio-temporal methods for detection (Jiang and Cooper 2010), algorithms for detection (Buckeridge et al. 2005; Li et al. 2012), automated surveillance (Cassa et al. 2006; Buckeridge 2007), and many more methods and instrumental mechanisms that involve the use of data management, information systems, and digital platforms. As highlighted by Houlihan and Whitworth (2019, p. 142), in comparison to previous events, *"outbreak response has made significant and impressive progress involving a wider range of disciplines, embracing modern technology, and recognising the importance of research during and in between outbreaks"*. Hence, we can argue in a positive perspective that disruption provides a chance for innovation; a chance for the trial of those innovative interventions, digital-based platforms, community enhancement, and multi-sectoral management. Moreover, such disruptive situations provide an opportunity to create synergies in management and operations, identify the city's weaknesses and deficiencies, and find means of resilience. It is almost like conducting a large scale SWOT analysis of the city's resilience, only that there is no analysis of such kind but strategic (and usually immediate) actions through preparedness, planning, and responsiveness.

Finally, while in most cases we cannot simply stop a disease outbreak from happening, we have to ensure we are resilient enough against any possible adversity. In the first steps, we can increase the effectiveness of our detection procedures, enhance precautionary measures, data collection, observation measures, and prepare a reflective and feasible action plan. This preparedness can happen through resilience enhancement, and it should be comprehensive enough to include multiple aspects of our cities and communities. In doing so, we could boost the effectiveness of our responsiveness in disruptive disease outbreak events. There may be some losses and there may be some necessary adjustments or rearrangements. But above all, it is better to have a bit of loss at first than to deal with longer-term adversity.

References

100 Resilient Cities. (2018). *What is urban resilience?* Retrieved February 2, 2020, from https://www.100resilientcities.org/resources/.

Abdulkareem, M., & Elkadi, H. (2018). From engineering to evolutionary, an overarching approach in identifying the resilience of urban design to flood. *International Journal of Disaster Risk Reduction, 28,* 176–190.

Acuti, D., Bellucci, M., & Manetti, G. (2020). Company disclosures concerning the resilience of cities from the Sustainable Development Goals (SDGs) perspective, *Cities, 99,* 102608 (In press).

Adger, W. N. (2000). Social and ecological resilience: Are they related? *Progress in Human Geography, 24*(3), 347–364.

Adger, W. N., Brown, K., Nelson, D. R., Berkes, F., Eakin, H., Folke, C., et al. (2011). Resilience implications of policy responses to climate change. *WIREs Climate Change, 2*(5), 757–766.

Admiraal, H., & Cornaro, A. (2018a). *Underground spaces unveiled: Planning and creating the cities of the future.* London, UK: ICE Publishing.

Admiraal, H., & Cornaro, A. (2018b). Future cities, resilient cities—The role of underground space in achieving urban resilience. *Underground Space*, from 2019 Tongji University and Tongji University Press. Production and hosting by Elsevier B.V., pp. 2467–9674.

AECOM. (2020). *Risk management & resilience.* Retrieved February 26, 2020, from https://www. aecom.com/services/risk-management-resilience/.

Ahern, J. F. (2011). From fail-safe to safe-to-fail: Sustainability and resilience in the new urban world. *Landscape and Urban Planning, 100*(4), 341–343.

Al-Abri, S. S., Kurup, P. J., Al Manji, A., Al Kindi, H., et al. (2020). Control of the 2018–2019 dengue fever outbreak in Oman: A country previously without local transmission. *International Journal of Infectious Diseases, 90*, 97–103.

Alberta Municipal Affairs. (2015). *Building community resilience through asset management: A handbook and toolkit for Alberta municipalities.* Retrieved February 26, 2020, from http://www. municipalaffairs.alberta.ca/documents/ms/2015-11-17_Handbook_-_FINAL.PDF.

Alberti, M., & Marzluff, J. M. (2004). Ecological resilience in urban ecosystems: Linking urban patterns to human and ecological functions. *Urban Ecosystems, 7*, 241.

Alcamo, J., Hassan, R., Scholes, R., Ash, N., Carpenter, S., Pingali, P., et al. (2003). *Ecosystems and human well-being: A framework for assessment (Millennium ecosystem assessment).* Washington, DC: Island Press.

Alexander, D. E. (2013). Resilience and disaster risk reduction: An etymological journey. *Natural Hazards and Earth Systems Sciences, 13*(11), 2707–2716.

Algan, Y. (2013). *Public policy and trust*, from Organisation for Economic Co-operation and Development (OECD) workshop on "joint learning for an OECD Trust strategy". Retrieved February 27, 2020, from https://www.oecd.org/governance/publicationsdocuments/articles/6/.

Almklov, P., Antonsen, S., & Fenstad, J. (2012). Organizational challenges regarding risk management in critical infrastructures. In P. Hokstad, I. B. Utne, & J. Vatn (Eds.), *Risk and interdependencies in critical infrastructures: A guideline for analysis* (pp. 211–226). London: Springer.

Alterman, R. (1988). Adaptive planning. *Cognitive Science, 12*, 393–421.

Alwidyan, M. T., Trainor, J. E., & Bissell, R. A. (2020). Responding to natural disasters vs. disease outbreaks: Do emergency medical service providers have different views? *International Journal of Disaster Risk Reduction, 44*, 101440.

Amadeo, K. (2020). *SARS, Ebola, Coronavirus: How disease outbreaks affect the economy*, the balance news material. Retrieved February 27, 2020, from https://www.thebalance.com/corona virus-plague-ebola-economic-impact-4795744.

Ambat, A. S., Zubair, S. M., Prasad, N., Pundir, P., Rajwar, E., Patil, D. S., et al. (2019). Nipah virus: A review on epidemiological characteristics and outbreaks to inform public health decision making. *Journal of Infection and Public Health, 12*(5), 634–639.

Anderies, J. M. (2014). Embedding built environments in social-ecological systems: Resilience-based design principles. *Building Research and Information, 42*(2), 130–142.

Anderies, J. M., Folke, C., Walker, B., & Ostrom, E. (2013). Aligning key concepts for global change policy: Robustness, resilience, and sustainability. *Ecology and Society, 18* (2: 8), 1–16.

Anderson, K. L. (2009). Norovirus outbreak management in a resident-directed care environment. *Geriatric Nursing, 30*(5), 318–328.

Angeon, V., & Bates, S. (2015). Reviewing composite vulnerability and resilience indexes: A sustainable approach and application. *World Development, 72*, 140–162.

Ansumana, R., Keitell, S., Roberts, G. M. T., Ntoumi, F., Peterson, E., Ippolito, G., et al. (2017). Impact of infectious disease epidemics on tuberculosis diagnostic, management, and prevention services: experiences and lessons from the 2014–2015 Ebola virus disease outbreak in West Africa. *International Journal of Infectious Diseases, 56*, 101–104.

Argyroudis, S. A., Mitoulis, S. A., Hofer, L., Zanini, M. A., Tubaldi, E., & Frangopol, D. M. (2020). Resilience assessment framework for critical infrastructure in a multi-hazard environment: Case study on transport assets. *Science of The Total Environment, 714,* 136854.

Asian Development Bank (ADB). (2014). *Urban climate change resilience: A synopsis.* Manila: Asian Development Bank.

Aswi, A., Cramb, S., Duncan, E., Hu, W., White, G., & Mengersen, K. (2020). Climate variability and dengue fever in Makassar, Indonesia: Bayesian spatio-temporal modelling. *Epidemiology,* 100335 (In press).

Athar, M. N., Khalid, M. A., Ahmad, A. N., Bashir, N., Baqai, H. Z., Ahmad, M., et al. (2005). Crimean-Congo hemorrhagic fever outbreak in Rawalpindi, Pakistan, February 2002: Contact tracing and risk assessment. *The American Journal of Tropical Medicine and Hygiene, 72*(4), 471–473.

Atkinson, P. A., Martin, C. R., & Rankin, J. (2009). Resilience revisited. *Journal of Psychiatric and Mental Health Nursing, 16*(2), 137–145.

Awalime, D. K., Bernard, B., Davies-Teye, K., Vanotoo, L. A., Owoo, N. S., & Nketiah-Amponsah, E. (2017). Economic evaluation of 2014 cholera outbreak in Ghana: a household cost analysis. *Health Economics Review, 7*(45), 1–8.

Ban, K. (2012). *The future we want.* Article from The New York Times. Retrieved from February, 23, 2020.

Berg, P. G., & Nycander, G. (1997). Sustainable neighbourhoods—A qualitative model for resource management in communities. *Landscape and Urban Planning, 39*(2–3), 117–135.

Beutels, P., Jia, N., Zhou, Q.-Y., Smith, R., Cao, W-C., & De Vlas, S. J. (2009). The economic impact of SARS in Beijing, China. *Tropical Medicine and International Health, 14*(1, Special Issue: SARS in China), 85–91.

Biosecurity Incident National Communication Network. (2019). *Media arrangements during the outbreak of an agricultural pest or disease.* A document produced by the Australian Government. Retrieved February 27, 2020, https://www.outbreak.gov.au/sites/default/files/documents/media-guidelines-outbreak.pdf.

Bloom, D. E., & Canning, D. (2006). Epidemics and economics. PGDA Working Paper No. 9. Retrieved February 27, 2020, from https://cdn1.sph.harvard.edu/wp-content/uploads/sites/1288/2013/10/BLOOM_CANNINGWP9.2006.pdf.

Bloom, D. E., & Mahal, A. S. (1997). Does the AIDS epidemic threaten economic growth? *Journal of Econometrics, 77*(1), 105–124.

Bloom, D. E., Canning, D., & Sevilla, J. P. (2004). The effect of health on economic growth: A production function approach. *World Development, 32*(1), 1–13.

Boin, A., & McConnell, A. (2007). Preparing for critical infrastructure breakdowns: The limits of crisis management and the need for resilience. *Journal of Contingencies and Crisis Management, 15*(1), 50–59.

Bosher, L. (2008). *Hazards and the built environment: Attaining built-in resilience.* Abingdon: Taylor & Francis.

Bosher, L. (2014). Built-in resilience through disaster risk reduction: Operational issues. *Building Research and Information, 42*(2), 240–254.

Bosher, L. S., & Dainty, A. R. (2011). Disaster risk reduction and 'built-in' resilience: Towards overarching principles for construction practice. *Disasters, 35*(1), 1–18.

Bosher, L., Dainty, A., Carillo, P., & Glass, J. (2007). Built-in resilience to disasters: A preemptive approach. *Eng Constr Archit Management, 14*(5), 434–446.

Boyd, E., Nykvjist, B., Borgström, S., & Stacewicz, I. (2015). Anticipatory governance for social-ecological resilience. *Ambio, 44*(Suppl 1), 149–161.

Brassett, J., & Vaughan-Williams, N. (2015). Security and the performative politics of resilience: Critical infrastructure protection and humanitarian emergency preparedness. *Security Dialogue, 46*(1), 32–50.

Briggs, R., Schulter, M., Briggs, D., Bohensky, E. L., et al. (2012). Toward principles for enhancing the resilience of ecosystem services. *Annual Review of Environment and Resources, 37,* 421–448.

Briggs, R., Schluter, M., & Schoon, M. L. (2015). *Principles of building resilience: Sustaining ecosystem services in social-ecological system*. Cambridge: Cambridge University Press.

Brunetta, G., Caldarice, O., & Tollin, N. (Eds.). (2019). *Urban resilience for risk and adaptation governance: Theory and practice.*, Resilient cities: Re-thinking Urban transformation Singapore: Springer.

Buckeridge, D. L. (2007). Outbreak detection through automated surveillance: A review of the determinants of detection. *Journal of Biomedical Informatics, 40*(4), 370–379.

Buckeridge, D. L., Burkom, H., Campbell, M., Hogan, W. R., & Moore, A. W. (2005). Algorithms for rapid outbreak detection: A research synthesis. *Journal of Biomedical Informatics, 38*(2), 99–113.

Burns, T. (2013). *Trust and education*, from Organisation for Economic Co-operation and Development (OECD) workshop on "joint learning for an OECD Trust strategy". Retrieved February, 27, 2020, from: https://www.oecd.org/governance/publicationsdocuments/articles/6/.

Bush, J., & Doyon, A. (2019). Building urban resilience with nature-based solutions: How can urban planning contribute? *Cities, 95,* 102483.

Cariolet, J.-M., Vuillet, M., & Diab, Y. (2019). Mapping urban resilience to disasters—A review. *Sustainable Cities and Society, 51,* 101746.

Cassa, C. A., Grannis, S. J., Overhage, J. M., & Mandi, K. D. (2006). A context-sensitive approach to anonymizing spatial surveillance data: Impact on outbreak detection. *Journal of the American Medical Informatics Association, 13*(2), 160–165.

Cassady, J. D., Higgins, C., Manizer, H. M., Seys, S. A., Sarisky, J., Callahan, M., et al. (2006). Beyond compliance: Environmental health problem solving, interagency collaboration, and risk assessment to prevent waterborne disease outbreaks. *Journal of Epidemiology and Community Health, 60*(8), 649.

Centers for Disease Control and Prevention (CDC). (2016). *Cost of the Ebola epidemic*. Retrieved February 27, 2020, from https://www.cdc.gov/vhf/ebola/pdf/cost-ebola-infographic.pdf.

Centers for Disease Control and Prevention (CDC). (2018). *Manual and tools, resources for emergency health professionals*, under 'Crisis and Emergency Risk Communication'. Retrieved February 27, 2020, from https://emergency.cdc.gov/cerc/resources/index.asp.

Charles-Smith, L. E., Reynolds, T. L., Cameron, M. A., Conway, M., Lau, E. H. Y., Olsen, J. M., et al. (2015). Using social media for actionable disease surveillance and outbreak management: A systematic literature review. *PLoS ONE, 10*(10), e0139701.

Chaturverdi, K., Matheus, A., Nguyen, S. H., & Kolbe, T. H. (2019). Securing spatial data infrastructures for distributed smart city applications and services. *Future Generation Computer Systems, 101,* 723–736.

Chelleri, L. (2012). From «The Resilient City» to urban resilience: A review essay on understanding and integrating the resilience perspective for urban systems documents. *d'Anàlisi Geogràfica, 58.*

Chelleri, L., & Olazabal, M. (Eds.). (2012). *Multidisciplinary perspectives on urban resilience: A workshop report* (p. 100). Basque centre for Climate Change (BC3), Bilbao, Spain.

Chen, K., Zhang, Y., Zhan, Y., Fan S., & Si, W. (2020, February 20). *How China can address threats to food and nutrition security from the Coronavirus outbreak*, professional article on IFPRI Blog. Retrieved February 29, 2020, from https://www.ifpri.org/blog/how-china-can-address-threats-food-and-neutrition-security-coronovirus-outbreak.

Cheshmehzangi, A. (2016). Multi-spatial environmental performance evaluation towards integrated urban design: A procedural approach with computational simulations. *Journal of Cleaner Production, 139,* 1085–1093.

Cheshmehzangi, A. (2020). *Comprehensive urban resilience for the city of Ningbo* (in Chinese: 宁波市城市综合抗灾弹性框架). Report submitted to local government units in February 2020, Ningbo, China.

Cheshmehzangi, A., & Dawodu, A. (2018). *Sustainable urban development in the age of climate change—People: The cure or curse*. Singapore: Palgrave Macmillan.

Chester, T. L. S., Taylor, M., Sandhu, J., Forsting, S., Ellis, A., Stirling, R., & Galanis, E. (2011). Use of web forum and an online questionnaire in the detection and investigation of an outbreak. *Online Journal of Public Health Informatics, 3*(1), ojphi.v3i1.3506.

Chirambo, R. M., Songolo, P., Masaninga, F., & Kazembe, L. N. (2019). Mumps outbreak in Copperbelt province, Zambia: Epidemiological characteristics. *Clinical Epidemiology and Global Health, 7*(3), 325–330.

Choi, J. S., & Kim, K. M. (2016). Crisis prevention and management by infection control nurses during the Middle East respiratory coronavirus outbreak in Korea. *American Journal of Infection Control, 44*(4), 480–481.

Cities Development Initiative for Asia (CDIA). (2010). *City infrastructure investment programming & prioritisation toolkit: User manual.* Retrieved February 28, 2020, from https://cdia.asia/wp-content/uploads/2014/09/CDIA-toolkit-project-programming-prioritization_2010.pdf.

Coaffee, J. (2013). Towards next-generation urban resilience in planning practice: From securitization to integrated place making. *Planning Practice & Research, 28*(3). Deconstructing planning and resilience: Lessons in translating theory to practice, pp. 323–339.

Coaffee, J., & Bosher, L. (2008). Integrating counter-terrorist resilience into sustainability. *Proceedings of ICE, 161*(2), 75–83.

Coaffee, J., & Clarke, J. (2017). Critical infrastructure lifelines and the politics of anthropocentric resilience. *Resilience, 5*(3), 161–181.

Cohen, J. (2018). *Congo's new Ebola outbreak is hitting health care workers hard.* Online article. Retrieved February 28, 2020, from https://www.sciencemag.org/news/2018/08/congo-s-new-ebola-outbreak-hitting-health-care-workers-hard.

Collinson, S., Khan, K., & Heffernan, J. M. (2015). The effects of media reports on disease spread and important public health measurements. *PLoS One, 10*(11), e0141423.

Condon, B. J., & Sinha, T. (2010, April). Who is that masked person: The use of face masks on Mexico City public transportation during the Influenza A (H1N1) outbreak. *Health Policy, 95*(1), 50–56.

Covello, V. (2002). *Risk and crisis communication: 77 questions commonly asked by journalists during a crisis.* New York: Center for Risk Communication. Retrieved February 27, 2020, from http://www.nwcphp.org/docs/pdf/journalist.pdf.

Coyle, D. (2013). *Why is trust in government important today?* From Organisation for Economic Co-operation and Development (OECD) workshop on "joint learning for an OECD Trust strategy". Retrieved February 27, 2020, from https://www.oecd.org/governance/publicationsdocuments/articles/6/.

Crowe, P. R., Foley, K., & Collier, M. J. (2016). Operationalizing urban resilience through a framework for adaptive co-management and design: Five experiments in urban planning practice and policy. *Environmental Science and Policy, 62*, 112–119.

Curley, M., & Thomas, N. (2004). Human security and public health in Southeast Asia: The SARS outbreak. *Australian Journal of International Affairs, 58*(1), 17–32.

Cutter, S. L., Burton, C. G., & Emrich, C. T. (2010). Disaster resilience indicators for benchmarking baseline conditions. *Journal of homeland security and emergency management, 7*(1).

da Silva, J., Kernaghan, S., & Luque, A. (2012). A systems approach to meeting the challenges of urban climate change. *International Journal of Urban Sustainable Development.* https://doi.org/10.1080/19463138.2012.718279.

Danial, J., Ballard-Smith, S., Horsburgh, C., Crombie, C., Ovens, A., Templeton, K. E., et al. (2016). Lessons learned from a prolonged and costly norovirus outbreak at a Scottish medicine of the elderly hospital: Case study. *Journal of Hospital Infection, 93*(2), 127–134.

Davidson, K., Phuong Nguyen, T. M., Beilin, R., & Briggs, J. (2019). The emerging addition of resilience as a component of sustainability in urban policy. *Cities, 92*, 1–9.

Davoudi, S. (2014). Climate change, securitisation of nature, and resilient urbanism. *Environment Planning C, 32*, 360–375.

Davoudi, S., Shaw, K., Haider, L. J., Quinlan, A. E., Peterson, G. D., Wilkinson, C., et al. (2012). Resilience: a bridging concept or a dead end? "Reframing" resilience: Challenges for planning theory and practice interacting traps: Resilience assessment of a pasture management system in northern Afghanistan urban resilience. *Planning Theory and Practice, 13*(2), 299–333.

De Bruijne, M., & Van Eeten, M. (2007). Systems that should have failed: Critical infrastructure protection in an institutionally fragmented Environment. *Journal of Contingencies and Crisis Management, 15*(1), 18–29.

de Groot, M., Ogris, N., & Kobler, A. (2018). The effects of a large-scale ice storm event on the drivers of bark beetle outbreaks and associated management practice. *Forest Ecology and Management, 408*, 195–201.

Deng, W., & Cheshmehzangi, A. (2018). *Eco-development in China: Cities, communities, and buildings*. Singapore: Palgrave Macmillan.

Doménech-Sánchez, A., Laso, E., Pérez, M. J., & Berrocal, C. I. (2020). Efficient management of a norovirus outbreak causing gastroenteritis in two hotels in Spain, 2014 (in Spanish: Control de un brote de gastroenteritis causado por norovirus en 2 hoteles en España, 2014). *Enfermedades Infecciosas y Microbiología Clínica* (In press).

Dyakova, M., Hamelmann, C., Bellis, M. A., Besnier E., Grey, C. N. B., Ashton, K., Schwappach, A., & Clar, C. (2017). *Investment for health and well-being: Review of social return on investment from public health policies to support implementing the Sustainable Development Goals by building on Health 2020*. Health Evidence Network synthesis report 51, by World Health Organisation (WHO) regional office for Europe.

Elston, J. W. T., Moosa, A. J., Moses, F., Walker, G., Dotta, N., Waldman, R. J., et al. (2016). Impact of the Ebola outbreak on health systems and population health in Sierra Leone. *Journal of Public Health, 38*(4), 673–678.

Eraydin, A., & Taşan-Kok, T. (Eds.). (2013). *Resilience thinking in urban planning* (Vol. 106)., GeoJournal Library Dordrecht: Springer.

Ernstson, H., van der Leeuw, S. E., Redman, C. L., Meffert, D. J., Davis, G., Alfsen, C., et al. (2010). Urban transitions: On urban resilience and human-dominated ecosystems. *Ambio, 39*(8), 531–545.

Etinay, N., Egbu, C., & Murray, V. (2018). Building urban resilience for disaster risk management and disaster risk reduction. *Procedia Engineering, 212*, 575–582.

European Commission. (2006). *European Programme for Critical Infrastructure Protection (EPCIP)*, COM 786 final. Brussels: European Commission.

Fabbricatti, K., & Biancamano, P. F. (2019). Circular economy and resilience thinking for historic urban landscape regeneration: The Case of Torre Annunziata, Naples. *Sustainability, 11*(3391), 1–29.

Fan, V. Y., Jamison, D. T., & Summers, L. H. (2017). Pandemic risk: How large are the expected losses? *Bulletin of the World Health Organization*, Article ID BLT.17.1995881.

Fekete, A., & Fiedrich, F. (Eds.). (2018). *Urban disaster resilience and security*. The Urban Book Series. Singapore: Springer.

Fiksel, J. (2003). Designing resilient, sustainable systems. *Environment Science Technology, 37*(23), 5330–5339.

Firestone, S. M., Christley, R., M., Ward, M. P., & Dhang, N. K. (2012). Adding the spatial dimension to the social network analysis of an epidemic: Investigation of the 2007 outbreak of equine influenza in Australia. *Preventive Veterinary Medicine, 106*(2), 123–135.

Folke, C., Carpenter, S., Elmqvist, T., Gunderson, L., Holling, C. S., & Walker, B. (2002). Resilience and sustainable development: Building adaptive capacity in a world of transformations. *Ambio, 31*(5), 437–440.

Forland, F., De Carvalho Gomes, H., Nokleby, H., Escriva, A., Coulombier, D., Giesecke, J., et al. (2012). Applicability of evidence-based practice in public health: Risk assessment on Q fever under an ongoing outbreak. *Eurosurveillance, 17*(3), 1–6.

Ghebrehewet, S., Torrtington, D., Farmer, S., Kearney, J., Blissett, D., McLeod, H., et al. (2016). The economic cost of measles: Healthcare, public health and societal costs of the 2012–13 outbreak in Merseyside, UK. *Vaccine, 34*(15), 1823–1831.

Global Health Security Agenda. (2017). Retrieved February 2020, from https://www.ghsagenda.org.

Godefroy, R., Chaud, P., Ninove, L., Dina, J., Decoppet, A., Casha, P., et al. (2018). Measles outbreak in a French Roma community in the Provence-Alpes-Côte d'Azur region, France, May to July 2017. *International Journal of Infectious Diseases, 76,* 97–101.

Gomes Ribeiro, P., & Gonçalves, L. A. P. J. (2019). Urban resilience: A conceptual framework. *Sustainable Cities and Society, 50,* 101625.

Grafe, C., J., Staes, C. J., Kawamoto, K., Samore, M. H., & Evans, S. (2018). State-level adoption of national guidelines for norovirus outbreaks in health care settings. *American Journal of Infection Control, 46*(10), 1084–1091.

Guillier, L., Danan, C., Bergis, H., Deligenette-Muller, M.-L., Granier, S., Rudelle, S., et al. (2013). Use of quantitative microbial risk assessment when investigating foodborne illness outbreaks: The example of a monophasic Salmonella Typhimurium 4,5,12:i:—Outbreak implicating beef burgersm. *International Journal of Food Microbiology, 166*(3), 471–478.

Haase, D., Fratzeskaki, N., & Elmqvist, T. (2014). Ecosystem services in urban landscapes: Practical applications and governance implications. *Ambio, 43*(4), 407–412.

Hajer, M., & Dassen T. (2014). *Smart cities: Illustrating the task for 21st-century Urban planning <<original: Slimme Steden: de opgave voor de 21e-eeuwse stedenbouw in beeld>>*, Nai010, Rotterdam, the Netherlands (In Dutch).

Hall, A. J., Vinje, J., Lopman, B., Park, G. W., Yen, C., Gregoricus, N., et al. (2011). Updated norovirus outbreak management and disease prevention guidelines. *Morbidity and Mortality Weekly Report: Recommendations and Reports, 60*(3), 1–15.

Han, M. S., Chung, S. M., Kim, E. J., Lee, C. J., Yun, K. W., Choe, P. G., Kim, N. J., & Choi, E. H. (2020). Successful control of norovirus outbreak in a pediatric ward with multi-bed rooms, *American Journal of Infection Control, 48*(3), 297–303.

Hassani, H., Yeganegi, M. R., Silva, E. S., & Ghodsi, F. (2019). Risk management, signal processing and econometrics: A new tool for forecasting the risk of disease outbreaks. *Journal of Theoretical Biology, 467,* 57–62.

Hassler, U., & Kohler, N. (2014). The ideal of resilient systems and questions of continuity. *Building Research and Information, 42*(2), 158–167.

Heinzlef, C., Becue, V., & Serre, D. (2019). Operationalizing urban resilience to floods in embanked territories—Application in Avignon. *Provence Alpes Côte d'azur region, Safety Science, 118,* 181–193.

Hepburn, S. L. (2017). Chapter Five—Strengthening informal supports to promote behavioral health of youth with intellectual and/or developmental disabilities in rural communities. *International Review of Research in Developmental Disabilities, 53,* 203–234.

Hiller, U., Mankertz, A., Köneke, N., & Wicker, S. (2019). Hospital outbreak of measles—Evaluation and costs of 10 occupational cases among healthcare worker in Germany, February to March 2017. *Vaccine, 37*(14), 1905–1909.

Hogg, D., Kingham, S., Wilson, T. M., & Ardagh, M. (2016). Spatio-temporal variation of mood and anxiety symptom treatments in Christchurch in the context of the 2010/11 Canterbury earthquake sequence. *Spatial and Spatio-temporal Epidemiology, 19,* 91–102.

Hopkins, R. (2009). *Resilience thinking.* London, UK: Resurgence.

Hossain, L., Kam, D., Kong, F., & Wigand, R. T. (2016). Social media in Ebola outbreak. *Epidemiology and Infection, 144*(10), 2136–2143.

Houlihan, C., & Whitworth, J. (2019). Outbreak science: Recent progress in the detection and response to outbreaks of infectious diseases. *Clinical Medicine, 19*(2), 140–144.

Huang, L., Wu, J., & Yan, L. (2015). Defining and measuring urban sustainability: A review of indicators. *Landscape Ecology, 30,* 1175–1193.

Huber, C., Finelli, L., & Stevens, W. (2018). The economic and social burden of the 2014 Ebola outbreak in West Africa. *The Journal of Infectious Diseases, 218*(5), S698–S704.

Huck, A., & Monstadt, J. (2019). Urban and infrastructure resilience: Diverging concepts and the need for cross-boundary learning. *Environmental Science & Policy, 100*, 211–220.

Huitt, W. (2004). *Maslow's hierarchy of needs, educational psychology interactive.* Retrieved February 23, 2020, from http://ways-ahead.net/meditation/037-2-Maslow.pdf.

ICLEI webpage. *Local governments for sustainability: Cities for climate protection programme.* Retrieved February 20, 2020, from www.iclei.org.

Institute for Social and Environmental Transition (ISET), MercyCorps, URDI and CCROM SEAP-IPB. (2010). *Vulnerability and adaptation assessment to climate change in Bandar Lampung City.* Retrieved February 24, 2020, from http://www.acccrn.org/sites/default/files/documents/ACCCRN_lampung_ENG_26APRIL2010.pdf.

Intergovernmental Panel on Climate Change (IPCC). (2007). *Climate change 2007: Synthesis report.* Geneva, Switzerland: UN-IPCC.

Interior Health website. *Active facility outbreaks.* Retrieved February 29, 2020, from https://www.interiorhealth.ca/YourEnvironemnt/CommunicableDiseaseControl/Outbreaks/Pages/default.aspc.

Jardine, A. K. S., & Tsang, A. H. C. (1973). *Maintenance, replacement, and reliability: Theory and applications* (2nd ed., 2013 by CRC Publishers), Lanham, USA: Pitman Publishing.

Jiang, X., & Cooper, G. F. (2010). A Bayesian spatio-temporal method for disease outbreak detection. *Journal of American Medical Informatics Association, 17*(4), 462–471.

Jiang, L., He, S., & Zhou, H. (2020). Spatio-temporal characteristics and convergence trends of PM2.5 pollution: A case study of cities of air pollution transmission channel in Beijing-Tianjin-Hebei region, China. *Journal of Cleaner Production*, 120631 (In press).

Jin, H. K., & Kim, J.-C. (2008). The effects of the BSE outbreak on the security values of US agribusiness and food processing firms. *Applied Economics, 40*(3), 357–372.

Johnston, C. P., Qiu, H., Ticehurst, J. R., Dickson, C., et al. (2007). Outbreak management and implication of a nosocomial norovirus outbreak. *Clinical Infectious Diseases, 45*(5), 534–540.

Jung, Y., Jang, H., & Matthews, K. R. (2014). Effect of the food production chain from farm practices to vegetable processing on outbreak incidence. *Microbial Biotechnology, 7*(6), Retrieved February 29, 2020, from https://sfamjournals.onlinelibrary.wiley.com/doi/full/10.1111/1751-7915.12178.

Kabisch, S., Koch, F., & Gawel, E. (Eds.). (2018). *Urban transformations: Sustainable urban development through resource efficiency, quality of life, and resilience* (Vol. 10). Future City. Singapore: Springer.

Kahn, P., & Kinsolving, A. (2010). *Methods and apparatus to provide outbreak notifications based on historical location data.* US Patent, 7, pp. 705–723.

Katko, T. S., Juuti, P. S., & Schwartz, K. (Eds.). (2013). *Water services management and governance: Lessons for sustainable future.* London: IWA Publishing.

Kershaw, T. (2018). *Climate change resilience in urban environments.* IOP Expanding Physics, Bristol, UK: IOP Publishing.

Klein, R. J. T., Nicholls, R. J., & Thomalla, F. (2003). Resilience to natural hazards: How useful is this concept? *Environmental Hazards, 5*(1), 35–45.

Klopp, J. M., & Petretta, D. L. (2017). The urban sustainable development goal: Indicators, complexity and the politics of measuring cities. *Cities, 63*, 92–97.

Knieling, J. (2016). *Climate adaptation in cities and regions.* Oxford: Wiley Blackwell.

Kombugabe-Dixson, A. F., de Ville, N. S. E., Trundle, A., & McEvoy, D. (2019). Environmental change, urbanisation, and socio-ecological resilience in the Pacific: Community narratives from Port Vila, Vanuatu. *Ecosystem Services, 39*, 100973.

Korte, S., Pettke, A., Kossow, A., Mellmann, A., Willems, S., & Kipp, F. (2016). Norovirus outbreak management: How much cohorting is necessary? *Journal of Hospital Infection, 92*(3), 259–262.

Kott, A., & Limaye, R. J. (2016). Delivering risk information in a dynamic information environment: Framing and authoritative voice in Centers for Disease Control (CDC) and primetime broadcast news media communications during the 2014 Ebola outbreak. *Social Science and Medicine, 169*, 42–49.

Kruk, M., E., Myers, M., Varpilah, S., T., & Dahn, B. T. (2015). What is a resilient health system? Lessons from Ebola. *The Lancet, 385*(9980), 1910–1912.

Kurup, K. K., John, D., Ponnaiah, M., & George, T. (2019). Use of systematic epidemiological methods in outbreak investigations from India, 2008–2016: A systematic review. *Clinical Epidemiology and Global Health, 7*(4), 648–653.

Leichenko, R. (2011). Climate change and urban resilience. *Current Opinion in Environmental Sustainability, 3*(3), 164–168.

Levin-Rector, A., Nivin, B., Yeung, A., Fine, A., D., & Greene, S. K. (2015). Building-level analyses to prospectively detect influenza outbreaks in long-term care facilities: New York City, 2013–2014. *American Journal of Infection Control, 43*(8), 839–843.

Levy, B., Edholm, C., Gaoue, O., Kaodera-Shava, R., Kgosimore, M., Lenhart, S., et al. (2017). Modeling the role of public health education in Ebola virus disease outbreaks in Sudan. *Infectious Disease Modelling, 2*(3), 323–340.

Li, Z., Lai, S., Buckeridge, D. L., Zhang, H., Lan, Y., & Yang, W. (2012). Adjusting outbreak detection algorithms for surveillance during epidemic and non-epidemic periods. *Journal of the American Medical Informatics Association, 19*(e1), e51–e53.

Liao, K.-H., Le Anh, T., & Van Nguyen, K. (2016). Urban design principles for flood resilience: Learning from the ecological wisdom of living with floods in the Vietnamese Mekong Delta. *Landscape and Urban Planning, 155,* 69–78.

Linkov, I., Eisenberg, D. A., Bates, M. E., Chang, D., Convertino, M., Allen, J. H., et al. (2013). Measurable resilience for actionable policy. *Environmental Science and Technology, 47,* 10108–10110.

Linkov, I., Bridges, T., Creutzig, F., Decker, J., et al. (2014). Changing the resilience paradigm. *Nature Climate Change, 4*(6), 407.

Lister, N.-M. (2016). *From reactive to proactive resilience: Designing the new sustainability.* Retrieved February 23, 2020, from http://www.thenatureofcities.com/2016/03/15/from-reactive-to-proactive-resilience-designing-the-new-sustainability/.

Lodeiro-Colatosti, A., Reischl, U., Holzmann, T., Hernández-Pereira, C. E., Rísquez, A., & Paniz-Mondolfi, A. E. (2018). Diphtheria outbreak in Amerindian communities, Wonken, Venezuela, 2016–2017. *Emerging Infectious Diseases, 24*(7), 1340–1344.

Lowe, J., Guager, P., Harmon, K., Zhang, J., Connor, J., Yeske, P., et al. (2014). Role of transportation in spread of Porcine epidemic diarrhea virus infection, United States. *Emerging Infectious Diseases, 20*(5), 872–874.

Lowe, C. F., Leung, V., Karakas, L., Merrick, L., Lawson, T., Romney, M. G., Ritchie, G., Payne, M., & Providence Health Care Infection Prevention and Control Team. (2019). Targeted management of influenza A/B outbreaks incorporating the cobas® Influenza A/B & RSV into the virology laboratory. *Journal of Hospital Infection, 101*(1), 38–41.

Lundberg, J., & Moberg, F. (2003). Mobile link organisms and ecosystem functioning: Implications for ecosystem resilience and management. *Ecosystems, 6*(1), 87–98.

Luo, Y. (2013). Assess the crisis management effort following the outbreak of Assess the crisis management effort following the outbreak of the subprime crisis. *Lingnan Journal of Banking, Finance and Economics, 4,* 2012/13 Academic Year Issue, Article 5, 1–8.

Lurie, N., & Fermont, A. (2009). Building bridges between health care and public health: A critical piece of the health reform infrastructure. *The Journal of the American Medical Association (JAMA), 302*(1), 84–86.

Luyten, J., & Beutels, P. (2009). Costing infectious disease outbreaks for economic evaluation: A review for hepatitis A. *Pharmacoeconomics, 27*(5), 379–389.

Olazabal, M. Chelleri, L., Waters, J., & Kunath, A. (2012). *Urban resilience: towards an integrated approach.* Conference Paper presented at the 1st International Conference on Urban Sustainability and Resilience, London, UK, ISSN 2051-1361.

Ma, L., Hong, H., Chen, K., Tu, S., Zhang, Y., & Liao, L. (2019). Green growth efficiency of Chinese cities and its spatio-temporal pattern. *Resources, Conservation and Recycling, 146,* 441–451.

Manyena, S. B. (2006). The concept of resilience revisited. *Disasters, 30*(4), 433–450.

Manyena, S. B., O'Brien, G., O'Keefe, P., & Rose, J. (2011). Disaster resilience: A bounce back or bounce forward ability? *Journal of Local Environment, 16*(5), 417–424.

Marana, P., Eden, C., Eriksson, H., Grimes, C., Hernantes, J., Howick, S., Labaka, L., Latinos, V., Lindner, R., Majchrzak, T. A., Pyrko, I., Radianti, J., Rankin, A. Sakurai, M., Sarriegi, J. M., & Serrano, N. (2019). Towards a resilience management guideline—Cities as a starting point for societal resilience. *Sustainable Cities and Society, 48,* 101531.

Martin, V. (2020). What impact could the Coronavirus epidemic have on agriculture and food security? Online article on *China Daily*. Retrieved February 29, 2020, from https://www.chinad aily.com/cn/a/202002/24/WS5e53af6fa310128217279e6d.html.

Marvin, S., & Medd, W. (2006). Metabolisms of obecity: Flows of fat through bodies, cities, and sewers. *Environmental Planning A, 38*(2), 313–324.

Maslow, A. (1971). *The farther reaches of human nature.* New York, NY: Viking Press.

McPhearson, T., Hamstead, Z. A., & Kremer, P. (2014). Urban ecosystem services for resilience planning and managemnet in New York City. *Ambio, 43,* 502–515.

Meerow, S., & Newell, J. P. (2016). Urban resilience for whom, what, when, where, and why? *Urban Geography, 40*(3). Geographic Perspectives on Urban Sustainability, 1–21.

Mercy Corps (2019). *Chapter 4: How does Ebola affect the economy?* Retrieved February 27, 2020, from https://www.mercycorps.org/blog/ebola-outbreaks-africa-guide/chapter-4.

Mian, J. (2018). *Putting the resilience building blocks in place—Every time.* A session presentation at the Institute of Asset Management's (IAM) Annual Conference 2018, Birmingham, UK. Retrieved February 25, 2020, from https://www.resilienceshift.org/asset-management-resilience/.

Minucci, G. (2012). In L. Chelleri, & M. Olazabal (Eds.), *Multidisciplinary perspectives on urban resilience: A workshop report.* Basque Centre for Climate Change (BC3), Bilbao, Spain.

Mitchell, T., & Harris, K. (2012). Resilience: A risk management approach, background note, Overseas Development Institute (ODI). Retrieved January 25, 2020, from https://www.odi.org/sites/odi.org.uk/files/odi-assets/publications-opinion-files/7552.pdf.

Mitchell, A., Gottfried, J., Barthel, M., & Shearer, E. (2016). *The modern news consumer: News attitudes and practices in the digital era.* Washington, DC: The Pew Research Center. Retrieved February 26, 2020, from http://www.journalism.org/2016/07/07/the-modern-news-consumer/.

MMI Engineering webpage. *Asset resilience.* Retrieved February 26, 2020, from https://www.mmi engineering.com/services/resilience/.

Moench, M., & Tyler, S. (Eds.). (2011). *Catalyzing urban climate resilience: Applying resilience concepts to planning practice in the ACCCRN program (2009–2011).* Bangkok, Thailand: ISET-Boulder.

Moffatt, S. (2014). Resilience and competing temporalities in cities. *Building Research and Information, 42*(2), 202–220.

Monstadt, J. (2009). Conceptualizing the political ecology of urban infrastructures: Insights from technology and urban studies. *Environmental Planning A, 41*(8), 1924–1942.

Monstadt, J., & Schmidt, M. (2019). Urban resilience in the making? The governance of critical infrastructures in German cities. *Urban Studies, 56*(11), 2353–2371.

Monteiro, A., Carvalho, V., Velho, S., & Sousa, C. (2012). Assessing and monitoring urban resilience using COPD in Porto. *Science of the Total Environment, 414,* 113–119.

Morgan, O. (2019). How decision makers can use quantitative approaches to guide outbreak responses. *Philosophical Transactions B, 374*(1776), 20180365.

Moser, C. (1998). The asset vulnerability framework: Reassessing urban poverty reduction strategies. *World Development, 26*(1), 1–19.

New South Wales (NSW) webpage. Section Justice: Office of Emergency Management, Infrastructure Resilience: Asset Management. Retrieved February 26, 2020, from https://www.emerge ncy.nsw.gov.au/Pages/emergency-management/local-government/nsw-critical-infrastructure-res ilience-strategy/resilience-priority-three-provide/infrastructure-resilience-asset-management. aspx.

Nunes, D. M., Pinherio, M. D., & Tomé, A. (2019). Does a review of Urban resilience allow for the support of an evolutionary concept? *Journal of Environmental Management, 244,* 422–430.

Ogie, R. I., Shukla, N., Sedlar, F., & Holderness, T. (2017). Optimal placement of water-level sensors to facilitate data-driven management of hydrological infrastructure assets in coastal mega-cities of developing nations. *Sustainable Cities and Society, 35,* 385–395.

Ohuabunwo, C., Ameh, C., Oduyebo, O., Ahumibe, A., Mutiu, B., et al. (2016). Clinical profile and containment of the Ebola virus disease outbreak in two large West African cities, Nigeria, July–September 2014. *International Journal of Infectious Diseases, 53,* 23–29.

Organisation for Economic Co-operation and Development (OECD). (2019). *Trust in Government.* Retrieved February 27, 2020, from http://www.oecd.org/gov/trust-in-government.htm.

Orsi, A., Butera, F., Piazza, M. F., Schenone, S., Canepa, P., Caligiuri, P., Arcuri, C., Bruzzone, B., Zoli, D., Mela, M., Sticchi, L., Ansaldi, F., & Icardi, G. (2019). Analysis of a 3-months measles outbreak in western Liguria, Italy: Are hospital safe and healthcare workers reliable? *Journal of Infection and Public Health* (In press).

Parpia, A. S., Skrip, L. A., Nsoesie, E. O., Ngwa, M. C., Abah Abah, A., S., Galvani, A. P., & Ndeffo-Mbah, M., L. (2019). Spatio-temporal dynamics of measles outbreaks in Csameroon. *Annals of Epidemiology* (In press).

Pascapurnama, D. N., Murakami, A., Chagan-Yasutan, H., Hattori, T., Sasaki, H., & Egawa, S. (2018). Integrated health education in disaster risk reduction: Lesson learned from disease outbreak following natural disasters in Indonesia. *International Journal of Disaster Risk Reduction, 29,* 94–102.

Pickett, S. T. A., Cadenasso, M. L., & Grove, J. M. (2004). Resilient cities: Meaning, models, and metaphor for integrating the ecological, socio-economic, and planning realms. *Landscape and Urban Planning, 69*(4), 369–384.

Pickett, S. T. A., Cadenasso, M. L., & McGrath, B. (Eds.). (2013). *Resilience in ecology and Urban design: Linking theory and practice for sustainable cities.* Future City 3, Singapore: Springer.

Pires Maciel, A. L., de Assis, D. B., Madalosso, G., & Padoveze, M. C. (2014). Evaluating the quality of outbreak reports on health care-associated infections in São Paulo, Brazil, during 2000–2010 using the ORION statement findings and recommendations. *American Journal of Infection Control, 42*(4), e47–e53.

Prager, F., Wei, D., & Rose, A. (2017). Total economic consequences of an influenza outbreak in the United States. *Risk Analysis, 37*(1), 4–19.

Qiu, W., Chu, C., Mao, A., & Wu, J. (2018). The impacts on health, society, and economy of SARS and H7N9 outbreaks in China: A case comparison study. *Hindawi, Journal of Environmental and Public Health, 2018,* Article ID 2710185, 7p.

Quah, S. R., & Lee, H.-P. (2004). Crisis prevention and management during SARS outbreak, Singapore. *Emerging Infectious Diseases, 10*(2), 364.

Rasoulkhani, K., Mostafavi, A., Cole, J., & Sharvelle, S. (2019). Resilience-based infrastructure planning and asset management: Study of dual and singular water distribution infrastructure performance using a simulation approach. *Sustainable Cities and Society, 48,* 101577.

Rauch, S., Jasny, E., Schmidt, K. E., & Petsch, B. (2018). New vaccine technologies to combat outbreak situations. *Frontiers in immunology, 9,* 1963.

Register, R. (2014). Much better than climate change adaptation. In S. Lehmann (Ed.), *Low carbon cities: Transforming urban systems* (pp. 75–84). Abingdon: Routledge.

Renner, R. (2018). *Urban being—Anatomy & identity of the city.* Salenstein, Switzerland: Niggli.

Resilience Alliance. (2017). *Resilience.* Retrieved February 23, 2020, from https://www.resalliance.org/resilience.

Reyes-Castro, P. A., Harris, R. B., Brown, H. E., Christopherson, G. L., & Ernst, K. C. (2017). Spatio-temporal and neighborhood characteristics of two dengue outbreaks in two arid cities of Mexico. *Acta Tropica, 167,* 174–182.

Reynolds, B., & Quinn Crouse, S. (2008). Effective communication during an influenza pandemic: The value of using a crisis and emergency risk communication framework. *Health Promot Pract, 9*(4 Suppl), 13S–17S.

Reynolds, B., & Seeger, M. (2014). *Crisis and emergency risk communications manual* (2014 ed.). Retrieved February, 26, 2020, from https://emergency.cdc.gov/cerc/manual/index.asp.

Richardson, E. T., Kelly, J. D., Sesay, O., Drasher, M. D., Desai, I. K., Frankfurter, R., et al. (2017). The symbolic violence of 'outbreak': A mixed-methods, quasi-experimental impact evaluation of social protection on Ebola survivor wellbeing. *Social Science and Medicine, 195,* 77–82.

Risa, K. J., McAndrew, J. M., & Muder, R. R. (2009). Influenza outbreak management on a locked behavioral health unit. *American Journal of Infection Control, 37*(1), 76–78.

Rizkalla, C., Arroyo, A., Zerzan, J., O'Keefe, M., Okereke, M., Dickman, E., Drapkins, J., & Marshall, J. (2020). Urban emergency department response to measles outbreak. *Annals of Emergency Medicine* (In press).

Roberts, D. (2010). Prioritizing climate change adaptation and local level resilience in Durban, South Africa. *Environment and Urbanization, 22*(2), 397–413.

Rodrigue, J.-P., Luke, T., & Osterholm, M. (2020). Transportation and Pandemics. In Rodrigue, J.-P. (Ed.), *The geography of transport systems* (5th ed.). New York: Routledge, Chapter 11. Retrieved February 28, 2020, from https://transportgeography.org/?page_id=8869.

Rosa, J. J., & Rosa, R. D. (2011). *Pedagogy in the age of media control: Language deception and digital democracy.* International Academic Publishers: Peter Lang Inc.

Sanchez, A. X., van der Heijden, J., & Osmond, P. (2018). The city politics of an urban age: Urban resilience, conceptualisations and policies. *Palgrave Communications, 4*(25), 1–12.

Sandman, P. (1989). *Hazard versus outrage in the public perception of risk.* In: V. T. Covello, D. B. McCallum, & M. T. Pavlova (Eds.), *Effective risk communications* (Vol. 4, pp. 45–49). Contemporary issues in risk analysis. New York: Plenum Press.

Santos, J., May, L., & Haimar, A. (2013). Risk-based input-output analysis of influenza epidemic consequences on interdependent work-force sectors. *Risk Analysis, 33*(9), 1620–1635.

Schewenius, M., McPhearson, T., & Elmqvist, T. (2014). Opportunities for increasing resilience and sustainability of urban social-ecological systems: Insights from the URBES and the cities and biodiversity outlook projects. *Ambio, 43*(4), 434–444.

Schoch-Spana, M., Watson, C., Ravi, S., Meyer, D., Pechta, L. E., Rose, D. A., Lubell, K. M., Podgornik, M. N., & Sell, T. K. (2020). Vector control in Zika-affected communities: Local views on community engagement and public health ethics during outbreaks. *Preventive Medicine Reports,* 101059 (In press).

Schuster, J., & Newland, J. G. (2015). Management of the 2014 Enterovirus 68 outbreak at a Pediatric Tertiary Care Center. *Clinical Therapeutics, 37*(11), 2411–2418.

Sharifi, A. (2019). Urban form resilience: A meso-scale analysis. *Cities, 93,* 238–252.

Sharpe, W. F. (1992). Asset allocation: Management style and performance measurement. *The Journal of Portfolio Management, 18*(2), 7–19.

Shears, P., & Garavan, C. (2020). The 2018/19 Ebola epidemic the Democratic Republic of the Congo (DRC): Epidemiology, outbreak control, and conflict. *Infection Prevention in Practice, 2*(1), 100038.

Siemens, Arup, & RPA. (2013). *Toolkit for resilient cities: Infrastructure, technology and urban planning,* 60 p. document. Retrieved January 24, 2020, from https://assets.new.siemens.com/siemens/assets/public.1543066657.641ee2256c5a0d5919d1aa3094a701f6ec9c3f90.toolkit-for-resilient-cities.pdf.

Singh, R. B. (Ed.). (2015). *Urban development challenges, risks and resilience in Asian mega cities.* Singapore: Springer.

Singh, R. B., Srinagesh, B., & Anand, S. (Eds.). (2020). *Urban health risk and resilience in Asian cities.* Singapore: Springer.

Singleton, C. D., Fey, R., & Appleby, C. (2000). Media management of a community outbreak of meningococcal meningitis. *Communicable Disease and Public Health, 3*(4), 267–270.

Smart Water Magazine. (2019). Integrated management of public services: The basis for a real smart city. Article from 25 April 2019. Retrieved February 26, 2020, from www.smartwatermagazine.com.

Smith, R. D. (2006). Responding to global infectious disease outbreaks: Lessons from SARS on the role of risk perception, communication and management. *Social Science and Medicine, 63*(12), 3113–3123.

Smith, A., & Stirling, A. (2010). The politics of social-ecological resilience and sustainable sociotechnical transitions. *Ecology and Society, 15*(1), 11–25.

Smith, K. M., Machalaba, C. C., Seifman, R., Feferholtz, Y., & Karesh, W. B. (2019). Infectious disease and economics: The case for considering multi-sectoral impacts. *One Health, 7,* 100080.

Srikantiah, P., Bodager, D., Toth, B., Kass-Hout, T., Hammond, R., Stenzel, S., et al. (2005). Web-based investigation of multistate salmonellosis outbreak. *Emerging Infectious Diseases, 11*(4), 610.

Sun, H., Zhen, F., Lobsang, T., & Li, Z. (2019). Spatial characteristics of Urban life resilience from the perspective of supply and demand: A case study of Nanjing, China. *Habitat International, 88,* 101983.

Tabibian, M., & Movahed, S. (2016). Towards resilient and sustainable cities: A conceptual framework. *Scientia Iranica, Transactions A: Civil Engineering, 23*(5), 2081–2093.

Tainter, J. A., & Taylor, T. G. (2014). Complexity, problem-solving, sustainability and resilience. *Building Research and Information, 42*(2), 168–181.

Taori, S. K., Khonyongwa, K., Hayden, I., Dushyanthie, G., Athukorala, A., Letters, A., et al. (2019). *Candida auris* outbreak: Mortality, interventions and cost of sustaining control. *Journal of Infection, 79*(6), 601–611.

The Rockefeller Foundation and Arup. (2014). *City Resilience Framework,* Retrieved February 2, 2020, from https://www.rockefellerfoundation.org/report/city-resilience-framework/.

The State Council (2020). *China takes measures to ensure transport for epidemic control.* Xinhua Report on 11 February 2020, China's official webpage. Retrieved February 28, 2020, from http://english.www.gov.cn/statecouncil/ministries/202002/11/content_WS5e426afd c6d0595e03c206c6.html.

The World Bank. (2012). *Building urban resilience: Principles, tools, and practice, specifically for managing the risks of disasters in East Asia and the Pacific.* Retrieved February 20, 2020, from https://www.gfdrr.org/sites/default/files/publication/EAP_handbook_principles_t ools_practice_web.pdf.

The World Bank Group. (2014). The economic impact of the 2014 Ebola epidemic: Short and medium term estimates for West Africa. Retrieved February 27, 2020, from http://documents.worldbank.org/curated/en/524521468141287875/pdf/912190WP0see0a0 0070385314B00PUBLIC0.pdf.

The World Organisation for Animal Health (OIE). (2007). *Prevention and control of animal diseases worldwide: Economic analysis—Prevention versus outbreak costs.* Retrieved February 27, 2020, from https://www.oie.int/doc/ged/D4309.PDF.

Thomson, D. K., Muriel, P., Russel, D., Osborne, P., Bromley, A., Rowland, N., et al. (2003). Economic costs of the foot and mouth disease outbreak in the United Kingdom in 2001. *Revue scientifique et technique (International Office of Epizootics), 21*(3), 675–687.

Thomson-Dyck, K., Mayer, B., Anderson, K. F., & Galaskiewicz, J. (2016). *Brining people back in: Crisis planning and response embedded in social contexts.* In Y. Yamagata, & H. Maruyama (Eds.), *Urban resilience.* Advanced sciences and technologies for security applications. Singapore: Springer.

Tom-Aba, D., Olaleye, A., et al. (2015). Innovative technological approach to Ebola virus disease outbreak response in Nigeria using the open data kit and form hub technology. *PLoS ONE, 10*(6), e0131000.

Tumpey, A. J., Daigle, D., & Nowak, G. (2018). *Communicating during an outbreak or public health investigation.* Online source Retrieved February 24, 2020, from https://www.cdc.gov/eis/ field-epi-manual/chapters/Communicating-Investigation.html.

UN Habitat. (2012). *Resilience.* Retrieved February 23, 2020, from https://unhabitat.org/urban-the mes/resilience/.

UN Habitat. (2018). *City Resilience Profiling Tool (CRPT).* Retrieved January 28, 2020, from http:// urbanresiliencehub.org/wp-content/uploads/2018/02/CRPT-Guide.pdf.

Ung, C. O. L. (2020). Community pharmacist in public health emergencies: Quick to action against the coronavirus 2019-nCoV outbreak. *Research in Social and Administrative Pharmacy* (In press).

United Nation Economic Commission for Africa (UN ECA). (2014). *Socio-economic impacts of the Ebola virus disease on Africa.* Retrieved February 26, 2020, from https://reliefweb.int/sites/reliefweb.int/files/resources/eca_ebola_report_final_eng.pdf.

United Nation's International Strategy for Disaster Reduction (UN/ISDR). (2014). *Disaster resilience scorecard for cities.* Working Document. United Nations International Strategy for Disaster Risk Reduction (UNISDR). Retrieved February 28, 2020, from http://www.unisdr.org/2014/campaign-cities/Resilience%20Scorecard%20V1.5.pdf.

United Nation's International Strategy for Disaster Reduction (UN/ISDR). (2017). *Disaster resilience scorecard for cities.* Preliminary level assessment. Retrieved February 28, 2020, from https://www.unisdr.org/campaign/resilientcities/assets/toolkit/Scorecard/UNDRR_Disaster%20resilience%20%20scorecard%20for%20cities_Preliminary_English.pdf.

Urban Resilience webpage. Retrieved February 22, 2020, from https://urbanresilience.net/.

US Department of Transportation. (2013). Risk-based transportation asset management: Building resilience into transportation assets, Report 5: Managing external threats through risk-based asset management, by Federal Highway Administration. Retrieved February 1, 2020, from https://www.fhwa.dot.gov/asset/pubs/hif13018.pdf.

Vale, L. J. (2014). The politics of resilient cities: Whose resilience and whose city? *Building Research and Information, 42*(2), 191–201.

van der Heijden, J. (2014). *Governance for urban sustainability and resilience: Responding to climate change and the role of the built environment.* Cheltenham: Edward Elgar Publishers.

Vogel, B., & Henstra, D. (2015). Studying local climate adaptation: A heuristic research framework for comparative policy analysis. *Global Environmental Change, 31,* 110–120.

Wang, C., Yu, E., Xu, B., Wang, W., Li, L., Zhang, W., et al. (2012). Epidemiological and clinical characteristics of the outbreak of 2009 pandemic influenza A (H1N1) at a middle school in Luoyang, China. *Public Health, 126*(4), 289–294.

Wang, Y., Wang, X., Li, H., Dong, Y., Liu, Q., & Shi, X. (2020). A multi-service differentiation traffic management strategy in SDN cloud data center. *Computer Networks, 171,* 107143.

Washington, H. (2015). *Demystifying sustainability: Towards real solutions.* London: Routledge/Earthscan.

Waters, J. J. (2012). Reconsidering resilience in rapidly urbanising areas. In L. Chelleri, & M. Olazabal (Eds.), *Multidisciplinary perspectives on Urban resilience: A workshop report* (pp. 59–65). Basque Centre for Climate Change (BC3), Bilbao, Spain.

Watson, C. (1993). *Public communication in the management of an outbreak of infectious disease* (Vol. 4, No. 9, p. 99, 102). A document as part of Health Department of Western Australia.

Westley, F., et al. (2011). Tipping toward sustainability: Emerging pathways of transformation. *AMBIO: A Journal of the Human Environment, 40,* 762.

Wilson, S., Pearson, L. J., Kashima, Y., Lusher, D., & Pearson, C. (2013). Separating adaptive maintenance (resilience) and transformative capacity of social-ecological systems. *Ecology and Society, 18*(1), 1–11.

Woolf, S. H. (2013). Public health and medical care systems. In S. H. Woolf & L. Aron (Eds.), *US health in international perspective: Shorter lives, poorer health.* Washington, DC, USA: National Academies Press.

World Economic Forum. (2019). *Outbreak readiness and business impact: Protecting lives and livelihoods across the global economy.* Retrieved February 27, 2020, from http://www3.weforum.org/docs/WEF%20HGHI_Outbreak_Readiness_Business_Impact.pdf.

World Health Organization (WHO). (2008). *Outbreak communication planning guide.* Retrieved February 27, 2020, from http://www.who.int/ihr/elibrary/WHOOutbreakCommsPlanngGuide.pdf.

World Health Organization (WHO). (2009). *WHO guide to identifying the economic consequences of disease and injury.* Retrieved February 27, 2020, from https://www.who.int/choice/publications/d_economic_impact_guide.pdf?ua=1.

World Health Organisation (WHO). (2018). *Communicating risk in public health emergencies. A WHO guideline for emergency risk communication (ERC) policy and practice,* 78 p. document. Retrieved February 27, 2020, from https://www.who.int/risk-communication/guidance/download/en/.

Wu, Q. S., Wang, X., Liu, J-Y., Chen, Y-F., Zhou, Q., Wang, Y., et al. (2019). Varicella outbreak trends in school settings during the voluntary single-dose vaccine era from 2006 to 2017 in Shanghai, China. *International Journal of Infectious Diseases, 89,* 72–78.

Yamagata, Y., & Sharifi, A. (Eds.). (2018). *Resilience-oriented Urban planning: Theoretical and empirical insight.* Lecture notes in energy. Singapore: Springer.

Yi, Y., Zhang, Z., & Gan, C. (2019). The outbreak threshold of information diffusion over social–physical networks. *Physica A: Statistical Mechanics and its Applications* (Vol. 526, p. 121128).

Zautra, A. J., Hall, J. S., & Murray, K. E. (2010). *Resilience: A new definition of health for people and communities.* New York: The Guilford Press.

Zeng, D., Chen, H., Lynch, C., Eidson, M., & Gotham, I. (2005). Infectious disease informatics and outbreak detection. *Medical Informatics,* part of Integrated Series in Information Systems (ISIS) book series (Vol. 8, pp. 359–395).

Zheng, Y., Xie, X.-L., Lin, C. Z., Wang, M., & He, X. J. (2018). Development as adaptation: Framing and measuring urban resilience in Beijing. *Advances in Climate Change Research, 9*(4), 234–242.

Chapter 4
Responsiveness Through City Management

4.1 Enhancing City Management: How Should Cities React?

A larger part of what we do in the case of an outbreak is centered on our responsiveness to the event. Following from the concluding remarks of Chap. 3, we progress from the preparedness planning into responsiveness planning, through the enhancement of city management. This is complementary and practical to the overarching concept of urban resilience thinking (Eraydin and Taşan-Kok 2013; Elmqvist 2014; Harrison et al. 2014; 100 Resilient Cities and EY 2017b, b; Sellberg et al. 2018; Cheshmehzangi 2020a). The practical suggestions of this chapter are aimed to enhance matters of connectivity (Huck et al. 2020) from multiple perspectives, enhance the process of recovery (Kenney and Phibbs 2014), and improve the societal wellbeing. In order to achieve them, the city needs to react to the conditions of the outbreak progression. Hence, under the main question of 'how should cities react', this chapter addresses the importance of responsiveness through city management.

To start with, responsiveness is the central factor in city management scenarios, helping to reduce challenges and respond to potential vulnerabilities caused by the outbreak event. Apart from the race for rapid development of vaccines or treatment methods, the city itself should be prepared to comprehensively respond to any types of shortages and deficiencies—both quantitatively and qualitatively—such as in supplies, services, operations and management of the city (as the four main dimensions covered in Chap. 3). Thus, we have to maintain coordination of an array of factors, as well as the coordination between our assessment of the situation, development of risk perception, and socio-economic responses. Hence, information is vital and we have to assess and respond to the gathered information as promptly as possible. This should be processed through assessment, validation, and reflection. The next step for us is to ensure cities can react effectively and efficiently on such occasions. In general, we have already covered that we deal with a range of adversities caused by emergencies, crisis or even disasters. To address these adequately, we should provide the necessary support to enhance city management. This is the ultimate practice of urban resilience, and how it can help the city to react vigorously with adequate city management. It also requires the implementation of multiple resilience measures and the strategic action plan that fully reflects on the dynamism and complexity of outbreak progression.

An example of effective management is the one proposed by the World Health Organisation (WHO) (2020, online source), which proposes a set of "*plans and strategies for the safe reactivation of essential health services and longer-term health system functions*". In this example, the proposed priority areas are towards recovery measures (particularly for early recovery), including: (1) infection prevention and control, and patient safety, (2) National and regional surveillance systems, (3) essential packages of services, and (4) building towards a fit-for-purpose workforce. Through such measures, WHO (ibid) also suggest the applicability of cross-cutting technical areas, as shown in the followings (extracted and modified from the WHO online source):

- **People-centered services**, which is focused on methods of community engagement, redesign of the healthcare delivery system, addressing primary needs, and coordination between the associated delivery systems. This includes co-benefit planning between multiple stakeholders (i.e. related facilities, health workers, and communities). It is also about community engagement and empowerment to ensure all primary services are addresses and health structures are sustained or enhanced. Through this, we have to address issues of discrimination and strengthen the social values of the society with the right resources and in a responsive manner.
- **Institutional health partnerships**, which is primarily based on the immediate needs of the health institution services and development of health workforce capacity. As part of service delivery and support, it is vital to provide necessary tools, resources, and techniques to enhance the safety of infected patients and develop larger scare healthcare partnerships.
- **Information and communication technology**, which is aimed to strengthen surveillance and health information systems as well as to integrate available and gathered information and communication technologies solutions. The focus is not only on the utilisation of ICT platforms but on how through them we can refine our procedures of health services. Some of these can happen through an integrated surveillance system for better control and monitory. On the other hand, some ICT tools/instruments can help to develop essential health services, to support our physical facilities, and to reinforce workforce capacity.
- **Giving patients and workers the supplies they need**, which is primarily focused on monitory of availability and quality of necessary supplies, such as essential

health services, relevant medicine, and medical supplies. This requires reinforcement of regulatory systems, quality assurance systems, and procurement systems to ensure sustaining an adequate supply chain management. Apart from the availability of the named products, it is also essential to monitor both factors of accessibility (to products) and quality (of products). Right channels should be in place to assure such supplies are provided promptly.

- **Health financing**, which is a policy-based factor that focuses primarily on mobilisation and reallocation of necessary financial resources. This is important to healthcare providers as it increases the variety of essential factors, such as operations, the effectiveness of systems, capacity, and quality of specific health services. These sources should be arranged adequately to ensure having a resilient service delivery system, through which certain funding mechanisms may be required and should be acquired. This factor also requires public finance management to formalize channels of funding mechanisms and excel in the procedures of enhanced service delivery availability and health financing reforms. It is also suggested to cover the informal sector to ensure specific services are provided to specific stakeholders or groups of people.
- **Learning for the future**, which is similar to what we have identified as reflectiveness. Any new knowledge is valuable that can potentially improve our future responsiveness. The suggestions are focused on the employment of in-built processes to capture the pros and cons of each activity. It is important to utilise the multitude of technical resources, global examples, and lessons that gain more support to reduce uncertainties of the situation. Through national-level recovery plans, we have to integrate resources of multiple sources and aim to strengthen our health systems and other related systems for better responsiveness.

The overall arrangement of this specific example is defined as the summary of six technical areas and specific factors for outbreak management and recovery support for health system services through an integrated management model (WHO 2020). This example includes four primary factors of 'infection, prevention and control (IPC), and patient safety', 'surveillance', 'essential package of services', 'health workforce'. All these four factors together address six very important technical areas of (1) people-centered services, (2) partnerships, (3) information, communication, and technology, (4) supply chain, (5) health financing, and (6) knowledge harvesting.

As a small city-country, Singapore has a prolonged response phase of multiple stages. As explained in Chap. 2, the phases of the action plan are different from the phases of outbreak progression. However, they have certain overlaps depending on how the procedures are handled. For action plans, there is a range of responses that are relevant to our six phases of outbreak progression. For Singapore, their plan includes three stages of alert, containment, and mitigation (MoH 2014, p. 8), as shown here:

1. ALERT: The disease is mainly overseas and the response is to detect and minimise importation of disease. This requires border control measures and may require measures to try to stop the spread from individual cases or resultant clusters if they are imported into Singapore.

2. CONTAINMENT: The disease has arrived in Singapore and the primary response is to stop or limit the spread of the disease as much as possible. This requires extensive contact tracing and quarantine measures.

3. MITIGATION: The disease is spreading widely through the community, and measures to try to stop its spread are no longer effective. The response is to reduce the overall impact of the disease in the community. This requires an overall activation of business continuity plans, surge capacity for healthcare and essential services, and community-based public health measures.

As described above, the details of three possible disease response phases in Singapore—Alert, Containment and Mitigation—all describe three main action plans to minimise importation, limit spread, reduce impact respectively (ibid). Based on this method, the local epidemic is expected to peak at the mitigation response phase, which then requires slowing down and then towards the recovery of the disease outbreak. Singapore's methods of reaction are similar to what has been proposed in this book in six phases, but with three methods of minimising importation, limiting transmission, and reducing impact. Hence, in order to have a better overview of responsiveness, it is important to evaluate how the city and its management should react in each phase of the outbreak event.

As demonstrated in the previous chapter, the backbone of preparedness is city resilience (Argonne National Laboratory 2012; Ebisudani et al. 2017; Cheshmehzangi 2020a, c), which should be enhanced through multiple sectors, in multiple urban systems, for and by multiple stakeholders/actors, and at multiple levels. The multiplicity of resilience is an important fact of consideration, one that requires comprehensive planning as proposed in Chap. 3. Following from key discussions of preparedness planning and urban resilience enhancement, we ought to evaluate how the city should react throughout the outbreak progression. Therefore, this chapter serves as a follow-up to Chap. 3, but from the narrative of responsiveness. At first, we evaluate all target groups of the conceptualised urban resilience framework across all six phases of the outbreak event. In this evaluation, we then provide a brief understanding of actions and reactions that are needed for city management enhancement. In doing so, we lead the discussion into certain aspects of responsiveness from the perspective of city management.

4.2 Evaluating the Management Aspects Across Six Phases of Outbreak Events

In this section, we explore all five management aspects (or defined as target groups), namely asset management, services management, healthcare management, media management, and economic management (Sects. 4.2.1–4.2.5 respectively), and evaluate their responsiveness across all six phases of outbreak progression (see Sect. 2.2 and Fig. 2.1). These phases are numbered 1 to 6, matching the six steps of outbreak progression, in the order of: (1) Identification phase, (2) Response and Containment phase, (3) Transmission phase, (4) Transition phase, (5) Recovery phase, and

Table 4.1 The hierarchical understanding of three primary considerations for each management target group and during all six phases of outbreak progression—Legend for text: GM (General Management), RM (Risk Management), and CMM (Control and Monitory Management); Legend for colour: Red (high priority—top position), Orange (medium priority—medium position), and Yellow (lower priority—bottom position) for each target group

'MANAGEMENT' target group		Phase 1	Phase 2	Phase 3	Phase 4	Phase 5	Phase 6
1	Asset management	GM	GM	GM	CMM	GM	GM
		RM	RM	CMM	GM	CMM	RM
		CMM	CMM	RM	RM	RM	CMM
2	Services management	GM	RM	GM	CMM	GM	GM
		RM	GM	CMM	GM	CMM	CMM
		CMM	CMM	RM	RM	RM	RM
3	Healthcare management	RM	RM	CMM	RM	CMM	GM
		CMM	CMM	GM	CMM	RM	CMM
		GM	GM	RM	GM	GM	RM
4	Media management	CMM	RM	RM	CMM	CMM	GM
		RM	CMM	CMM	RM	RM	CMM
		GM	GM	GM	GM	GM	RM
5	Economic management	GM	RM	CMM	RM	CMM	GM
		RM	GM	RM	CMM	GM	RM
		CMM	CMM	GM	GM	RM	CMM

Source The Author's own

(6) Post-recovery phase. This thorough evaluation is narrated based on the three main management considerations of risk management (RM), control and monitory management (CMM), and general management (GM). Table 4.1 demonstrates the hierarchical understanding of all considerations and their level of importance in this matrix of management target groups and six phases of outbreak progression. The following five sub-sections shed light on matters of responsiveness for these management target groups. The target-based and phase-based suggestions are given in the form of practical recommendations.

4.2.1 Evaluating Asset Management (Under Operational Dimension)

For asset management, in five out of six phases, the priority is on general management, which represents the importance of general management demand and operations in the city. While risk management is considerably lower in the recovery phase, it is important that thorough risk management is conducted in the earlier phases (refer to Table 4.1).

Phase 1—In this phase, the primary assets of the city should be cross-checked at multiple sectors and levels. It is important that this procedure includes both the quantity and quality of particular assets, such as critical infrastructures (European Commission 2006). In other words, the governance of critical infrastructures should be resilient itself (Monstadt and Schmidt 2019). At this stage of outbreak progression, human resource management is important and the city authorities should plan for possible reallocation of certain resources. This includes, in particular, the provision of extended support to disease detection units and investigators. Early detection of disease and any relevant information about it, would certainly help to equip the healthcare units efficiently and appropriately. Early detection can also enhance the arrangement of specific emergency assets and precautionary products. Moreover, early risk management is highly recommended in this phase.

Phase 2—By now, supply and demand measures become crucial. The city's primary assets should be preserved and increased in numbers (when possible). These include the preparedness of emergency units as well as keeping the critical infrastructures safe and operational. At this point, the city authorities should assess the operationalisation of their assets and the institutions supporting them. Mobile assets, such as transportation and distribution units, should receive careful control and monitory to ensure their healthy operations are maintained with a range of enhanced precautionary measures. The city needs to maintain its continuous risk management of primary assets to foresee any potential shortages.

Phase 3—In this phase, it is recommended to develop a system-based approach that enhances control and monitory measures of primary assets in each sector. Some assets, such as transportation, can run in a one-way system; meaning that they could be utilised to import essential goods, mostly of medical products and medicine. Critical infrastructures, as primary assets of the city, need to be kept safe and secure with some restricted operations. More importantly, it is recommended to increase the level of control and monitory measures to ensure the smooth operation of primary assets in the city.

Phase 4—The integration of resilience-based infrastructure planning with asset management becomes a crucial response in the transition phase. The quality and quantity of assets should be assessed thoroughly through enhanced control and monitory management measures. There may be some tangible shifts in healthcare systems, either fully equipped or facing shortages—all depending on how the conditions are at this particular phase. It is recommended to prioritise control and monitory of assets, respond to issues of resource management, and provide a revised plan for the overall operationalisation of the city's primary assets (Marana et al. 2019). In addition, resources should be allocated to places in need.

Phase 5—As the recovery starts to progress gradually, the city should reflect on the last two phases of outbreak progression to ensure all primary assets and fully operational through a robust general management plan. The potential risk reduction at this phase indicates a chance for increasing control and monitory management measures. These should be kept either at the level of the previous phase, or should be narrowed down to certain areas that require further attention. The healthcare system plays a major part as a primary asset, and its general management for speeding the process towards recovery is essential. It is recommended to not relax any of the control and monitory measures until full recovery is achieved.

Phase 6—The first step in the post-recovery phase is to provide general management of all primary assets. The need to assess and respond to losses is essential, providing a chance for the city authorities to provide certain levels of subsidies to support businesses and to gradually improve their revenues. Through this process, the city authorities have to review losses in financial assets and propose for a short to midterm recovery plans. The review of critical infrastructures is highly recommended and certain relocations of resources may be needed in order to address the needs of the society. It is important to develop a reflective report to evaluate the alterations of asset management during the process. Moreover, it is critical to highlight the primary shortages and/or deficiencies in the assets and resources of the city.

4.2.2 Evaluating Services Management (Under Operational Dimension)

For services management, the priority consideration changes during the outbreak progression, precisely starting with general management and also into the post-recovery phase with enhanced general management. The risks of response and containment phase need to be carefully managed before we can impose a range of enhanced control and monitory management measures from the third phase onwards. This also reflects on a reduced risk management requirement from the transition phase, while it is advisable to have a robust risk management plan developed in the first two phases (refer to Table 4.1).

Phase 1—In this phase, the general management of services should be maintained as a high priority. There may not be major changes in many sectors, but the city requires developing a multi-services management co-creation model that is emerged from an effective multi-service differentiation plan (Wang et al. 2020a). Enhanced risk management should run in parallel to this process, providing an early plan for all production units across various sectors. Similar to asset management, there is a need for early resource assessment to maintain smooth operations of main urban services. It is highly recommended to boost the capacity of emergency services and primary public services, to identity any shortages at the early stage of outbreak progression.

Phase 2—Through our response plan, risk management takes a high priority with continuous support from the general management of services. The gradual change in the status of control and monitory measures starts to transpire towards the end of this particular phase. The city would benefit from early multi-management support through an integrated approach, one that could assess and monitor the city's primary productions, respond to their risks, and have immediate plans to react in the time of need. It is recommended to develop a robust risk management plan at this stage, which can be durable (with only some minor flexibilities through reflective assessment) until the end of the outbreak progression.

Phase 3—From this phase onwards, services management should help to enhance the resilience of society. Also, it is recommended to develop a set of guidelines for the control and monitory measures, which should be done in a progressive approach. In this phase, it is recommended to develop a comprehensive services management system to consolidate all management objectives in the city's general management plan. We also expect them to deploy crucial services (Linkov et al. 2014), which can support the emergency and medical services at most. In this process, the developed co-creation management model should be utilised in full capacity, with careful operations of primary services, and limited operations of secondary services. It is recommended to maintain primary productions but the city should review the secondary productions, those that may then be restricted or temporarily stopped.

Phase 4—In this phase, the already established co-creation approach should be maintained while the city authorities are expected to increase the capacity of control and monitory management. As a priority, control and monitory measures should be across all sectors and systems. In this process, it is vital to regularly monitor the updates and foresee any unexpected issues. It is beneficial for the city to develop a multi-scenario strategic plan, which can be reflective enough on the actual conditions and operations of all services in the city. It is recommended to make certain adjustments to the general management of services and prioritise primary services.

Phase 5—An enhanced general management is expected to be in place at this stage, which should have already matured throughout the outbreak progression. We anticipate the general management plans to be more reflective in this phase. We also expect them to readjust some of the operations, and start normalisation of primary productions and services. It is advisable to maintain careful control and monitory management to ensure that general management is adaptive and reflective of difficulties in the recovery process. It is highly recommended to boost public services and manage their operations through regular monitory measures.

Phase 6—The hierarchy of management considerations remains the same as in the previous phase. In addition, we expect the city to develop a network-based system of multi-sectoral services that can put together an integrated model of the post-recovery plan for all services, and specifically for public services. While in practice, this model may not be feasible, it is then recommended to develop a network-based system that can suggest the individual plans of each sector and how they may support one another as well as what support they may require. The city authorities should reflect on losses of primary services and help to support the gradual operations of secondary services. The need for financial support is not essential in the case of a short outbreak, but in longer events, there may be some financial needs of public services in particular.

4.2.3 Evaluating Healthcare Management (Under Operational/Institutional Dimensions)

For healthcare management, the priority consideration is mostly either risk management or control and monitory management. There is only one expectation in the post-recovery phase, when the general management is expected to support the recovery plans more effectively. As the main target group under the management dimension, it is expected the city authorities prioritise healthcare management and support its uninterrupted operations throughout the outbreak progression. The risk management plan should be developed and implemented in the first two phases, while it is then required before full recovery is achieved (refer to Table 4.1).

Phase 1—As the hit on the healthcare system is expected to be significant, the management of healthcare both as an institute and its operations, ought to be integrated with an immediate risk management plan. This requires to be accompanied by intensified control and monitory management to ensure the necessary support and precautionary measures are in place from the inception. The early management and assessment of the healthcare system and services should include a comprehensive reporting of multiple aspects, such as services (including emergency services), resources, facilities, logistics, and equipment. Risk management should be implemented in this phase and develop added safety measures that respond to the nature of the infectious disease. It is vital to prepare specific units or additional support in the time of need.

Phase 2—The hierarchy of management considerations remains the same as in the previous phase, with risk management remaining as a priority. The healthcare system and services should remain on alert and have full responsiveness through quantity and quality of resources, facilities, equipment, detection tools, and emergency units. The other non-outbreak health issues should be addressed adequately, with minimised risks to larger groups of people. It is suggested to maintain a high level of control and minatory management in order to avoid any early and/or rapid cases of community transmission. The immediate response through risk management is to reduce shockwaves of the outbreak and address any vulnerabilities of the healthcare system and services at this phase of outbreak progression. A reflective approach from the inception helps to develop a set of precautionary measures, early control, and better monitory of outbreak progression (if and when it occurs).

Phase 3—In this phase, healthcare management should increase its control and monitory measures to the highest level. The system and its services would be under the highest risk and require immediate attention through systematic management of all operations. It is also suggested to provide and manage separate facilities and temporary units of healthcare in multiple locations, ideally in close proximity to main hospitals and healthcare facilities or to be allocated in each community/district (if feasible and depending on the size of the city). It is highly recommended to monitor the procedures of patient delivery and treatment. It is required to have adequate training of emergency medical staff and those teams involved in the treatment procedures. A highly controlled procedure should be developed and immediately implemented to reduce the risk of widespread transmission from those specific locations. This is also meant to provide detailed information about the procedures and disseminate them appropriately. People should be informed not to rush to the main hospitals or designated healthcare services. Main healthcare infrastructures should be kept safe and secure in order to avoid their potential conversion into unwanted hotspots of the outbreak. If this occurs, we may risk the impact on healthcare workforces and weaken the regular operationalisation of major facilities and infrastructures.

Phase 4—As the risks continue to overshadow the regular operations of the healthcare system and services, it is vital to have a revised high-risk management plan in this phase. It is highly recommended to avoid the reduction of any control and monitory measures, and the city authorities should do regular checks on such measures and procedures. In order to avoid any misconceptions, it is important to adhere to structured and formal procedures. This should be widespread not only to healthcare workforces but also to the general public. Effectiveness of treatment measures are crucial, hence it is suggested to expand the temporary facilities and their continued management with extra care and careful monitory. It is also recommended to remain in the alert status as the situation of outbreak progression can reverse at any point.

Phase 5—It is anticipated that treatment procedures are managed well in both the quality and quantity of resources and facilities. Careful resource management is required to ensure no shortages, and should instead plan for faster recovery. The procedures should be maintained with the highest level of control and monitory management. In combination with this plan, it is recommended to have regular checks and revise the risk management plan in a reflective approach. The recovery can only be reachable through the enhancement of healthcare services and systems, and their management through a range of enhanced safety, increased support, and multiple resources (e.g. human power, equipment, medical products, etc.). It is highly recommended to continuously adhere to structured and formal procedures. In addition, a contingency plan is required to avoid any further issues before full recovery is achieved.

Phase 6—Healthcare management plays a major part in restoring the health and wellbeing of society. While risk management is reduced to the lowest level during the outbreak progression, it is anticipated the operations of the healthcare system and services go back to general management processes. A contingency plan is no longer needed, but it is essential to keep some of the precautionary measures in place. It takes time for the society to go back to their original mental status, hence we have to manage certain needs of the society to overcome some of the earlier distress and anxiety caused by the outbreak. We anticipate having a management plan for potentially populated counselling services (or other similar facilities). In addition, it is important to empower the communities to support the post-recovery activities of healthcare services and systems. Hence, a co-management plan would help to reflect on some of the adversities and reduce the vulnerabilities as fast as possible.

4.2.4 Evaluating Media Management (Under Institutional Dimension)

For media management, priority consideration is mostly a combination of control and monitory management and risk management. This should remain in place before the general management can retake its position in the post-recovery phase. As a crucial target group that can have adverse effects on societal wellbeing, it is important to increase the control and monitory of media management. This should be done through reliable sources and official channels of communication. Transparency and trust should be kept in the best possible way, as they can simply dictate how society may behave in the later phases of outbreak progression. Media management also affects people's perception of the disease, people's knowledge of precautionary measures, people's level of compliance, and people's support. As an institution that disseminates information (such as updates, monitory checks, etc.), knowledge (for response), and education, media management remains as an effective target group during the whole process of outbreak progression (refer to Table 4.1).

Phase 1—In the first phase, the media is probably silent but its management is crucial. Hence, media management should operate with extra care and robust control and monitory management. It is vital to establish official channels (if not already in place), and make them operational as the main official and trustworthy means of communication. The reliability of the source is very important and must be recognised from inception. It is recommended to structure and define the communication channel, provide a rightful resource of information, ensure full accessibility to all, and have clear and positive announcements. The tone of communication should be reassuring, supportive, and constructive. More importantly, all communications should avoid manipulative viewpoints or politically-tailored announcements. There should be no announcement of repudiation nature and the communication should effactually reflect on the preparedness and responsiveness of the authorities. In this phase, it is important to start notifying the general public through gradual and soothing announcements, so that necessary precautionary measures are in place, and the society is aware and involved from inception.

Phase 2—The response and containment phase is a crucial phase for the media and its management. In this phase, information pours in from multiple sources and it requires to be managed in the best possible way. Hence, the establishment of a robust structure in the previous phase can help to overcome any of the sudden shockwaves. The ripple effect of unreliable information and false communications could last for a long time and could weaken the reliability of official sources. It is important the response announcements are strong, confident, and informative in order for them to be effective and trustworthy. In the context, where governmental trust is fragile, it is recommended that official authorities help to reassure society through sincere support and increased reliability. In this phase, the highest risk management of information should be carefully conducted in line with the control and monitory management of the media. On particular occasions, certain regulations may be needed to reduce conflicts, implement an action plan against false information/news, and tackle any potential belligerence behaviour. It is highly recommended that media is managed transparently and for the benefit of the society's health and wellbeing.

Phase 3—The hierarchy of management considerations remains the same as in the previous phase, while we anticipate a higher risk level for false news and information. Similar to the response and containment phase, all information should be disseminated through the right sources, and are made available to all in the most transparent way. More importantly, the media should be utilised and managed to provide an educational recommendation to the general public. It should be the primary source of new information, precautionary education, necessary health training, and professional and valid viewpoints from relevant officials, medical officials, and relevant scientists and academics. While possibilities are high, it is advisable not to make the announcements as political means.

Phase 4—In this phase, media management should be conducted in a way that it can reduce any uncertainties or any false information that was previously disseminated from other sources. As early as this phase, media should be utilised and managed to provide factual reflections of the outbreak progression. The updates should highlight regular control and monitory measures and positively showcase the main progress and success stories. The progress towards recovery should be managed cautiously; and in response, the broadcasting of any updates should be done officially and reflectively. In this stage, any conflicts or mistrusts could be amended if the progress is positive and if relevant authorities have shown reliability and responsibility to combat any adversities from the outbreak progression. It is advisable to prioritise control and monitory management of media at this stage.

Phase 5—The hierarchy of management considerations remains the same as in the previous phase, with control and monitory management remaining as the priority. While general risks are expected to reduce by this stage, it is still important to keep up the risk management procedures. This approach would help to maintain consistency in the flow of information and updates. Media should be managed proactively at this stage and should be responsive to specifics of the outbreak and the societal wellbeing. It is recommended to manage the media towards appropriate on-site responses (Singleton et al. 2000) and involve the communities as much as possible. A healthy involvement of the general public in media can support the wider dissemination of the right knowledge and updates, and continue the control and monitory of the situation until completion of the recovery phase.

Phase 6—For the first time throughout the outbreak progression, general management of media becomes a priority consideration at this phase. Hence, media management could effectively enhance various tools of media communication to guarantee the inclusiveness of all factors, stakeholders, and different groups of society. Risks are expected to be the lowest of all time, but it is still recommended to keep up with control and monitory measures. In the post-recovery phase, media management should also take a reflective position, addressing critical aspects of continuous education, support, and reclamation progression. It is highly recommended to detect any false updates and utilise the media with factual information to reverse some of the distress and anxiety caused by the outbreak event.

4.2.5 *Evaluating Economic Management (Under Institutional Dimension)*

For economic management, the priority consideration alters in each phase as it reflects on the dynamic nature of this important target group. The economic impacts are often worryingly high and only become visible from the third or fourth phase of outbreak progression. The economic consequences are generally long-term, multi-sectoral, multi-spatial, and usually beyond the boundaries of the city (Amadeo 2020). They could sometimes last for years after the outbreak. Managing the economy, based on its significant losses, is crucial to help the revitalisation of society after the outbreak. Hence, it is inevitable that the city and the society would face certain levels of economic losses; some that can be considered as socio-economic losses. To reduce the vulnerabilities and the level of losses, we have to utilise combined techniques and methods of economic management in each phase (refer to Table 4.1).

Phase 1—In the first phase of outbreak progression, there are little or no impacts on the economic management of the city. Nevertheless, this provides an opportunity to assess the conditions and propose necessary action plans for the next phase. Depending on the conditions of the outbreak, it is possible to increase the production of certain sectors and increase the capacity of stocks for necessary products. In this phase, general management should focus on the stability of financial and economic sectors, which can then be utilised as preliminary risk management reports. It is recommended that the initial assessment, conducted through multi-sectoral risk management, should develop a scenario-based action plan. Specific control measures are not yet required but vulnerabilities from uncontrolled economic management should be carefully assessed.

Phase 2—In this phase, the first wave of economic impacts are perceived on a small scale. It is required to put risk management as a priority in order to develop realistic measures and support the general management of the city's economy. Measures of economic stability in connection with various networks should be assessed and put in place for effective operations. This includes maintenance of economic development plans beyond the city boundaries, and in multiple networks of regional, national, sub-regional, and global. It is recommended to develop a multi-sectoral precautionary plan and to create co-benefits through enhanced risk management. In parallel to these general management plans, it is recommended to develop sectoral coping strategies and identify potential shortfalls. Financial support could only start for primary sectors and those that are prioritised during the outbreak progression. This is a context-specific selection which can differ based on the city's economic status, import/export patterns, and financial characteristics. In addition, it is recommended for the staff of secondary businesses an enterprises, as well as the finance sector, to start working from home through available digital platforms. Early public-private partnerships should develop as early as this phase. These should be utilised in the case of emergencies, and for later production shifts and necessary support to particular sectors or vulnerable groups.

Phase 3—The main wave of economic impacts are perceived by now, and impacts will become widespread and beyond the city's boundaries. The scale and intensity of impacts depend on the political or economic status of the city (Cheshmehzangi 2020b) and how it plays a part in multi-scalar trade (i.e. from regional to global). For the city, however, the impacts become tangible as we face temporary closure of businesses and productions. The situation will increase the risk of redundancies and the permanent closure of businesses. The financial sector of the city should respond to potential deficiencies and propose a multi-system adjustment that has the capacity to rearrange some of the primary operations and productions. While larger businesses can temporarily cope with issues of the outbreak, small and medium-sized enterprises and local communities would suffer heavily. The financial support should then be prioritised for those smaller businesses. Another unfortunate example is the construction sector, which may stop all its on-site operations. For this specific sector, we recommend to only operate with emergency maintenance projects. In this regard, the city authorities should provide sufficient leverage to increase cooperation between multiple sectors, and particularly from larger enterprises to smaller businesses. For sectors that face more impactful adversities and may not be able to develop a coping strategy, it is recommended that responsible authorities provide necessary subsidies and financial support. Moreover, it is advisable to prioritise control and monitory management plans. In addition, it is recommended for the staff of selected primary businesses and enterprises to start working from home through available digital platforms. Public-private partnerships should be strengthened and prepared for any potential production shifts or support. These should be prioritised for the support of healthcare systems and services. If needed, the production shifts could help to temporarily adjust the supply and demand chains.

Phase 4—By now, the city has (possibly) gone through large-scale economic losses. The reduction in functionality of economic operations becomes perceptible, and the city has to respond to the needs of local economies and small to medium-sized enterprises. This should be implemented through a reflective approach to measure the subsistence level of each sector. Such an approach would help to better assess the costing and demands of the city productions (Luyten and Beutels 2009), which could then carefully move into gradual production and functional status. Hence, the city can develop a short-term contingency plan and increase the synergies or capacity for co-benefit and co-management between different sectors. In this phase, while we anticipate a reduction in instabilities, it is suggested to have a combined priority plan for risk management, and control and monitory management of the city's economy, businesses, enterprises, and organisations, as well as their operations.

Phase 5—While risks are relatively reduced in this phase, control and monitory management should remain as the priority. The earlier risk management approach should have highlighted the level of economic losses across multiple sectors and in different categories and sizes of businesses and enterprises. The city may face high fiscal impacts, some of which hit the society the most like the impacts on local economics, job losses, self-employment instabilities, and business closures. The city should have a contingency plan to support those in need. In practice, it is feasible to conduct this planning through a sector-based approach than a generic one. More importantly, the inclusive general management approach should be emerging here, which should ultimately put the vulnerable sectors and businesses as the priority.

Phase 6—In this phase, the authorities have to respond to three overarching factors: (1) to reduce the economic burdens; (2) to regain the face of the city/community; and (3) to improve public policy. Post-recovery management still has some risks that require careful management. This is primarily due to the long-lasting impacts of the economic burdens (e.g. on businesses, enterprises, and society) that require time to recover (Huber et al. 2018). It usually takes a while before the situation is improved before it can return to its original status. The city has to cope with large-scale and multi-sectoral losses, and the recovery strategies should first address the vulnerable sectors/areas/groups. To regain the face of the city or community, the city has to reflect on specific losses caused by deteriorated external relations, the decline in tourism, a decline in production, etc. Most important of all, the city should develop an all-inclusive public policy strategy (Algan 2013), which should include the involvement of all stakeholders, particularly its communities. Steps taken from the post-recovery phase onwards may need more time to regenerate the city and its economic stability. Hence, it is recommended to take steps accurately and gradually, and make necessary readjustments to institutional structures, financial sector, and any associated economic production. The latter should again be reflective and inclusive to ensure it includes the role of all goods, capital, labour and services for the sole purpose of the economic regeneration of the city.

4.3 Evaluating the Provision Aspects Across Six Phases of Outbreak Events

In the same order as demonstrated in the previous section, we also explore all five provision aspects or target groups, namely transportation services, safety and security services, social services, food supply, and amenities supply (Sects. 4.3.1–4.3.5 respectively). In the same order as in Sect. 4.2, we evaluate their responsiveness across all six phases of outbreak progression. This thorough evaluation is narrated based on the three fundamental considerations of risk assessment (RA), adaptive maintenance (AM), and prioritisation plan (PP). Table 4.2 demonstrates the hierarchical understanding of all considerations and their level of importance in this matrix of provision target groups and six phases of outbreak progression. The following five sub-sections shed light on matters of responsiveness for these provision target groups. Similar to Sect. 4.2, the suggestions are given in the form of practical recommendations.

Table 4.2 The hierarchical understanding of three primary considerations for each provision target group and during all six phases of outbreak progression—Legend for text: RA (Risk Assessment), AM (Adaptive Maintenance), and PP (Prioritisation Plan); Legend for colour: Red (high priority—top position), Orange (medium priority—medium position), and Yellow (lower priority—bottom position) for each target group

'PROVISION' target group		Phase 1	Phase 2	Phase 3	Phase 4	Phase 5	Phase 6
1	Transportation services	RA	RA	RA	RA	RA	AM
		AM	AM	PP	PP	AM	RA
		PP	PP	AM	AM	PP	PP
2	Safety and security services	PP	PP	RA	RA	AM	AM
		AM	AM	AM	AM	PP	PP
		RA	RA	PP	PP	RA	RA
3	Social services	RA	PP	PP	AM	AM	AM
		PP	RA	AM	RA	RA	PP
		AM	AM	RA	PP	PP	RA
4	Food supply	PP	RA	RA	RA	AM	AM
		RA	PP	PP	PP	RA	PP
		AM	AM	AM	AM	PP	RA
5	Amenities supply	PP	RA	RA	RA	AM	AM
		RA	PP	PP	PP	RA	PP
		AM	AM	AM	AM	PP	RA

Source The Author's own

4.3.1 Evaluating Transportation Services (Under Services Dimension)

For transportation services, the priority consideration is mostly risk assessment with the only exception being adaptive maintenance at the post-recovery phase. The expected high(er) risks are based on the fact that transportation services are extremely dynamic and provide services for various purposes. The multiplicity of transportation services poses a high risk for disease transmission from person to person, and location to location. Progressive adaptive maintenance, with some flexibility in operational analysis, is identified as effective strategic planning in the first two phases. This approach could help better responsiveness in the later transportation rearrangements. For instance, this approach can help to implement action plans for later restrictions, precautionary measures, and monitory of transportation operations. It is highly recommended to keep both transportation services and their associated infrastructures safe and secure throughout the outbreak progression, even if this may depend on making significant alterations to the services. The mid-stage enhanced prioritisation plan should be in correlation with the established adaptive maintenance plans. Moreover, it is recommended to conduct an early asset and resource assessment for all transportation services, and develop a cross-sectoral strategic plan. In doing so, we can provide reallocation of assets to emergency services, medical services, and support of certain provisions to the city and its communities; particularly in the time of need (refer to Table 4.2).

Phase 1—The consistency in risk assessment should be maintained from phase one of outbreak progression. The initial phase is not necessarily of high risk, but the assessment of any risk before the speeding process of the outbreak is critical. In this phase, the city authorities should develop scenario-based and cross-sectoral adaptive maintenance plans. The need to facilitate support for emergency transportation services is crucial and should be addressed as part of initially planning. Through a preliminary sector-based risk assessment, it is recommended to suggest adaptive measures and potential changes to operations in the next two phases. Prioritisation plan is not yet essential, even though the priority should be given to emergency transportation services. The dynamic characteristic of transportation services is not yet a major threat, but precautionary measures should be advised to related stakeholders.

Phase 2—In comparison to the previous phase, we can see only some minor changes in the response and containment phase. Risk assessment stands high as the priority while adaptive management should be enhanced to be implemented towards the end of this phase. The prioritisation plan should start with some restructuring of multi-sectoral transportation services before the major upheaval of the outbreak progression. In this phase, we anticipate the city authorities to increase their attention on transportation services, and implement a range of precautionary measures. In a responsive approach, the city authorities should start to utilise redundant transportation services for the delivery of primary goods and products. The city's transportation services are expected to have early measures in place in order to reduce the intensity of the transmission period. In addition, early disinfection procedures should be introduced and implemented.

Phase 3—In this phase, transportation services are at the highest risk. Hence, top-down interventions are essential to overcome any shortages, and to reduce (or limit) the secondary services. In this phase, the highest level of risk assessment should be conducted in parallel with a systematic prioritisation plan. The city's multiple transportation services are likely to boost the speed, scale, and intensity of transmission. This is somehow inevitable unless all transportation networks are immediately stopped, allowing only emergency services to deal with patient delivery and treatments. While this may not be feasible, it is highly recommended to impose the highest possible restrictions on transportation services. To the largest possible scale, this restriction should be applied to public transportation networks, which should ultimately minimise the population density of transportation hubs/nodes. While it is not feasible to close off communities, it is recommended to limit the number of access points in and out of communities or neighbourhoods. Moreover, the transportation services suffer at most through long hours of operation, high frequency, and high density of users. These should be assessed promptly and adaptive measures should be in place at the earliest possible time. Through these assessments, the city should prioritise only certain primary transportation services, such as emergency, medical, primary logistical, and private modes of transportation. It is advisable to either reduce or temporarily stop the operations of public transportation services and shared transportation systems. Potentially, this can develop in the form of a density control plan, which could then include possible restriction measures. The city should also impose a careful monitory plan for intra-city and inter-city transportation systems, and limit some of the secondary logistical mobility. It is also recommended to generate a data-based mobility pattern of transportation services. This approach would create a better overview of transmission patterns through spatio-temporal analysis. In doing so, we could provide a valuable dataset to analyse the intensity and scale of transmission. In addition, regular disinfection procedures should be increased in frequency.

Phase 4—The hierarchy of primary considerations remains the same as in the previous phase, with risk assessment continuing as the priority. By now, the prioritisation plan should be effective, the main public transportation systems should be stopped or only partly operational. The city authorities should empower the communities and their representatives (such as community management teams, or committees, etc.) in order to carefully monitor people's mobility in and out of their communities. The use of the digital platform can be useful to create an information-based data collection mechanism, one that can store mobility patterns, detect any unusual mobility, and record any newly infected cases (as well as their mobility). The city should temporarily stop the operations of secondary logistics, selective primary logistics (which should be cross-checked and selected through a robust prioritisation plan), and all public transportation of the city. The main transportation hubs should continue with their control measures and have a full monitory of all mobility patterns of their territory. It is recommended to reduce the frequency and intensity of intra-city and inter-city transportation systems, and put adaptive measures on private transportation. While emergency services are expected to remain as the priority, the city should limit transportation across all other sectors. In this phase, continuous limitations on transportation networks and services would help to get closer to the recovery phase. In addition, regular disinfection procedures should be maintained, but could only reduce their frequency.

Phase 5—In this phase, we anticipate further adaptive maintenance while risk assessment remains the priority. The normalisation of selected primary transportation systems could gradually happen, allowing for minimised frequency of operations, limited hours of operation, secured mobility, and point-to-point mobility plans. Through an enhanced adaptive maintenance plan, the city should start making new adjustments to restructure the operations of transportation services across all sectors. In doing so, careful monitory should be maintained to avoid any issues that can potentially cause the spread of the disease again. Through comprehensive risk management, regular checks and disinfection procedures should continue. The regularity of these procedures is not required to be consistent across multiple phases of the outbreak progression. However, this should be decided separately and a thorough risk assessment of each phase. Furthermore, the city should allow mobility of private transportation, but should also implement monitory measures for all mobility patterns. It is advisable to review and revise any priorities at this stage, and allow for a gradual normalisation of all transportation networks and services. Lastly, it is recommended to put more restrictions on those people traveling back to their workplace, particularly from outside the city or from other infected zones. These safety and monitory measures should help to develop a range of regulations on multi-spatial mobility checks, necessary precautionary measures, and the possibility of quarantine controls.

Phase 6—In this phase, we anticipate careful adaptive maintenance as the priority and in order to respond to some of the shortfalls during the earlier phase. While we anticipate minimal risks, it is recommended to keep risk assessment as the second priority consideration of this phase. The revised adaptive maintenance plan should start with the rearrangement of primary transportation resources and prioritise the private sector. In a strategic approach, the city should maintain its logistical operations and provide the necessary support to vulnerable sectors. Emergency transportation services could relax their intense operations and public transportation networks should gradually restore their normal operations. It is still advisable to continue with monitory checks but at a lower level. Moreover, it is recommended to make gradual rearrangements rather than rapid normalisation. The procedures should be embedded in the situation's progression and this should be maintained by an enhanced adaptive maintenance plan. It is also advisable to have a gradual progression to achieve the optimal level of operations for secondary transportation services.

4.3.2 Evaluating Safety and Security Services (Under Services Dimension)

For safety and security services, the priority consideration changes in a progressive way, i.e. prioritisation plan in the first two phases, risk assessment in the middle two phases, and adaptive maintenance in the last two phases. Throughout the outbreak progression, careful planning is required to ensure flexibility is provided to all services under this category. By responding to the role of co-benefits resiliency actions (Siemens, Arup, and RPA 2013), the city's safety and security should be maintained through primary services; and those are required to be available, receptive, and operational. Safety and security situations may differ from one case to another, and they also differ from one phase to another. This is reflective of how such services are arranged and rearranged throughout the outbreak progression. The correlation between such services and the safety of the city and its communities is a major factor for further consideration into the practice of city management (refer to Table 4.2).

Phase 1—In this phase, safety and security services should be on alert but with minimal risk assessment. Most of their responsiveness should address the assessment of priorities, impacting, in particular, the support that is needed for selective emergency services, specifically related to health and medical services. An important aspect of this phase is to reflect on the developing uncertainties of the outbreak and monitor the safety of services' operations. It is highly recommended that the city authorities work with relevant management teams of safety and security authorities at the earliest possible time, and develop two adaptive maintenance plans: the first one is required to be sector-based looking into specific needs and challenges of each sector—i.e. in the form of a contingency plan with adaptive measures; and the second one is required to be multi-sectoral with possible integration measures and co-creation of services' operations. These two plans should be modified and readjusted according to the needs of each phase. The safety and security services should develop a procedural plan to promptly deal with potential disorders in the society and to reduce the impact of disruptions caused by them.

Phase 2—Safety becomes a major priority in this phase, as any reaction can direct the possibilities of the outbreak. With enhanced safety, we can create an opportunity for early containment or potential speed recovery. In this phase, more attention should be given to emergency units under the safety services. The adaptive maintenance plan must reflect on the outbreak progression with added resources and workforces. The provision of services is essential, and the safety of them is more so. Through a more responsive reaction, the provision of small and isolated temporary treatment facilities is beneficial, such as the use of containers, or any other approach of quick assembly and operations. This approach needs careful feasibility assessment with rapid procedures in a short time. Further reactions should include enhancement of safety services, as well as preparedness for security units to combat any unrest or unexpected issues raised directly and indirectly from the outbreak event. The announcements should be clear and sufficient enough for the reassurance of society. Moreover, it is advisable to start developing security services at multiple spatial levels of the city, mostly to be managed at the community level and to monitor security from a community-based approach.

Phase 3—While safety issues remain as the priority, the city needs to conduct a comprehensive risk assessment of the outbreak progression. This phase is quite crucial for city management as responses are under surveillance by the society, media, and outside communities. Any misconduct would turn into a tool against the authorities and if so, this can cause damage to outbreak control plans. As a result, the provision of safety and security services should be developed in line with a step-by-step risk assessment plan that is all-inclusive and reflective. It is recommended to consider the provision of large scale temporary treatment locations/facilities, through which we can boost the capacity of safety and emergency units. All operations should be carefully monitored and safety measures should be adapted and imposed to the highest level. Moreover, the city management should reflect on any shortages, and seek external support in the time of need. The involvement of regional and national support is essential, and the involvement of relevant international organisations should be maintained officially and healthily. If necessary, it is advisable to consider national and international support and increase the capacity of support teams and on-site emergency units. Moreover, it is highly recommended to implement security services at multiple spatial levels of the city, mostly to be managed at the community level and to monitor security from a community-based approach. This approach should develop from the earlier phase, and to be fully implemented and monitored in this phase.

Phase 4—The hierarchy of primary considerations remains the same as in the previous phase, which includes a high priority for risk assessment plans. It is expected the adaptive maintenance measures are well-developed by now, and they should be fully implemented in the practice of city management. It is advisable to continue with the earlier established prioritisation plan (from phase 2) and only make minor adjustments if required. There are still potential risks in this phase particularly that they can lead to the treatment of societal distress and anxiety. Any issues of shortfall or shortages could potentially become a backfire. Therefore, it is recommended to address those as prompt as possible and maintain the security of the society. The involvement of community representatives and community management teams would be helpful to support the adaptive maintenance plans and ensuring the society's health and wellbeing is maintained inclusively.

Phase 5—In this phase, we anticipate the growing progress of adaptive maintenance plans in the practice of city management. This becomes a priority while a comprehensive risk assessment is still needed but no longer as a high priority. The review and revision of prioritisation plan are essential at this stage, to ensure new measures are developed and implemented across all relevant sectors. This planning should reflect on adaptive measures put in place through careful monitory of the outbreak progression. In addition, it is recommended to increase the involvement of communities in order to develop their support for a full recovery. In a reflective approach, it is necessary to respond effectively to the reallocation of units between multiple units of safety and security services, and ensure they have adequate resources and facilities.

Phase 6—The hierarchy of primary considerations remains the same as in the previous phase, which includes a high priority for adaptive maintenance. In this phase, we anticipate the early normalisation of safety and security services and their operations. This should effectively be reflective of the outbreak progression and societal needs, as the authorities should allow some flexibility in the early operations of those relevant services. It is recommended to keep the community involvement so that all responses are reflective of the real conditions/situations. Hence, it is important to prioritise what needs to be addressed in a hierarchical setting, by also creating a possibility for specific services to address the needs of vulnerable stakeholders and communities. Lastly, it is advisable to continue with cross-sectoral plans in order to provide support internally and externally at the time of need.

4.3.3 Evaluating Social Services (Under Services/Supplies Dimensions)

For social services, the priority consideration also changes progressively, starting with a comprehensive risk assessment, towards two phases of prioritisation plan, and three last phases of adaptive maintenance. Social services are a crucial part of the provision indicator category, addressing both dimensions of services and supplies. Hence, their issues should be addressed carefully throughout the early phases, as they progressively develop to become a major of the post-recovery phase. In reality, social services help to sustain the provision of certain services and also provide support to necessary systems, such as healthcare systems. Their positions are usually embedded in the socio-economic structure of the city; hence, they are recognised as both services and supplies that maintain the societal wellbeing. In order to sustain the smooth operations of primary social services, it is essential to have an early preparedness plan and to provide necessary financial support (refer to Table 4.2).

Phase 1—In this phase, we expect risk assessment to be the priority. This is due to unforeseen impacts on some of the primary operations of social services at both community and city scales. These impacts become more perceptible in later phases. Hence, it is important to develop a thorough prioritisation plan in the first phase. The categorisation of primary and secondary social services would help to ease the later decision-making processes, in terms of adaptive measures and imposing restrictions on selective services and their operations. The risk assessment plan should include a concise adjustment plan for vulnerable groups of society. It is also recommended to involve relevant stakeholders in the process of assessment, ensuring that their conditions and needs are considered from inception. The impacts are yet minimal but careful planning would reduce the later impacts on primary social services.

Phase 2—It is expected that by now the prioritisation plan is fully implemented in line with a continuous risk assessment. The response and containment phase may potentially bring some uncertainties to both primary and secondary social services. Hence, the involvement of relevant stakeholders, ideally the representatives of social services, is essential to strengthening the contingency plans. Through an enhanced prioritisation plan, it is recommended to assemble relevant information about the needs and potential shortfalls of all social services. This assessment-based approach should verify what are the priorities? And how each priority is dealt with responsively but differently? Hence, it is advisable to make a scenario-based assessment for primary social services, as it is likely they remain in full or partial operation.

Phase 3—The priority consideration of this phase remains the same as in the previous phase, meaning that we have to maintain a continuous prioritisation plan for the support of decision making at the city management level. It is encouraged to start with the development of an adaptive maintenance plan that requires to be implemented and practiced from the next phase onwards. This adaptive maintenance plan should propose certain restrictions, monitory measures, and even policy guidelines. The readjustments made from this proposal are expected to get implemented by the city authorities (if not higher levels) who should then take a reflective approach in the assessment of the outbreak progression and the impacts on all social services. The city authorities (or even national-level authorities) should develop and implement monitory measures on price control and hoarding activities associated with primary products (particularly food, sanitation, and anti-infection products, medical products, and medicine). Early adaptive maintenance should be structured to address any of the shortfalls in primary social services. It is advisable that, through a comprehensive prioritisation plan, secondary social services have limited or no operations at this phase.

Phase 4—By now, we expect adaptive maintenance plans to be fully developed and implemented. This should occur through an all-inclusive reflective approach, representing the realities of the situation and responding to them promptly. It is highly recommended to keep vulnerable groups as the priority while conducting a revised risk assessment of the outbreak progression. It is also recommended to maintain accessibility to primary social services, and sustain their partial operations. The relevant social services should be consulted by the city authorities, and necessary financial support should be provided to them inconsistent with the already established prioritisation plan. The consultation should address any deficiencies and suggest potential methods of securing external support (in emergency cases). Social services should implement the adaptive maintenance plan and work closely with the city management team. This co-planning mechanism should help to provide necessary supplies and services at the time of need.

Phase 5—The hierarchy of primary considerations remains the same as in the previous phase, which includes a high priority for adaptive maintenance in the combination with continuous risk assessment. As the city progresses in the recovery phase, the adaptive maintenance plan should develop and expand to include secondary social services. In doing so, the suggestions are to maintain restrictions, consultations, and monitory measures. By bringing back the gradual operations of secondary social services, the city authorities can start lifting some of the high-level restrictions. Nevertheless, it is still advisable to limit the operations of all social services and ensure priorities are given to certain groups and stakeholders (i.e. as suggested by the earlier prioritisation plan). Most important of all, it is recommended to start working on the redevelopment of societal resilience that should be implemented through social services and by the end of this phase.

Phase 6—In this phase, adaptive maintenance remains the priority. As the general outbreak risks are expected to reduce by now, we expect the city management team to develop a new prioritisation plan. This approach should help to reflect on the realities of social services, particularly that they may have faced a period of decline during the whole outbreak progression. Depending on the length of the disease outbreak, it is vital to assess the conditions of both primary and secondary social services. The situations can change over a longer period of the outbreak than those sudden and quick events. As a result, it is advisable to allocate new financial support to revitalise the full operations of all social services as early as possible. In doing so, the city could safeguard the faster process of societal revitalisation through the power of social services. It is also advisable to put risk assessment as a lower priority, and in its place focus more on adaptive measures, which could be utilised to nurture a large array of social services. The whole process could benefit from close and consistent consultation with social services and the communities, endowing their positions in the post-recovery phase.

4.3.4 Evaluating Food Supply (Under Supplies Dimension)

For food supply, the priority consideration also changes progressively, starting with an early prioritisation plan, towards three phases of risk assessment, and two last phases of adaptive measures. As discussed earlier in Chap. 3, food supply is separated from other amenities supply in a category of its own. This is mainly due to its significance and the multiple factors it embraces for urban resilience and city management. To secure food supply as a major and regular provision, the city authorities have to respond to issues associated with its four factors of safety, production, distribution, and provision. It is vital to keep consistency in all four factors in order to sustain food supply for the city and its communities (refer to Table 4.2).

Phase 1—At the earliest possible time, it is vital to develop a robust prioritisation plan for food supply. This can help to reduce any vulnerabilities, such as food safety, stock shortages, and delivery complications. This planning should be conducted at a time when there is little risk and no visible impact on food supply. Hence, it is important to keep records of multiple factors associated with food supply to ensure relevant authorities and stakeholders are involved in the initial risk assessment. It is recommended to keep monitory of internal and external food production and food distribution networks. This should be provided in a form of report or guideline, enabling the stakeholders to have an overview of the overall operations and patterns of food supply to the city and its communities. Moreover, an early evaluation of public-private partnerships (PPPs) in food supply and its provision would be advantageous to overcome later issues of shortages. These early PPPs would help the later production shifts, if needed.

Phase 2—In this phase, we anticipate the priority to change into risk assessment; an arrangement that remains the same until the end of the transition phase. It is suggested to conduct risk assessment progressively and reflectively, enabling the opportunity for direct consultation with food suppliers, food distribution centers/factories/enterprises, and food safety officers. A holistic risk assessment also sheds light on any issues or deficiencies in the whole cycle of the food supply. It is recommended to conduct these assessments in the procedural approach, which should then provide a reflective narrative for issues of each phase. It is highly recommended to keep an eye on any unexpected changes, including issues of food safety in particular. An adaptive maintenance plan should start developing gradually and in line with the conditions of the outbreak progression. It is suggested to introduce and practice rationing measures from this phase until the stabilisation of the food supply chains. In doing so, we can partially control issues of panic-buying and ensure all people have access to essential goods.

Phase 3—In this phase, we expect to see further challenges, such as potential shortages and disruptions in delivery and distribution services. We expect larger supermarkets to reduce operations or have temporary closures. To avoid this, it is important to conduct an early risk assessment and sustain the operations of the main supermarkets of each community. If their closure and reduced production are inevitable due to conditions of the disease spread, it is recommended to then allocate food supply at the community level. With reinforced monitory measures, it is advisable to give priority to local shops and smaller supermarkets, which are accessible within a walkable (or cyclable for larger cities) catchment zone. By empowering those micro-scale food supply networks, we increase the level of provision as well as the level of community accessibility at a small scale. In return, this approach may help to reduce the need to commute for long distances and by vehicular transportation modes. The most perceptible impact, which is temporary, is the reduction of a variety of products. Hence, the city needs to rely more on local and regional food production, and compensate for any shortages through regional support. Other recommendations in this phase, include the availability of an online food delivery system, which should remain operational with careful safety and monitory measures. The contactless food supply helps to maintain the safety of this important provision. More important of all, there is a need for regulatory measures and their monitory for healthy food supply, and methods of distributions in the city. If needed, it is suggested to tighten ration measures and provide new mechanisms for food delivery to vulnerable groups and essential workforces.

Phase 4—The hierarchy of primary considerations remains the same as in the previous phase, which includes a high priority for risk assessment in combination with a continuous prioritisation plan. In this phase, careful measures should be in place to monitor food production and food delivery systems. The safety of all operations is essential and should be reported through an all-inclusive risk management plan. It is recommended for the city authorities to work closely with those involved stakeholders and develop an adaptive maintenance plan, which can then be utilised and implemented in the new two phases of outbreak progression. It is also highly recommended to keep necessary reinforcements and restrictions in place, which would help to avoid any further disruptions.

Phase 5—As we progress into the recovery phase, the priority consideration should be changed to adaptive maintenance. This should have been gradually developed throughout the last one or two phases of the outbreak progression. By implementing the adaptive maintenance in the practice of city management, we can provide the necessary support to those shortages that may have been experienced in the last two phases. It is advisable to implement this plan in combination with a reflective risk assessment. Further recommendations are to continue with consultations, regulatory reinforcements, and micro-level support. Selective restrictions can be eased as the progress continues with some signs of improvement. The online food delivery system should become more active and food safety measures should remain as the priority. Moreover, food import and export can start their gradual operations under careful safety and monitory measures.

Phase 6—The main difference in this phase is to develop a new prioritisation plan, while adaptive maintenance remains the priority. This new planning approach should be all-inclusive and respond to all four factors of the food supply. The regulatory measures can ease restrictions but should put food safety and production as their priorities. Depending on the nature of the disease, there may be some adjustments for the local and regional food production. This can intensify some of the existing regulations or push for new policies that address healthier food production. While it is recommended to continue with regular risk assessment plans, it is advisable to have more adaptive measures in place. Moreover, it is recommended to continue with consultations and inform the general public with the progress.

4.3.5 Evaluating Amenities Supply (Under Supplies Dimension)

For amenities supply, the priority consideration is also progressive, and the same pattern is recommended as the one described for food supply. In the first phase, we start with a comprehensive prioritisation plan across multiple amenities, which is then followed by a progressive and responsive risk assessment in the next three phases. Similar to the food supply, the last two phases would require to implement a robust adaptive maintenance plan and make up for any losses or shortages. While amenities supply covers a larger number of services, systems, and supplies, it is important to have both cross-sectoral and sector-based measures in place. Each particular urban system should be assessed according to their conditions, needs, and risks through the continuous risk assessment of amenities supply (refer to Table 4.2). The following practical recommendations are all-inclusive and based on four main categories of amenities as mentioned in Chap. 3 (see Sect. 3.5.5).

Phase 1—In this phase, the development of an early prioritisation plan is highly recommended. Hence, it is suggested as the priority consideration, addressing the complexity of services across all four categories. Nevertheless, we anticipate no impacts on these services at this stage. The city authorities could start liaising with relevant stakeholders of multiple amenities and could conduct an early risk assessment. This can be done through a sector-based approach, which could then provide a briefing report document to the authorities. While there are no major adjustments needed at this stage, it is vital to proceed with the development of precautionary measures.

Phase 2—The impacts on amenities supply are relatively similar to phase one, or are expected to be very minimal. We anticipate the first tangible impacts on basic amenities, through which the prioritisation plan should have raised safety and monitory measures of those specific amenities. From this phase, the provision and full operation of basic amenities should be secured. Safety measures should be in place and on alert as the city prepares its responses. Also, preventive measures should be in place as the city aims to reach containment. In this phase, the city authorities should involve the relevant stakeholders in a procedural approach of continuous participation, through which healthy consultation should be maintained. In return, all amenities should work in line with official recommendations and ensure their operations are safe and any shortages are addressed promptly. It is advisable to involve trustworthy community representatives to ensure those amenities are responding to the needs of society progressively. It is recommended to start working on the adaptive measures and identity replacements for the provision of secondary amenities.

Phase 3—In order to keep the orderly operations of the basic amenities, risk assessment remains the priority in this phase. It is expected that by now, secondary amenities have already reduced their operations or are temporarily closed. The basic amenities in form of supply (e.g. water, energy, internet, etc.) should be placed as the priority, while those in form of facilities and infrastructures (e.g. elderly health care units, roads, hospitals, etc.) should increase their level of safety and monitory measures. For the latter, the city authorities may reduce or limit access and mobility to and from them. While essential needs should be maintained, it is recommended to avoid cross-sectoral decision making. It is also recommended to impose restrictions on secondary amenities, and only allow essential operations to continue. In this phase, closure of public facilities is highly recommended from the start as they have a higher risk of disease spread. This also applies to larger groups of entertainment facilities, education buildings/zones, restaurants, public buildings, and some restrictions on public places with multiple access points. The operation of some of these amenities can be replaced by online provisions, such as for education, entertainment, online food delivery, etc. This approach reduces the chance of overcrowded environments, mobility, and large scale transportation. It is recommended to also minimise the provision of mobile amenities, while the opportunity for the provision of virtual amenities can be strengthened.

Phase 4—In this phase, the priority remains a risk assessment. This approach should include a reflective prioritisation plan that prioritises both basic and virtual amenities. It is recommended to keep the provisions and operations of amenities at the same level as in phase 3, and in order to reduce risks of the disease transmission. It is vital also to continue with the provision of basic and virtual amenities, while the same restrictions or limitations are applied to the other two categories of amenities (namely secondary and mobile amenities). The provision of basic and virtual amenities should be maintained in a well-ordered manner as they can also compensate for some of the provisions of secondary and mobile amenities. Moreover, it is recommended to increase the level of community consultation to continue with the provision of basic needs that reflect on societal wellbeing and quality of life. In addition, it is advisable to keep high measures of safety and monitory across all amenities and sectors.

Phase 5—In this phase, the priority changes and remains as adaptive maintenance for the next two phases. For instance, the utilisation of our current digital technologies can help to replace some of the operations through digital and virtual platforms. This should have already developed in the last two phases, and should be extended and encouraged more widely. In this regard, adaptive maintenance should develop in the form of adaptive planning (Alterman 1988), which suggests relevant readjustments when and where needed. It is highly recommended to keep risk assessment of individual sectors or amenities in place, in order to avoid generic decision making. While some flexibilities could be allowed in this phase, it is recommended to keep basic amenities as the priority. In a gradual pace, the city authorities can relax some of the restrictions or limitation measures on less risky secondary amenities. For instance, it is advisable to discourage the provision of secondary amenities that encourage populated environments, closed crowded environments, and a high rate of mobility. These should happen through a careful density control plan or partial lockdown measures. It is important to closely monitor any mobility and travels from other infected locations. This requires some temporary regulations for the enhanced safety measures, mobility checks, community-level monitory, and even having a possibility of self-isolation and quarantine controls. In the form of immediate regulations, these should be carefully monitored and get readjusted in a reflective approach.

Phase 6—In this phase, we anticipate gradual progress in adaptive planning measures that were implemented in the previous phase. The priority remains for basic and virtual amenities, their provisions and operations. A revised prioritisation plan should be developed in this phase to allow the gradual operations of all other amenities in a timely manner. It is advisable to continue with sector-based assessment and planning to ensure generic decisions are not made for all sectors. It is recommended to continue with risk assessment of individual sectors, and provide accurate data and guidelines to the city management team and relevant authorities. The main suggestions are to approach post-recovery with attentiveness and at a gradual pace, to reflect on the needs of the secondary amenities, to provide necessary help to compensate for any losses or shortages (i.e. particularly for the secondary and mobile amenities), and to respond to the needs of society through mechanisms of enhanced engagement (Houlihan and Whitworth 2019) and support. Lastly, it is important to regain the position of amenities supply as the primary instrument for recreating and revitalising the societal wellbeing and quality of life.

4.4 City Management Reactions at the Time of Need

City management and governance structures follow their institutional arrangements (Raven et al. 2017; Dawodu et al. 2018; Yahia et al. 2019; Wang et al. 2020b), and they can differ from one context to another (Cheshmehzangi and Dawodu 2018). Depending on the nature of such arrangements and structures (of the city management), the city has to allocate duties to respective city authorities, bureaus, and any other responsible managerial units. But there are no universal models of city management structures that can be suggested as a guideline here. Therefore, in the early phase of outbreak progression, the city authorities need to provide clarity and identify the roles and duties of each unit. By clarifying the responsibilities of main actors and their secondary stakeholders through a holistic action plan, we increase the possibility of having higher quality responsiveness. In this systematic approach, the city authorities can establish formal channels of communication and inform the communities regarding such arrangements. This approach of formalisation must be kept throughout the outbreak progression. This must be developed in line with the involvement of communities through various means and opportunities. Hence, it is required to have a co-planning approach to city management and its responsive operations; one that provides adequate flows of precautionary measures, safety and monitory measures, and control action plans.

It is important to note that the foremost grounds of malpractice in city management are often associated with the ambiguity of duties for specific actors, the unpredicted miscommunication between the actors, and lack of institutional support and operations. All three factors are crucial in strategic planning approaches, as well as in the overall procedures of outbreak control and resilience enhancement of the city. Hence, city management should be conducted in a structured approach, which is reflective and reliable. In doing so, the city can become more resilient, and its communities shall be more supportive and proactive. The structured approach to city management should be processed systematically and through co-planning arrangements, such as a detailed systematic framework (see Chap. 3), a set of influential instruments, guidelines, or through comprehensive strategic planning. And in return, city management reactions should not be improvised, and nor should they be decided without continuous consultations. The city authorities should avoid random decision making and random responses to the outbreak progression. They should rather focus on structured management planning approaches. Thus, these approaches are expected to be all-inclusive, reflective, and progressive. Through a systematic structure of who is doing what, and what needs to be done, the city authorities can define the reactions clearly and allow for synergies in between the management of the city. Also through a scenario-based approach, we anticipate careful monitory of the progression. Hence, the city needs to be well-prepared through enhanced urban resilience, and ought to be responsive through well-structured city management.

To summarise, this chapter highlights the practicalities of city management, those that can only be achieved through comprehensive assessment, reflective decision making (da Silva et al. 2012; Siemens, Arup, and RPA 2013; Fabbricatti and Biancamano 2019), and responsive implementation. By understanding the dynamic conditions of the outbreak progression, this chapter delivers a step-by-step plan for all defined target groups under the overarching dimensions of operational, institutional, services, and supplies. Hence, what we conclude here is not necessarily representative of by-default procedures, but rather what needs to be taken forward and in a broader understanding of responsiveness through adequate city management. This ideology goes back to our earlier introductory discussions on the role of holistic planning (see Sect. 1.1), which should be characterised as predictive, prescriptive, and preventive. Hence, the city's responsiveness should also follow the same mindset. In doing so, the city authorities could provide effective and realistic pathways towards outbreak control and full containment. The city then needs to recover, and should progress to regain its position; but this time, through a much-enhanced resilience. It is only then that we know we can save the city in need.

References

100 Resilient Cities and EY (2017a). Resilience thinking: Rhetoric and reality. Retrieved March 3, 2020, from http://100resilientcities.org/resilience-thinking-rhetoric-reality/.

100 Resilient Cities and EY (2017b). Getting real about resilience: How cities can build resilience thinking into infrastructure projects. Retrieved March 3, 2020, from http://100resilientcities.org/wp-content/uploads/2019/02/EY_100RC_Getting-Real-About-Resilience_FINAL.pdf.

Algan, Y. (2013). *Public policy and trust*, from Organisation for Economic Co-operation and Development (OECD) workshop on "joint learning for an OECD Trust strategy". Retrieved February 27, 2020, from https://www.oecd.org/governance/publicationsdocuments/articles/6/.

Alterman, R. (1988). Adaptive planning. *Cognitive Science, 12,* 393–421.

Amadeo, K. (2020). *SARS, Ebola, coronavirus: How disease outbreaks affect the economy*, the balance news material. Retrieved February 27, 2020, from https://www.thebalance.com/corona virus-plague-ebola-economic-impact-4795744.

Argonne National Laboratory. (2012). *Resilience: Theory and applications*. Argonne National Laboratory: US Department of Energy, Office of Scientific and Technical Information. Retrieved February 29, 2020, from http://www.osti.gov/bridge/product.biblio.jsp?osti_id=1044521.

Cheshmehzangi, A. (2020a). *Comprehensive urban resilience for the city of Ningbo* (in Chinese: 宁波市城市综合抗灾弹性框架). Report submitted to local government units in February 2020, Ningbo, China.

Cheshmehzangi, A. (2020b). *Identity of city and city of identities*. Singapore: Springer.

Cheshmehzangi, A. (2020c). 10 Adaptive Measures for Public Places to face the COVID-19 Pandemic Outbreak. City & Society, Article ID: CISO_12282, https://doi.org/10.1111/CISO.12282.

Cheshmehzangi, A., & Dawodu, A. (2018). *Sustainable urban development in the age of climate change–people: The cure or curse*. Singapore: Palgrave Macmillan.

da Silva, J., Kernaghan, S., & Luque, A. (2012). A systems approach to meeting the challenges of urban climate change. *International Journal of Urban Sustainable Development*. https://doi.org/10.1080/19463138.2012.718279.

Dawodu, A., Cheshmehzangi, A., & Akinwolemiwa, B. (2018). The systematic selection of headline sustainable indicators for the development of future neighbourhood sustainability assessment tools for Africa. *Sustainable Cities and Society, 41*, 760–776.

Ebisudani, M., Kishimoto, S., Yamaguchi, H., Nakakubo, T., & Tokai, A. (2017). An integrated measurement framework of city resilience for preparedness: A case study for Japan. *Journal of Sustainable Development, 10*(6), 106–123.

Elmqvist, T. (2014). Urban resilience thinking. *Solutions, 5*(5), 26–30.

Eraydin, A., & Taşan-Kok, T. (Eds.). (2013). *Resilience thinking in urban planning*. Dordrecht: Springer. GeoJournal Library 106.

European Commission. (2006). *European programme for critical infrastructure protection (EPCIP)*, COM 786 final. Brussels: European Commission

Fabbricatti, K., & Biancamano, P. F. (2019). Circular economy and resilience thinking for historic urban landscape regeneration: The case of Torre Annunziata, Naples. *Sustainability, 11*(3391), 1–29.

Harrison, P., Robbins, K., Culwick, C., Humby, T.-L., La Mantia, C., Todes, A., et al. (2014). *Urban resilience thinking for municipalities*. University of the Witwatersrand, Gauteng City-Region Observatory. Retrieved March 3, 2020, from http://wiredspace.wits.ac.za/jspui/bitstream/10539/17082/1/URreport_1901MR.pdf.

Houlihan, C., & Whitworth, J. (2019). Outbreak science: Recent progress in the detection and response to outbreaks of infectious diseases. *Clinical Medicine, 19*(2), 140–144.

Huber, C., Finelli, L., & Stevens, W. (2018). The economic and social burden of the 2014 Ebola outbreak in West Africa. *The Journal of Infectious Diseases, 218*(supply 5), S698–S704.

Huck, A., Monstadt, J., & Driessen, P. (2020). Building urban and infrastructure resilience through connectivity: An institutional perspective on disaster risk management in Christchurch, New Zealand. *Cities, 98*, 102573.

Kenney, C., & Phibbs, S. (2014). Shakes, rattles and roll outs: The untold story of Māori Engagement with community recovery, social resilience and urban sustainability in Christchurch, New Zealand. *Procedia Economics and Finance, 18*, 754–762

Linkov, I., Bridges, T., Creutzig, F., Decker, J., et al. (2014). Changing the resilience paradigm. *Nature Climate Change, 4*(6), 407.

Luyten, J., & Beutels, P. (2009). Costing infectious disease outbreaks for economic evaluation: A review for hepatitis A. *Pharmacoeconomics, 27*(5), 379–89.

Marana, P., Eden, C., Eriksson, H., Grimes, C., Hernantes, J., Howick, S., et al. (2019). Towards a resilience management guideline—cities as a starting point for societal resilience. *Sustainable Cities and Society, 48*, 101531.

Ministry of Health (MoH). (2014). MoH Pandemic Readiness and Response Plan for Influenza and other Acute Respiratory Diseases (Revised April 2014), online 32 pages document. Retrieved

February 28, 2020, from https://www.moh.gov.sg/docs/librariesprovider5/diseases-updates/int erim-pandemic-plan-public-ver-_april-2014.pdf.

Monstadt, J., & Schmidt, M. (2019). Urban resilience in the making? The governance of critical infrastructures in German cities. *Urban Studies, 56*(11), 2353–2371.

Raven, R., Sengers, S., Spaeth, P., Xie, L., Cheshmehzangi A., & de Jong, M. (2017). Urban experimentation and institutional arrangements. *European Planning Studies, 27*(2), 258–281. Urban Experimentation & Sustainability Transitions.

Sellberg, M. M., Ryan, P., Borgström, S. T., Norström, A. V., & Pereston, G. D. (2018). From resilience thinking to resilience planning: Lessons from practice. *Journal of Environmental Management, 217,* 906–918.

Siemens, Arup, and RPA (2013). *Toolkit for resilient cities: Infrastructure, technology and urban planning*, 60 pages document. Retrieved January 24, 2020, from https://assets.new.siemens.com/ siemens/assets/public.1543066657.641ee2256c5a0d5919d1aa3094a701f6ec9c3f90.toolkit-for-resilient-cities.pdf.

Singleton, C. D., Fey, R., & Appleby, C. (2000). Media management of a community outbreak of meningococcal meningitis. *Communicable Disease and Public Health, 3*(4), 267–270.

Wang, H., Cheng, Z., & Zhu, D. (2020a). Striving for global cities with governance approach in transitional China: Case study of Shanghai. *Land Use Policy, 90,* 104288.

Wang, Y., Wang, X., Li, H., Dong, Y., Liu, Q., & Shi, X. (2020b). A multi-service differentiation traffic management strategy in SDN cloud data center. *Computer Networks, 171,* 107143.

World Health Organisation (WHO). (2020). *What is early recovery?*. Retrieved March 1, 2020, from https://www.who.int/csr/disease/ebola/health-systems-recovery/early-recovery/en/.

Yahia, N. B., Eljaoued, W., Saud, N. B. B., & Colombo-Palacios, R. (2019, in press). Towards sustainable collaborative networks for smart cities co-governance. *International Journal of Information Management,* 102037.

Chapter 5
Reflection on Disruptions: Managing the City in Need, Saving the City in Need

We will neglect our cities to our peril, for in neglecting them we neglect the nation.
—John F. Kennedy.

5.1 The Anatomy of the City in Disruptive Times

At the time when mostly negative news flow between countries, the reactions of national-level and city-level governments should reflect on their full preparedness and responsive management. This book has covered these aspects in Chaps. 3 and 4. We know by now how cities are affected, how they need to be prepared, and in which ways they have to react to disruptive disease outbreak events. At the city level, changes are progressive, sudden transformations are inevitable, and impacts are tangible. In reality, the city's anatomy changes as it has to deal with new pressures, adversities, and disruptions. In the many examples of scholarly work on the anatomy of the city, focused mostly on how cities operate and function (Ascher 2007; Lamorgese and Petrella 2016; Renner 2018), there are a few that look into aspects of smartness (Al-Nasrawi et al. 2017; Bibri 2019; Desjardins 2019), creative examples (Cohendet et al. 2010), context-specific explorations (Shultz 1968; Quirk 2013; Tillett 2018), etc. Some may look into specific factors of historical analysis (Stone and Zimansky 1992), data-driven approaches (Bibri 2019), mobility analysis (Gallotti and Barthelemy 2015; Jauregui-Fung et al. 2019), transportation and planning (Goldwyn 2018), or even economic crisis (Baumol 1967). But none has looked into the anatomy of the city due to transformations, and most importantly the ones that are sudden, disruptive, and possibly temporary. The changes occur during the period of outbreak progression are impactful on the city and its communities. There are many factors that indicate and trigger substantial changes in the overall anatomy of the city during an outbreak. Hence, it is important for us to demonstrate some of those changes and how they become disruptions for the city and its management.

A. Cheshmehzangi, *The City in Need*, https://doi.org/10.1007/978-981-15-5487-2_5

So far, we have covered a wide range of viewpoints and approaches to urban resilience thinking. We have also explored how urban resilience plays an important part in better preparedness against outbreak events. We argued this could happen through urban resilience enhancement, which should be systematic and progressive. Later, we looked into specifics of urban resilience measures from the perspective of strategic planning and action plan. These arguments then led to further elaboration on city management issues during all phases of the outbreak progression (see Chaps. 3 and 4). These issues were looked at from the multiplicity of responsiveness, reflectiveness, and integrated thinking. In this chapter, these arguments are illustrated through the narrative of a city example. This is done in a twofold: in the first half, through experience and how the impacts are developed, measured, and perceived; and in the second half, through a reflection on what occurs and what are the reactions. In the first section covering the 'experience', disruption factors are discussed from five perspectives of society, safety and monitory, adaptability and resilience, economy and governance. These are then followed by highlighting the realities in the second section on the matters of 'reflection', capturing and covering key factors under four resilience dimensions (from Chap. 3), namely operational, institutional, services, and supplies (Cheshmehzangi 2020a). In this chapter, we only cover selective matters that are relevant to the central theme of the book. Hence, they mainly represent an example of an outbreak and its disruptions. In sum, this chapter serves as a case study/example chapter, which is narrated through perceptions, reactions, and reflections.

5.2 An Experience: Narrative of a City Example

On 23rd of January 2020, one day before the official celebration of the Chinese New Year, the source location of COVID-19 disease was under full lockdown. This was the populated City of Wuhan, the capital of Hubei Province in Central China (officially estimated more than 11 million people for the City of Wuhan, and approximately 60 million for the province). Weeks after the start of the outbreak, the number of cases was soaring rapidly. The new transmission cases were increasing beyond the boundaries of Wuhan. It only took a few days that a disease outbreak turned into a major epidemic outbreak event. Several weeks later, it was eventually declared a pandemic event. On 30th of January 2020, the International Health Regulations Emergency Committee of the World Health Organisation, declared the outbreak a "*public health emergency of international concern*" (CDC 2020a). Within one month after the start of Wuhan's lockdown, there were similar measures in not only other cities or towns of the same province, but also across the country. These measures were then replicated by other countries, too. By then, some of the major industrial provinces of the country were under restricted lockdown. While China was facing the epidemic through heightened control and monitory of many cities and provinces during its busiest period, the disease was already globally widespread. On the verge of a major pandemic outbreak, by end of February 2020, there were records of infected cases in all continents (with an exception of Antarctica). Under the likely threat of further spread, the situation worsened from the 20th of February onwards. By then, the global attention shifted from a one-country problem to a large-scale global health emergency. The economic impacts became global, rippling on every single sector. The biggest hit on oil production with China's reduced demand (i.e. estimated at 20% reduction in the first weeks of the event) was of the earliest news about the economic impacts of the disease. From an economic perspective, the situation turned into a global economic crisis. Soon later, the hit was on all sectors, affecting production, trade, marketing, transportation, supply chains, construction, medical, and all other industries. Somehow, the world was no longer just aware, it awakened!

As per records, the early signs of the disease cases were detected in the last days of December 2019. Assuming that it could get contained in that early phase of uncertainties, early containment could simply not happen. Conversely, the first sign of the novel coronavirus was detected earlier, and was acknowledged on the 10th of December, at least two or three weeks before it was an evident case of a novel disease. The official statement by the Chinese government to the World Health Organisation (WHO) confirmed the first diagnosed case was on the 8th of December 2019. Some unpublished data report traces of the disease as early as the 17th of November 2019. Nevertheless, without confirmed cases of human-to-human transmission and owing to a lack of knowledge about this novel disease, the official announcement of the outbreak happened later in January 2020. The source was then identified from a seafood merchant in the Hua'nan market of Wuhan, which was later closed as a result of this incident. But unfortunately, the transmission was inevitable, and community spread was already rapidly growing, from one case to another, and possibly from one location to another, too. By the 9th of January 2020, there were signs of the cluster of cases in Wuhan (WHO, an online source, 9 January 2020). The control was still not successful and containment could not happen. In the midst of uncertainties, the leaders of the epidemic response officially announced the outbreak on the 20th of January 2020. By then, the spread was partly out of hand. This made Wuhan become the original epicenter of this novel disease. Three days later, the city was under full lockdown. In some way, we can argue the disease outbreak happened at the wrong time and in the wrong place. It occurred just before the largest annual mobility of people (i.e. during the Chinese New Year) across the globe, and right in the most central node of the country, where the City of Wuhan is located. The City of Wuhan is the 42nd most populated city in the world, a first-tier city and one of the main populated urban hubs, which is well-connected to all other populated parts of the country. The city holds a large population of migrant workers and professionals, who had to travel back to be with their families and friends during the New Year celebrations. The city is a major transportation hub, extensively connected by railways and air travel to many major cities at nationwide, regional, and global levels. In particular, the city is a major economic hub connected to multiple regions and countries in East and Southeast Asia. There are direct cross-continental flights in and out of the city and on a regular basis. Thus, the inescapable mobility of the city and the province were unknowingly fuelling up the situation.

By the end of January 2020, it was clear that all Chinese provinces had confirmed cases, and the number of infected cases was already growing outside mainland China, in the region, and in many further locations like Australia, Europe, and North America. Soon after, the disease reached another parts of mainland Europe, North Africa, and the Middle East. The pattern of spread continued to reach more regions, countries, and locations, while also creating several emerging hotspots in different parts of the world. Those new hotspots caused further spread first in their immediate contexts and later exported the disease to other (further) locations. The whole pattern of spread represents an example of how fast a disease can spread through air travel and other modes of transportation; i.e. from one community to another, from one city to another, and so on. The speed of spread was extraordinarily fast, and in some cases occurred so sudden that some countries/cities did not have a chance to be well prepared in advance. The hit was clearly global. While mortality rate remained relatively low (in comparison to previous diseases of the same category), the severity of the disease and the growing numbers of person-to-person transmissions (or community transmission) remained a major worry for many people and governments. The disruptions became global, and many cities and communities struggled to cope and react to rapidly changing situations. There was little time to prepare and the struggle was clearly visible in unstructured announcements of governments in multiple regions.

The virus was initially phrased as a novel coronavirus before it was officially named severe acute respiratory syndrome coronavirus 2 (SARS-CoV-2) on the 12th of February 2020. The delay for such official naming was primarily caused by scientific studies on the virus, which should have collected enough data in order to identify and categorise it in the right way. At the same time, the disease associated with the virus was then renamed COVID-19. Also, this novel disease was officially recorded as the one never tracked in human cases before. While there remain many speculations about the actual source of the disease, to date it remains unclear how it actually developed and how it was transmitted to the first index case (i.e. human). At first, the precautionary guidelines were basic, mostly based on similar experiences of previous diseases or outbreaks (e.g. SARS-CoV and MERS-CoV). It took a while till the society recognised the seriousness of the situation, and by then uncertainties turned into anxiety. This was no longer the anxiety of China, but also became global distress. By the early days of March 2020, the disease already reached some unexpected places in four unreachable locations, like in Brazil, Iceland, New Zealand, and South Africa; hence, posing a larger threat to global health. Later, it reached more unexpected places, such as Greenland, Suriname, Madagascar, Papua New Guinea, etc.

It was as early as the 28th of January that we started to notice the gradual control measures of lockdown in some other cities outside China's Hubei Province. The steps were taken cautiously and gradually, but within just less than a week, many cities were under full control measures of cordon sanitaire and lockdown. The unprecedented public health measures affected millions of people, changed patterns of mobility, and had significant impacts on society. The City of Ningbo, a major port located in Zhejiang Province, in East China was one of these city examples. The unprecedented lockdown measures brought in significant control and monitory to all operations, to all mobility, and across all sectors. By the 7th of February 2020, the online food delivery systems were also affected. The society experienced many uncertainties, not knowing how long the situation would last. The city's reaction was essential and immediate support was necessary. The situation was no longer just a matter of pandemic outbreak; it was a major test for the city, and other cities and communities, to enhance their resilience and to manage the outbreak's impacts in the face of new adversities. With the gradual lockdown, the city had to increase its measures for safety, control, and monitory. With depopulated scenes in every part of the city apart from healthcare centers and hospitals, the society experienced a sudden shock; something the society was not used to before, something which will be remembered as a lifetime experience. A ghost-town effect was the highlight of many local and foreign media, something that was later on recognised as a collective approach by many stakeholders who played their part in speeding up the process of containment and reducing the possibility of community transmissions. The same scenes that were later seen and experienced in other countries, too. The city became a platform for innovation, where resilience enhancement was no longer a case of experimentation but merely implementation and success. With high risks of failure, the city had to reflect on the outbreak progression and face the adversities through new and temporary regulations, extended guidelines, enhanced monitory, new legislation, and community support. Lessons learned from this particular instance are invaluable and strategies that we can build upon are certainly important for managing the city in need, and ultimately saving the city in need.

5.2.1 The Experience of Disruptive Time

Disruptions are perceptible across the city as they hit every single sector. Through this, the city struggles to accommodate its daily operations. The overall operationalisation of the city and its urban systems are affected significantly through a sudden shock caused by a particular emergency. It is not yet recognised as a disaster situation unless it is not managed well. As identified by Ritchie et al. (2008) there can be certain aspects affecting the response. And if the response is affected by any means, then the whole process of containment is affected, too. Hence, the experience and impacts of the outbreak follow this major matter. As their findings indicate (ibid) the limited response is under influence of multiple factors, such as lack of preparedness which is common, speed of spread, and speed and severity of media coverage (particularly international media). The need for communication strategies is also highlighted as one of the effective methods to overcome issues of response, and help to overcome some of the disruptions.

In general, what we experience is not a common scenario. Society goes through a high level of uncertainties, with which it is hard to re-adjust day to day routines and the lifestyle that provides us comfort and security. The situation hits the vulnerable and low-income groups the most, with the highest impact on the urban poor often reside in crowded communities of impoverished infrastructures and unhealthy living environments. The risks can also increase drastically with the lack of hygiene or limited, and sometimes ineffective, precautionary measures. More than everything, the impacts on social infrastructure and socio-economic values of the city are of major concern to all stakeholders and communities. Businesses face decline and possible closure (even if temporary), and the society can only cope with all these adversities for a limited time. Hence, having resilience planning in place is ever essential. This impacts not only the governance and management of the city, but everything else that makes up the cityness of the city.

Society

It is noteworthy to observe how society perceives change and adapts to it in a process. The continuous flow of negative and false news impacts how society's perception of the disease can change at any time. Anything that gives a false alarm can be harmful and anything that reduces the society's awareness is even more so. Besides medical support and dealing with the treatment of the disease, a larger requirement is to maintain the mental stability of society. As disruptions impact people's daily lives, it is essential to be adaptive and positive. In order to cope with the transformative conditions of the outbreak, people should enhance their resilience both individually and collectively. The latter is vital to societal well-being as it helps to reunite the society against shared adversity.

In the race to finding the cure, often through the development of a novel vaccine, there seems to be more curiosity from the general public than the scientists' community. People get attracted to floating news and information that could reduce their anxiety, which is a natural self-calming effect needed in such events. Yet, collective measures are much more effective. For instance, society could play a major part in various ways, such as through voluntary support, engagement in community support, security enhancement, training, and education, etc. Only with a resilient mindset, one can overcome the negativities of the outbreak. But the ones affected directly from the disease would suffer the most. Hence, it is likely that maintaining continuous support to them, their pressures could be eased to some extent. Moreover, society's compliance with precautionary measures is highly important. It is only through a collective approach that the society can maintain large-scale self-isolation, and reduce the chances of community spread.

In our careful observations, we noticed the difference between urban and rural communities of the municipality. While we may wrongly assume rural communities are more vulnerable, they are in fact more resilient and self-sufficient, which are the exact characteristics of rural life. The slow pace of rural life remains unchanged (with the only impact on their tourism industry), while the city changes from a busy socio-economic hub into a static environment with limited operations. In other observations, we noticed the vulnerability of people who cannot adapt to sudden change as they may assume disruptions may last long. They may struggle to cope well, but healthier communications from the right sources could support them. Other examples are boredom, anxiety, and regular panic attacks that may affect the mental stability of society the most. We noticed examples of people enhancing their communities even through contactless activities, and provided care and support to those in need.

Approaches that include sharing correct knowledge and experiences could help to relax those individually-perceived adversities.

BOX 5.1 Notice for Preventing 2019 Novel Coronavirus from Spreading (A dual language document was released by Ningbo Science and Technology Bureau)

Recently, many areas have been affected by the 2019 Novel Coronavirus (2019-nCoV) pneumonia. The Chinese government has taken a series of rigorous measures to fully contain the spread of the epidemic (later became pandemic on 11 March 2020). Thus, we propose the following proposals to all foreigners and their employers in Ningbo:

1. **Everyday preventive actions**

Novel Coronavirus (2019-nCov) is a coronavirus, which is identified as the cause of the respiratory disease outbreak in Wuhan, China, and can be transmitted from person to person. Please follow these instructions to protect yourself:

(1) Restrict activities outside your home and avoid group activities such as parties. If you have to go out, you must wear a face mask correctly;
(2) Open windows and ensure good ventilation at home, wash hands frequently and disinfect regularly;
(3) Avoid close contact with anyone who has fever and/or cough;
(4) The consumption of raw or undercooked animal products should be avoided;
(5) Monitor body temperature of yourself and family members.

2. **What should you do if you feel unwell?**

(1) Contact your employers or friends in China and tell them about your status;
(2) According to your condition, visit medical institutions nearby in time, and try to avoid taking subway, bus and other means of public transportation;
(3) Tell doctors about your recent travels and living history especially in relation to epidemic-stricken areas, as well as the people in contact after the onset of the disease, and cooperate with doctors to carry out relevant investigations.

3. **Keep yourself reachable via phone or email**

Please keep the communication channels such as telephone or email unblocked during the epidemic/pandemic. The employer shall pay attention to the physical health and schedules of foreign employees. It is suggested that foreign employees who are not in Ningbo during the holidays may postpone their return.

4. About the extension of work permit service window

According to the arrangement of superior departments, the service time of the work permit service window for foreigners originally scheduled to open on January 31 may be postponed. After confirmation, we will give a further notice. For the applicants affected by this change, please contact desk receptionist to file for expedited service if necessary. We will arrange a green channel for urgent processing and coordinate with relevant departments.

5. Ningbo epidemic prevention and control contact number

Explanation Note: The document includes details of 10 Centers for Disease Control and Prevision, covering all urban and non-urban districts under the municipality of Ningbo (including: Ningbo (main central), Yuyao, Cixi, Xingshan, Fenghua, Ninghai, Yinzhou, Zhenhai, Jianbei, Haishu, and Beilun).

6. Appointed hospitals for 2019-nCoV treatment

Explanation Note: The document includes name and location of 15 medical institutions/hospitals whom accept patients for treatment, including five in the city centre area (Ningbo No. 1 Hospital, Ningbo No. 2 Hospital (or Hwa Mei Hospital, University of Chinese Academy of Sciences), Ningbo Medical Center Lihuili Hospital, The affiliated Hospital of Medical School of Ningbo University, and Ningbo Woman and Children's Hospital); as well as 10 other hospitals at district levels, including: Ningbo No. 9 Hospital (Jiangbei District), Ningbo Zhenhai District People's Hospital (Zhenhai), Ningbo Beilun District People's Hospital (Beilun), Ningbo Yinzhou District People's Hospital (Yinzhou), Ningbo Fenghua District people's Hospital (Fenghua), Yuyao People's Hospital (Yuyao), Cixi People Hospital (Cixi), Ninghai No.1 Hospital (Ninghai), Xiangshan No. 1 People's Hospital (Xiangshan).

Source Extracted from: http://kjj.ningbo.gov.cn/art/2020/1/29/art_9940_4 045698.html. Originally Developed from WHO's Novel Coronavirus (2019-nCoV) advice for the public (original source available from: https://www. who.int/emergencies/diseases/novel-coronavirus-2019/advice-for-public). Adapted from Ningbo Science and Technology Bureau (2020).

BOX 5.2 "A Letter to All Foreign Friends in Ningbo" (This English document is released by Ningbo Foreign Affairs Office)

On the occasion of Chinese Spring Festival, we extend our sincere greetings and best wishes to you all. In response to the outbreak of pneumonia caused by the novel coronavirus (2019-nCoV), Zhejiang Provincial Government has activated Level I public health emergency response since 23rd January. Attaching paramount importance to the situation, CPC Ningbo Committee and Ningbo Municipal People's Government have taken high-level prevention and control measures to curb the spread of the virus. For the health of you and your friends, you are kindly requested to follow the requirements of the law of the PRC on the prevention and control of infectious diseases and other relevant laws and regulations, and respond to the calls of the central and local governments by cooperating in the following practices:

I.   **Follow official updates and do not trust false and unverified information**: Please follow "nbfb0574" (the official Wechat account run by Ningbo Municipal People's Government) and other official media outlets for the most recent updates. Check out the list of designated hospitals in Ningbo, and guides for prevention and treatment. If you have any questions, please call 24-hour hotline 87680042. Avoid believing and spreading false or unverified information.

II. **Keep distance from infectious sources**: Please follow the official release of epidemic/pandemic information and avoid visiting epidemic areas or contacting anyone with cold or flu-like symptoms. Avoid close contact with wild animals or birds; eggs and meat must be thoroughly cooked before consumption. Avoid crowded places, parties and group activities. Correctly wear a mask (preferably a surgical mask) if you have to go out.

III. **Keep personal hygiene**: Keep good hygiene by washing your hands often with soap or alcohol-based sanitizer. Cover a cough or sneeze with a tissue or your flexed elbow. No spitting. Dispose of used masks appropriately. Maintain proper ventilation and keep rooms clean.

IV. **Join in our mass prevention and control efforts**: If you have travelled to the infected areas recently, or have contacted anyone who is from the epidemic areas, please immediately report to the community where you reside, cooperate with community or medical staff if they inquire about your condition, and stay at home under medical observation for 14 days.

V.   **Ask for hospital treatment immediately**: If you have a fever over 37.3°C, cough, fatigue or breathing difficulty, please wear surgical mask, avoid contacting anybody and immediately go to the designed hospitals for treatment. Please do not forget to tell the doctor your travel record.

Dear foreign friends, you have contributed to the economic and social development of Ningbo City. Ningbo is our home and we need to stay unified in solidarity to protect and defend our city in the time of difficulties.

Foreign Affairs Office, Ningbo Municipal People's Government, 30th January, 2020.

Note A later updates of the letter was also provided on 28 Feb 2020—*Extracted*

from original source http://english.ningbo.gov.cn/art/2020/2/1/art_931_104 1824.html.

BOX 5.3 Updated version: "A Letter to Foreign Nationals from Ningbo Government" (This English document is released by Ningbo Foreign Affairs Office)

Welcome (back) to Ningbo. In response to the outbreak of the COVID-19 epidemic (later pandemic on 11 March 2020), under the leadership of the Central Government of China, Ningbo Municipal People's Government has put the life and health of its people first and taken prompt action in a science-based and coordinated way. We have implemented the most comprehensive, rigorous and thorough measures of prevention and control, which have produced notable results. We have managed to maintain economic and social stability throughout the city, and currently businesses are re-opening and people returning to their normal life. However, we still face severe pressure from the epidemic as its cuts through borders and spread further afield. Whether you are living, working or travelling in Ningbo, we believe it is important to remind you of the following.

I.   **Cooperate in health checks for epidemic prevention and control**. On occasions (or in places) required by the government, please understand and cooperate with field staff in carrying out health checks. As arriving at your destination, please abide by the prevention and control measures implemented by Ningbo Municipal People's Government. If you come from countries or regions with large number of infected cases, or have contacted anyone who is from the epidemic areas, please immediately report to the community (hotel) where you reside, and try the best to self-quarantine at home for 14 days before engaging in other activities so as to stern possibilities of transmission.

II. **Step up self-protection**. Wear face masks and wash hands as often as needed. Maintain proper ventilation and keep a clean living and working environment. Avoid crowded places, parties or group activities. Have safe and healthy food. Stay optimistic and have proper exercise to improve your immune system.

III. **Ask for medical help immediately**. If you have a fever, cough, fatigue, breathing difficulty or other symptoms, please wear a mask and stay calm. Report promptly to your host or community (hotel), local health or immigration agencies, and go to the designated hospitals for treatment as advised.

IV. **Follow official updates and do not trust false and unverified information**. Please follow "http://weibo.com/nbfb" (the official Weibo account run by Ningbo Municipal People's Government) and other official media outlets for the most recent updates. Check out the list of designated hospitals in Ningbo, and guides for prevention and treatment. If you have any questions, please call 24-hour hotline 87680042. Avoid believing and spreading false or unverified information. You can also get into contact with Foreign Affairs office of Ningbo Municipal Government or further assistant.

Epidemic prevention and control needs the understanding, support, cooperation and participation of citizens as well as foreign nationals in Ningbo. Ningbo Municipal People's Government attaches great importance to the health and safety of foreign nationals in Ningbo. We have full determination, confidence and capability to win the battle against epidemic, and contribute to the international public health. Service Hotline of Ningbo Municipal Foreign Affairs Office: 89186517, 13805861661

Foreign Affairs Office, Ningbo Municipal People's Government, 28th February, 2020.

Extracted from third party source https://www.nottingham.edu.cn/en/Campus-Portal/News/Article.aspx?id=9e26c727-9fdf-413b-b3e7-1f38f275815d&language=en-GB.

In addition, four main areas of guidelines developed by the World Health Organization (WHO) (2020), specifically titled: "Coronavirus disease (COVID-19) advice for public", include precautionary measures, education of right and wrong information, safety measures of health workforces and preparation plans for work environments. Their four main areas are named 'when and how to use masks', 'myth-busters', 'advice for health workers', 'getting workplace ready'. Each area includes a set of guidelines, explained through text and several downloadable files in the form of posters. For instance, the myth-busters area is one of the most extended areas that include a range of factors and graphical posters. Those informative and educational posters include a range of facts regarding the disease, its transmission, and many issues that are identified as myths in particular contexts. Examples of these myths or wrong perceptions include: transmission of the disease through mosquito bites, methods of killing the virus, updates on the vaccine development (also include any misinformation or misconception), information on specific food consumption (especially those that are advised in different cultures), etc. It is important to note that as our knowledge of the disease develops, these posters and information also get upgraded. They are made available for the general public as well as for companies/enterprises to display them. The main intention is to raise awareness and educate the majority; a task that requires a lot of reinforcement and collective support.

Safety and Monitory

It is simply not adequate enough to merely depend on the placement of safety and monitory measures. More important than their placement, they should be regulated and the society also needs to comply with them. This sense of responsibility should be widespread and must be practiced throughout the communities. This is for the benefit of individuals, the wider community and those who are responsible for the delivery and implementation of those measures. Examples of these are collective reduction of unnecessary mobility, increased remote working, increased contactless activities, and full compliance with advisory and/or regulated measures. In addition, the safety measures must be taken seriously and any official updates should be followed accordingly. In a situation that management should be strengthened through its role of control, the city may require monitory of multiple aspects, such as entries and exits to public places and communities (or mobility record), temperature checks (or other health checks depending on the nature of the disease and its symptoms), regular checks on communities, more security, closure of multiple entries to public places, careful monitory of food safety, monitory of safe productions, monitory of business operations, regular monitory of product pricing, etc. With such a diversity of measures, community involvement is a necessity, ensuring that all aspects of safety and monitory are applied and conducted appropriately.

Furthermore, the implementation of robust management, and the presence of those managerial bodies, are certainly effective to increase the city's monitory and control measures. In our observations, we noticed examples of control on any sort of gathering, minimising human contact as much as possible. The extended limitations on the use of public places involve the presence of high security and even the relevant managerial teams who are responsible to regulate the order of uses and activities of places. In Ningbo, apart from high-security teams all over the main public places, there was a strong presence of other control measures at the community level. In addition to these, a new mobile application was developed to monitor people's health conditions and their mobility. This data-based and data-driven approach, regarded as 'Ningbo Health Code', was the only possible way to enter public places of different types. This mobile application was based on multiple colour codes reflecting on the health conditions of the individuals, and it was required for entry to any public and private premises (including housing compounds). The main intention was to enhance safety and monitory measures through the available technological and digital means. It was an instrument to do multiple things: detect, monitor, and record. Altogether, it was only through the multiplicity of effective measures that the city managed to speed up its containment process.

BOX 5.4 The Application Procedure for "Ningbo Health Code" (This English document was released by Ningbo government)

This code system registration is designed for people who are currently in Ningbo, and those who are coming to Ningbo. The simple guide (below) to obtaining your code should be followed once you arrive back in Ningbo, or in China if you return from overseas.

What is the "Health Code"?

The Ningbo Municipal Government released a code management system for people in and entering Ningbo. It is a self-report system that asks user to update their health information in the app. And it will be checked in transport hubs like metro and train station. It has color of "**Green, Yellow** and **Red**".

The **"Red Code"** including confirmed cases, suspected cases, asymptomatic infections, people in close contacts with confirmed cases, people under medical observation and quarantine, people who came to Ningbo from the hardest-hit areas of the epidemic, or have been to these areas and people from other specific areas. In principle, the "Red code" people who want to enter Ningbo will be advised to go back or delay their return. People with "Red code" who have already arrived in Ningbo will be required to have 14 days of centralized quarantine or strict home quarantine for medical observation.

The **"Yellow Code"** mainly include those people with fever, shortness of breath and respiratory symptoms, those who have close contact with patients with fever or respiratory symptoms, those who have close contact with confirmed cases or have confirmed cases in the same building where they live or live temporarily, and people from high-risk counties (cities, districts) within and outside the province. People with "Yellow code" will be subject to 7 days of home quarantine or centralized quarantine for medical observation according to the provisions.

The **"Green Code"** are identified as people other than "Red code" and "Yellow code". According to the decision from the Ningbo Leading Group for Epidemic Prevention and Control, people who have "Green code" are allowed to pass throughout the city after their temperature is taken. Checkpoints at all levels of the city shall not be allowed to refuse non-local license plates, non-local census register to entry Ningbo or entry into the village and residential areas. Employment and schooling of those people shall not be restricted and they have the freedom to rent houses.

How to apply for "Ningbo Health Code"?

- Step 1: open Alipay app* and then click "City Service";
- Step 2: make sure the city is "Ningbo" and then choose "Ningbo Health Code";
- Step 3: fill in the boxes.

How to use "Ningbo Health Code"?

In all the checkpoints that need to use "Health code", you can open Alipay, click on the "Pocket", "Credentials", and then click to check the "Health code" on the next page, the system will automatically display the user's exclusive "Health code" (red, yellow or green). According to Ningbo's epidemic control rules, users should take the initiative to show the code to staff to pass throughout the city (Ningbo Government Official Page 2020).

Note Alipay is a popular multi-functional mobile app, mainly used for payment.

(Adapted from main online source which was released and disseminated on official and social media platforms: https://mp.weixin.qq.com/s/tMk5tfXIX yxf_uaFObSwuQ).

Adaptability and Resilience

In line with safety and monitory measures, it is also important to implement adaptive and resilience measures. During the disruptive time, we cannot anticipate for business-as-usual practices or common daily routines. As it was expressed before, there are many cases of temporary closure, unavailability and limited operation of services, non-operational events and activities, redundant units/businesses/uses, and limited physical access. Hence, the city and its communities need to face the disruptions by increasing the adaptability in all sectors in the first instance. As previously discussed, this also provides a chance for innovation, where we can make better use of our technologies, information-based and data-based platforms, online facilities (or previously regarded as virtual amenities), digital platforms, etc. Through adaptability, we should not only remain adaptive but should also become creative in finding replacements or alternative methods for our daily operations. Now that we live in the age of digital technologies, this should no longer be a major challenge. With the availability of many digital networks, communication instruments, and advanced tools, we are able to develop smart-resilient opportunities (Cheshmehzangi 2017); a hybrid approach that is adaptive and integrative in the practices of adaptive planning and resilience enhancement.

The lockdown meant a major change of lifestyle for many people and its impact became widespread. Many companies and businesses continued with their online operations, education became virtual, and e-commerce was ever popular. The contactless city managed to keep some of its primary operations via digital and online platforms. This approach helped to ease some of the pressures, which helped to provide necessary services and supplies. In a positive turn, the ICTs supported secondary sectors that could not even partially operate. In addition, through the practices of resilience, the city had to go through some new practices. At the community level, adaptive measures were limited, which meant little chance for flexibility of practices and measures. For individuals and individual households, however, there were more chances to manoeuvre around the imposed adaptive measures. This became an opportunity for team-building exercises, extra care and support, efficient lifestyle, and replaced daily activities. Through these resilience measures, entertainment was redefined and people found a chance to learn new skills. In reality, the situation can either strengthen or worsen the methods of cooperation and integrative thinking. This fact depends on how we could effectively and efficiently manage to respond to the issues of adaptability, how we can strengthen the resilience, and how we can ultimately manage to cope with the disruptions.

In our observations, we noticed a tangible difference between people who are more individual than social (i.e. introvert vs extrovert people). This difference changes the way they could adapt to the changing situations of the outbreak. In both categories (individual and social groups), people choose their replacement activities differently. On one hand, people with individual mindsets tend to be more resilient, as they are used to pseudo-isolation situations or are generally less needy of social activities. On the other hand, socially active people (or people of the social personality type) may struggle to adapt themselves to less social activities and changes that affected their daily routine(s). The replacement activities could be healthier examples of fitness activities (both indoor and outdoor when possible), jogging, cycling, etc. Hence, adaptive thinking to minimise social contacts is not an easy task. But a variety of alternative activities should be planned to replace some of our everyday life's common activities. This requires good management of time and activities.

Economy

Apart from health impacts, one of the primary concerns is usually associated with the economic impacts of the outbreak. As far as they could get, there are traces of economic impacts in every sector. With small and medium-sized enterprises/businesses getting the biggest hit, the situation puts many secondary businesses and amenities/services in decline. At multiple levels of local, national, and international, we see many examples of economic losses; and in most cases, the situation is narrated as an economic crisis. While outbreak events may be somehow considered beneficial to environmental factors, they pose a significant threat to socio-economic values and particularly the economy of the city (and beyond). In the first two weeks of the COVID-19 outbreak, the reduction in China's fossil fuel consumption was equivalent to the whole oil consumption of Italy and the United Kingdom combined. This was not only due to lower fuel demand for transportation, but for other sectors too. An example of which is the construction sector, with nearly all its operations on halt. For a country with the highest number of construction projects and large scale development, the impact was perceptible on cities of various sizes. It took approximately six weeks for the construction sector to restart its operations at a much smaller scale, marking three weeks of delay after the Chinese New Year holidays. The economic impacts also include effects on the free movement of workforces such as labour forces. With the situation of lockdown, such movement was highly restricted at first and could only retain its regular flow from the early stages of the recovery phase.

Moreover, disruptions on the economy could create a range of imbalances in society, one of which is issues related to human resources. Hence, there are probable chances for increases in unemployment, business closure, redundancies, and special unpaid leave. With a lack of relevant legislation, society may not be able to cope with such added adversities of financial nature. A larger example of such imbalance is related to the tourism industry as well as the large scale and wide range of trade and marketing that are part of it. Undoubtedly, the tourism industry is affected significantly not only during the outbreak event but also during the post-recovery phase (and afterward). The stabilisation of the tourism industry for highly infected cities or regions may take a much longer time, months or even years after the outbreak. Hence, post-recovery management of the tourism industry, trades, and marketing is essential for the city's economic reclamation.

In our observations, we witnessed the depopulation effect in the main economic hubs of the city, both business and retail/commercial typologies. Known that the city cannot continue for long with such disruptive impacts on its economic activities, all industries have to safeguard the safety of their workforces and working environments. In all those weeks of many empty buildings, those affected industries have to develop their adaptive planning (Alterman 1988) and strategies. Depending on their main operations, we noticed different control and monitory measures across all sectors. For instance, primary services were mostly back to full operation after almost six to seven weeks, while secondary services were only partially operational by then. The city's step into economic revitalisation is crucial, but it simply is more important to do so in the safest possible way and a gradual process. The impacts are yet to be holistically assessed, and the full economic recovery is far ahead.

Governance

The main part of how the outbreak could be successfully managed comes under the role and duties of governance. It is clear that everyday governance cannot cope with such a transformative—but disruptive—period. Hence, it is urgently required to make governance as resilient as the city's management and provision factors. On the spur of the moment, decision making becomes more sensitive and processes adhere to new strategies. Governance should respond to multiple factors (see Chap. 4), such as the need for primary social services in the city as a main example of governance. The enforcement of new policies should be regulated and widespread, allowing little flexibility in control and monitory of the outbreak progression. Subsequently, the feasibility of practical implementation must be carefully assessed, inclusively developed, and formally announced. Hence, it is vital to equip governance of any level with the right instruments or tools for channelisation of all procedures in the formal process. The informality of responsiveness could harm governance, as it creates an unstructured approach to dealing with the situation, and it might also reduce the level of societal awareness or care. Therefore, the city governance ought to develop a group of stakeholders (Benson et al. 2016; Deng and Cheshmehzangi 2018; Hellsten et al. 2019), decision-making instruments (Ambat et al. 2019; Cheshmehzangi 2020a), adaptive measures (Monge et al. 2020; Shears and Garavan 2020), and formal procedures. Similarly, the informality of any of these factors would have adverse impacts on the control and containment plans. The society's full awareness of precautionary measures and guidelines is essential, and so is their engagement and involvement in strategic planning processes.

Furthermore, the need for adequate legislation and adjustment of institutional constellations (Cheshmehzangi and Dawodu 2018; Kang 2019) is evident in the process of decision making. The city governance should enhance cross-communication between: multiple actors of the city management teams, the society and various stakeholders, and with the authorities of different levels (i.e. both higher than the municipal level, and smaller levels under the municipality, such as districts, neighbourhoods, townships, etc.). Altogether, governance could provide added values beyond the position of the government. Hence, collaborative governance should merge into a network of key stakeholders, an all-inclusive approach that can assess the outbreak progress, verify the priorities, and make the right and reflective readjustments. This is not necessarily a top-down approach, but it should be driven and fully managed through the right top-down channels.

In our observations, we noticed regular changes; some more sudden than the others, and some in the form of updates. Eventually, the word 'Progress' meant a different thing in different contexts/countries/regions—one may just progress as the mean of improvement, and one may face the progression of the outbreak through increasing transmission and upsurge of the number of cases. To achieve the former, robust governance appeared to be essential and effective. This was evident through the role of governance in many Chinese cities, as also appraised by Bruce Aylward, head of the WHO-China Joint Mission on COVID-19, "*China is doing an unprecedented job in battling an unknown disease with rigor and innovation*" (reported by Zhao and Ma 2020). In other observations, two achievements of robust governance in China are appraisable, one is the formality of procedures, and the other is early control measures on transportation and mobility at multiple levels. Through formal procedures, the governance of the outbreak was effective, which helped to raised people's awareness and regulated collective compliance with precautionary measures. And through early control measures, the community transmissions were controlled, detected, and reduced in numbers.

BOX 5.5 "The provincial epidemic prevention and control supervision group visits Ningbo for supervision and inspection" (This English document is released by Ningbo Municipal Economic and Information Technology Bureau—Dated 13 February 2020)

In the afternoon of 11 February 2020, the provincial epidemic prevention and control supervision group visited Ningbo for supervision and inspection of the work on prevention and control of the COVID-19. Zhang Shifang, Director of the Ningbo Municipal Economic and Information Technology Bureau, made a report to the supervision group on the current support of materials for epidemic prevention and control, the difficulties in current material supply and demand, and the plan for the next step in Ningbo.

The supervision group fully affirmed the work that Ningbo has done and appreciated quick material support for epidemic prevention and control in Ningbo. Meanwhile, in view of the current difficulties in emergency materials supply, early return to work, labour recruitment, and transportation congestion and so on, four pieces of suggestion for the next step were put forward.

Firstly, strengthen examination and approval of enterprises' return to work.

It is suggested that we further understand the status-quo of enterprises, strengthen the work of providing "3 Services" for enterprises, and prevent any return to work without government approval.

Secondly, improve classified ability of rapid screening, such as nucleic acid detection.

It is suggested that we make full use of technology for nucleic acid detection and scientifically set the time for people under medical observation or home quarantine under the premise of ensuring safety.

Thirdly, provide guidance for enterprises to earnestly implement all requirements on epidemic prevention.

It is suggested that enterprises establish and improve daily cleaning, disinfection, ventilation and body temperature monitoring.

Fourthly, strengthen implementation of policies.

It is suggested that we strengthen publicity among and guidance for enterprises on policies, telling enterprises what to do and how, that we strengthen supervision and checking to make sure that relevant polices are complied with during the period of epidemic prevention and control.

Extracted from original source: http://www.nbec.gov.cn/art/2020/2/13/art_1907_4048855.html.

BOX 5.6 "Notice of "Twelve Regulations" for Epidemic Prevention and Control of Pneumonia Infection by the New Coronavirus in Ningbo" (This Chinese document is released by Office of the Leading Group of Epidemic Prevention and Control of Pneumonia Infection by the New Coronavirus, Ningbo—Dated 4 February 2020)

On February 4th, Ningbo's Leading Group of Epidemic Prevention and Control announced twelve newly enacted rules, known as the "Twelve Regulations", to further contain the epidemic and asks all citizens in Ningbo to abide by the rules. The translation of the announcement is below.

The battle against the pneumonia epidemic caused by 2019-nCov (later pandemic on 11 March 2020) now enters a key stage. In order to win this fight and to ensure the safety and health of our people, we've enacted twelve specific containment measures known as the "Twelve Regulations", based on previously issued requirements for epidemic prevention and control. The "Twelve Regulations" are as follows:

1. Anyone from outside Ningbo city, who either returns or visits the city, should report to the company/organization where they work, or the village/community residence quarter where they live, at the earliest possible opportunity.

2. All villages/community residence quarters should have their access controlled. People residing in a village (community residence quarter) are not allowed to go outside without a designated certificate or submitting to a temperature check. External visitors and vehicles are not allowed to enter. If without a designated property management unit, a community should control its own access. Any special circumstances should be recorded by property management staff.

3. It is strictly forbidden for a citizen to enter public places without wearing a mask. Each household (except those quarantined at home for medical observation) can send one family member out to purchase necessities every two days while others should stay at home unless they need epidemic prevention or control, need to see a doctor or have a work emergency.

4. It is strictly forbidden for people to gather or to visit friends and relatives. Anyone entering a farmers' market or a supermarket should wear a mask and submit to a temperature check. All areas for congregation within a village or community residence quarter should be closed down. Courier and mail services should be conducted in a non-contact manner.

5. Those quarantined at home for medical observation are strictly forbidden to leave their residence.

6. All centrally controlled air conditioning systems at companies/organizations or public places should be shut down. All elevators and other confined spaces should be disinfected on a daily basis.

7. Public places not related to people's livelihood should be closed down. Farmers' markets, supermarkets, pharmacies and other places should operate with reasonable arrangement of business hours, regular disinfection and all personnel should take temperature checks and wear masks.

8. Weddings should be suspended. Funerals should be simplified and recorded at the administrative departments of villages (community residence quarters), with the cadres managing the whole process

9.   Residents with a fever or cough should report to their villages (community residence quarters) at the earliest possible opportunity, and go to a designated fever clinic. It is strictly forbidden to conceal or fail to report any underlying health issues. Pharmacies should implement real-name registration for sales of fever or cough medicines and report these to the local health department at the first opportunity.
10. Landlords should carry out primary management responsibilities and not sign new leases with persons from areas with high infection rates. Tenants from areas with high infection rates should be instructed not to return to Ningbo in the near future.
11. It is strictly forbidden for companies/organizations to start or resume work before 9th February unless previously approved.
12. All citizens should abide by national, provincial and municipal regulations on epidemic prevention and control. It is strictly forbidden to spread rumors or misreport the epidemic situation. Violation of regulations shall be severely and promptly punished in accordance with the law.

The regulations shall be enforced immediately. The prohibitive measures above are effective until further notice according to the situation of epidemic prevention and control.

Leading Group of Epidemic Prevention and Control of Pneumonia Infection by the New Coronavirus, Ningbo

4 February, 2020.

Extracted from third party source: https://www.nottingham.edu.cn/en/Campus-Portal/News/Article.aspx?id=0a341fd2-a4c6-4d91-97a0-5cdd9deaa06c&language=en-GB.

Translated by the University of Nottingham Ningbo China, University working group team.

5.3   The Realities of the Experience

This section demonstrates a city example by representing some of the realities of the COVID-19 outbreak and their impacts on a real case. These are recorded to showcase the realities of the experience in the City of Ningbo, East China. The city was under a full lockdown situation for several weeks and the impacts were widespread. The city went from the status of emergency to low risk on 3rd of March 2020, with 157 infected cases at its peak. The rapid transformations were remarkable and the city managed to respond effectively. Hence, this pictorial section is an unpretentious reflection on the realities of the outbreak, including factors of high consideration, details of control and monitory approaches, measures of various kinds that were in place, and ultimately the ways the society managed to cope with such disruptive time (Figs. 5.1 to 5.103).

Fig. 5.1 The aerial view of Ningbo's southern areas (mostly with a combined urban layout of low to mid-rise housing and industrial areas), just one day after the lockdown of the original epicenter of the COVID-19 disease outbreak. At this time, the City Ningbo is still fully operational and people are preparing for the Chinese New Year celebrations. In the midst of early uncertainties, there are some growing concerns about disease transmission. People are more cautious with less outdoor activities and less mobility. The new conversation was to assess what is discussed in social media and the official media. Many discussions were around issues of the SARS outbreak and its experience several years ago. By this time, the majority wore facial masks, the news was developing fast, and there were temperature checkpoints in the main transportation hubs of the city. On 24th of January 2020, the city was under the shadow of its very own lockdown. *Source* Ali Cheshmehzangi

Fig. 5.2 As early as the 24th of January 2020, intensified health checks, high-level security, and empty flights were amongst the early expected scenes, reminding many people about the measures implemented during the SARS outbreak in 2002–2003. With many uncertainties and lack of knowledge about the disease, the immediate actions were to identify suspected cases travelling between cities, provinces, and countries. At this time, there was no sign of 'social distancing' or any of the later measures that became global. Only a few symptoms of the disease were identified and anyone with a temperature over 37.3 °C was not allowed to travel out of cities. While recapturing or remembering some of those earlier experiences, and not knowing the severity of the disease, most people started to use facial masks in the public. This was also recommended to all flight attendants, pilots, security workers, etc. Everything was moving fast and measures were just adding up one after another. Reactions were fast and there was no sign of inactions under the circumstances. Soon after, streets were empty not because of the Chinese New Year arrival but because of the fear of the unknown and hidden enemy. By then, the media coverage was continuous at the national level, and there were many reports of the outbreak progress in Wuhan and other cities across the country. I took a photo of a pilot wearing a facial mask, and told a few others this must be serious! *Source* Ali Cheshmehzangi

Fig. 5.3 The early closure of secondary gates/entrances of compounds from the 29th of January 2020, applied to residential compounds, educational compounds, and some public amenities. The temporary closure as shown in the image is basic but all residents complied with this temporary measure to ensure the security of their compound/working area/campus is maintained. Hence collective compliance is vital at any scale and in such disruptive event. *Source* Ali Cheshmehzangi

Fig. 5.4 The closure strategy soon applied to mixed use development of retail and businesses, and limiting access points. *Source* Ali Cheshmehzangi

Fig. 5.5 Early presence of three main services, namely security/safety, medical, and emergency, next to each other in front of main healthcare centers and hospitals to monitor entries and to provide immediate response. *Source* Ali Cheshmehzangi

Fig. 5.6 The gradual closure of secondary amenities, such as restaurants, cafés, pubs, bars, etc. to reduce the chance of gathering that pose higher risk of community transmission. *Source* Ali Cheshmehzangi

Fig. 5.7 The closure of public buildings and religious buildings, effective from 30th of January 2020. *Source* Ali Cheshmehzangi

Fig. 5.8 The early closure of larger and clustered retail areas, such as shopping centers/malls with only limited operations of their supermarkets, which were remained accessible with more control and monitory measures. *Source* Ali Cheshmehzangi

Fig. 5.9 The closure of underused public places with less visibility for control and monitory. *Source* Ali Cheshmehzangi

Fig. 5.10 The early closure of leisure and entertainment facilities or places that attract larger group of people or often clustered activities. *Source* Ali Cheshmehzangi

Fig. 5.11 The closure of secondary amenities for a longer period, the example of Ningbo Calligraphy and Painting Academy. *Source* Ali Cheshmehzangi

Fig. 5.12 The closure of educational buildings for a longer period, the example of primary school in Yinzhou District. *Source* Ali Cheshmehzangi

Fig. 5.13 The extended closure of all construction sites, regulated for any non-compliance activities, particularly that construction sector was affected significantly. *Source* Ali Cheshmehzangi

Fig. 5.14 The persistence of the slow(er) pace rural life with tangible impact on their tourism industry. *Source* Ali Cheshmehzangi

Fig. 5.15 Minimised tourism industry in rural and outskirt areas. *Source* Ali Cheshmehzangi

Fig. 5.16 Nearly empty touristic villages in outskirts. *Source* Ali Cheshmehzangi

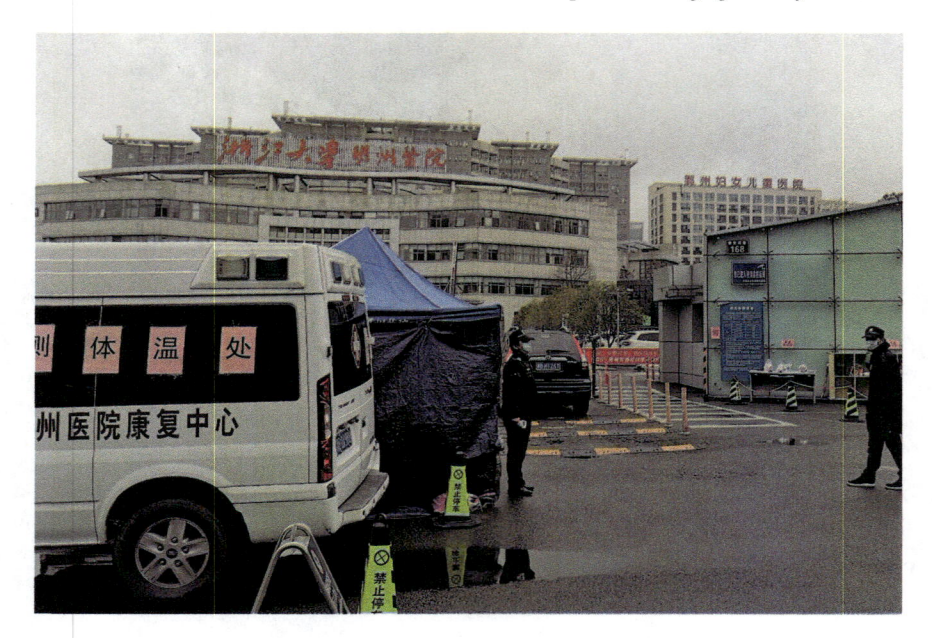

Fig. 5.17 Higher level of control and monitory measures for secondary hospitals during the transmission and transition phases, bearing in mind that only selective hospitals (15 in total) could accept infected patients across the city and its outer districts. *Source* Ali Cheshmehzangi

Fig. 5.18 Limited operations of secondary social services, such as the local tax office. *Source* Ali Cheshmehzangi

Fig. 5.19 The city authorities placed slogans with the main highlights of precautionary measures. In this example there are four protective reminders: *"put on the mask, check your temperature, wash your hands regularly, and no gathering"*. *Source* Ali Cheshmehzangi

Fig. 5.20 Empty and closed public places and secondary amenities across the city and its rural districts with a large impact on people's lifestyle. *Source* Ali Cheshmehzangi

Fig. 5.21 The closure of multiple access points to open public spaces in order to limit and centralise control and monitory of checkpoints. *Source* Ali Cheshmehzangi

Fig. 5.22 The city's many constriction sites were on halt for an extended period due to multiple facts, partly associated to limited resources and workforces or the need for larger mobility of workforces from other parts of the province and other provinces. *Source* Ali Cheshmehzangi

Fig. 5.23 Regular checks and control were implemented for hotels and resorts, ensuring the early detection of cases coming from others provinces or countries/regions. This was a time when hotel businesses are not doing well due to limited travel and significant impact on tourism. *Source* Ali Cheshmehzangi

Fig. 5.24 Regular disinfection procedures of main public places, public buildings, and public facilities, which was changed from a daily routine at first to several times a day. *Source* Ali Cheshmehzangi

Fig. 5.25 Societal disorders, regardless of their magnitude and impacts, became more visible throughout the transition and recovery phases. Approximately five minutes after the photo was taken, the police was on site to charge the cars. *Source* Ali Cheshmehzangi

Fig. 5.26 Another simple but unnecessary disorder that requires a particular a unit to get involved. That unit is perhaps non-operational, partial in their regular operations, or lack human resources. *Source* Ali Cheshmehzangi

Fig. 5.27 Designated food delivery and goods collection point beside the main entrance of one of the hospitals to help with contactless operations to maintain essential provisions to vulnerable and affected groups of the society. *Source* Ali Cheshmehzangi

Fig. 5.28 Mobile amenities only operate for those primary maintenance needs and operations, which were effectively regulated by relevant units. *Source* Ali Cheshmehzangi

Fig. 5.29 With limited resources and high risks of community transmission, construction works were limited to emergency and necessary construction jobs which could were mainly small scaled and required limited resources. *Source* Ali Cheshmehzangi

Fig. 5.30 With almost three weeks of delay, the construction work started with gradual and limited operations. Somehow, the construction noise brought life back to the city. *Source* Ali Cheshmehzangi

Fig. 5.31 Tangible decline in daily activities of one of the three Central Business Districts (CBD) of the city during the lockdown period and towards the end of transmission phase. *Source* Ali Cheshmehzangi

Fig. 5.32 Gradual operation of CBD area inclusive of financial sector and middle sized businesses immediately after the lockdown period and towards the end of transition phase. *Source* Ali Cheshmehzangi

Fig. 5.33 Gradual operation of retail and public services with limited hours of operations, limited workforces, and high safety measures. *Source* Ali Cheshmehzangi

Fig. 5.34 An example of checkpoint at the main entry node to rural areas, with restriction on non-local entry to specified areas. The set-up includes signage, a visible tent, disinfection products, record of documents, and community volunteers. *Source* Ali Cheshmehzangi

Fig. 5.35 Temporary but longer closure of shared bike hire facilities, which are normally with little control measures and high risks of person-to-person/community transmission. *Source* Ali Cheshmehzangi

Fig. 5.36 Ningbo faced a decline in the use of shared facilities. In fact, shared bikes and similar shared facilities may have less control or even add burden to disinfection teams. Hence, it is suggested to temporarily close or limit their operations in order to have a better monitory in place. *Source* Ali Cheshmehzangi

Fig. 5.37 Temporary closure of playground facilities is essential and they should be regularly disinfected alongside with all public facilities. *Source* Ali Cheshmehzangi

Fig. 5.38 A group of delivery workers stay in a designated zone nearby clustered retail and restaurant areas. *Source* Ali Cheshmehzangi

Fig. 5.39 High level of safety measures are essential for delivery workers as they may be in contact with more people during the day. In here, a delivery person is waiting for his next delivery job adjacent to a secondary retail area. *Source* Ali Cheshmehzangi

Fig. 5.40 The use of 'Ningbo Health Code' becomes compulsory and it takes 24 h to get confirmation after registration. In this instance, the entry to shopping center/mall was not granted and the security members are instructing the person with required procedures. *Source* Ali Cheshmehzangi

Fig. 5.41 Designated entry zone with multiple checks (health code and temperature) in underground parking of a shopping center/mall. *Source* Ali Cheshmehzangi

Fig. 5.42 A board providing instruction for 'Ningbo Health Code' at the entry of a shopping center/mall towards the end of transition phase. *Source* Ali Cheshmehzangi

Fig. 5.43 After a temporary half in operations, bust services became less frequent and often without any passengers. But they remained in operations for inter-city connection and reaching the rural/outer districts of the city. *Source* Ali Cheshmehzangi

Fig. 5.44 Metro/underground operations were back to normal after several weeks and people had to keep their distance inside the cabins. *Source* Ali Cheshmehzangi

Fig. 5.45 High level presence of security and monitory checks are added value to maintain safety of the public services. *Source* Ali Cheshmehzangi

Fig. 5.46 Regular cleaning and disinfection procedures play a big part in maintaining hygiene and safety of high risk transportation hubs. *Source* Ali Cheshmehzangi

Fig. 5.47 Temperature check alongside the confirmation of 'Ningbo Health Code' are compulsory for all people entering public buildings. *Source* Ali Cheshmehzangi

Fig. 5.48 The university campus imposed high level checks, and only allow entries for internal members who reside on campus and those internal and external members whom hold their daily permits with pre-approval at three managerial levels. *Source* Ali Cheshmehzangi

Fig. 5.49 The equipped security members check records (with relevant paper works) and measures temperature before allowing anyone to their premises. *Source* Ali Cheshmehzangi

Fig. 5.50 Fast food services only operate for delivery and pick-up purchases and temporary closed their entrances to public. *Source* Ali Cheshmehzangi

Fig. 5.51 Other retail services/amenities started limited operations after 3–4 weeks of no operations, and only operated with contactless services. *Source* Ali Cheshmehzangi

Fig. 5.52 Customer-workplace interactions were minimised through various ways, and online platforms were flooded with purchases and delivery requests. The retail units allocate designated space for either pick-up or delivery of their goods/products/food. *Source* Ali Cheshmehzangi

Fig. 5.53 Double checks imposed for populated supermarkets, to maintain safety and to avoid becoming hotspots for the community transmissions. *Source* Ali Cheshmehzangi

Fig. 5.54 People should keep their individual healthy activities such as jogging and cycling in unpopulated areas, while indoor fitness activities are also very effective. *Source* Ali Cheshmehzangi

Fig. 5.55 Limited camping is an easy way to spend hours with new activities which should be arranged in dispersed layout. In this instance, high safety and monitory checks were reinforced during the recovery phase. *Source* Ali Cheshmehzangi

Fig. 5.56 Gradual and dispersed family picnic activities can maintain the health and happiness of children in particular. This is recorded towards the end of the recovery phase. *Source* Ali Cheshmehzangi

Fig. 5.57 A major benefit of specialised hospitals is decentralisation of health system operations, which could provide safe services to special and non-infectious issues. This is an example of Orthopaedic Hospital in Ningbo. *Source* Ali Cheshmehzangi

Fig. 5.58 Throughout the outbreak response plan, the operation of secondary amenities were either forbidden or limited. *Source* Ali Cheshmehzangi

Fig. 5.59 Provision of primary products in local supermarkets. People should sustain a greater sense of social responsibility and avoid panic buying. People should remain considerate to their communities, while the provision of essential/primary products should be maintained throughout the outbreak progression. In Ningbo, some shortages of primary products were experienced for one to two weeks during the transmission phase. *Source* Ali Cheshmehzangi

Fig. 5.60 In the recovery phase, the city's primary operations were under the influence of rapidly intensifying global situations. Local supermarkets continued with high level of safety measures, as well as control and monitory of customers. This was essential to avoid reversing the containment at the city level. *Source* Ali Cheshmehzangi

Fig. 5.61 In the recovery phase, residential compounds preserved their security checks. In here, we see multiple aspects, from left to right: red banners for general guidelines, temporary tables for food collection, temporary goods collection point at micro-level, security checks (inclusive of temperature check), one secured access point to the compound, and rubbish bin next to the entrance for the disposal of any masks, gloves, etc. *Source* Ali Cheshmehzangi

Fig. 5.62 The gradual opening of larger scale institutes or businesses under full control and monitory were amongst the later signs of recovery phase. The actual date of reopening was postponed to a much later date. The reopening was subject to much-enhanced health checks, new prevention and control measures, new installations, and in-advance rehearsal procedures before the approval could be granted at multiple stages. *Source* Ali Cheshmehzangi

Fig. 5.63 In the recovery phase, closure of unnecessary public places were still in place, allowing for continuous practice of social distancing in populous areas or with high level of multiple functionalities. *Source* Ali Cheshmehzangi

Fig. 5.64 Early days of recovery phase meant return of popular local destinations for social activities, which then created road blockages and heavy traffic. *Source* Ali Cheshmehzangi

Fig. 5.65 Amid all monitory measures, large gathering was still disallowed and leisure activities were monitored by safety and security workforces. *Source* Ali Cheshmehzangi

Fig. 5.66 A new beginning for a city and its resident coming out of lockdown and reaching outbreak containment meant continuous caution, reviving the social life, and signs of positive progress. It became viral to say 'life goes on', which were meant to help the overall mental stability despite all adversities and disruptions. It was clear that many people were affected and more people became vulnerable during the outbreak. But under the shadow of the global emergency, people sensed a sign of hope to enjoy the outdoors more than any other time. With kites flying in blue skies, it simply felt like freedom. *Source* Ali Cheshmehzangi

Fig. 5.67 While considering protection measures, it is more than just the generalities but also details of certain aspects that require further attention. For instance, here we see examples of adhesive plastic wrap covering the area where we touch the most on a regular basis. They are disinfected regularly and are replaceable easily. Doing this is a simple measure but also effective not to damage the buttons due to regular disinfection procedures. *Source* Ali Cheshmehzangi

Fig. 5.68 Temporary electronic lock is placed on the residences' doors of people returning to China in the recovery phase. *Source* Ali Cheshmehzangi

Fig. 5.69 The officials put a temporary electronic lock on the doors of those in quarantine in their homes. The officials or allocated personnel provide and deliver the essentials such as water and food (three time a day). Also, there are three times medical checks per day for the allocated quarantine period during the recovery phase. All these measures highlight three main factors: measures to avoid reversing the outbreak progression, attention to details, and strong institutional arrangement through good governance. *Source* Ali Cheshmehzangi

Fig. 5.70 Upon the reopening of secondary amenities, such as restaurants, the waiting areas and procedures to dine inside the restaurants had to change. In some cases, you could only dine inside through pre-booking system with allocated time and place to sit. *Source* Ali Cheshmehzangi

Fig. 5.71 Similar to restaurants, most cafés/tea shops/coffee shops opened in a gradual pace and limited or closed their sitting-in facilities. *Source* Ali Cheshmehzangi

Fig. 5.72 In the recovery phase, some restaurants still delayed the restart of their operations or they were working on limited hours. Those that restarted their operations, closed their waiting areas, had a list to check and record details and temperature of whoever dined inside, and provided disinfection products for all users. *Source* Ali Cheshmehzangi

Fig. 5.73 This event was an opportunity to promote contactless food delivery in various modes—from leaving them in designated areas or at front of gates/properties (a and b), to online delivery of packages of essential food supply. *Source* Ali Cheshmehzangi

Fig. 5.74 Apart from the approved green health code, permission to enter all public premises was dependant on temperature checks and records. In China, this was set at 37.3 °C, while in some other countries this was slightly higher. *Source* Ali Cheshmehzangi

Fig. 5.75 Scenes of empty shopping malls became global. In Ningbo, many places remained depopulated or closed during the recovery phase. Many secondary services/amenities remained closed in light of business losses and lack of customers. *Source* Ali Cheshmehzangi

Fig. 5.76 All centrally-controlled air conditioning systems remained non-operational throughout the outbreak. The unpopular and/or smaller shops and restaurants remained closed even during the recovery phase. Their closure was based on several factors, including lack of customers, lack of workforces, and significant business losses. Smaller businesses simply struggle to cope with their economic loss. The permanent closure of such businesses is expected globally as they are expected to be more vulnerable. *Source* Ali Cheshmehzangi

Fig. 5.77 Despite earlier signs of re-opening public entertainment amenities, the national-level decision decided to continue with cinema closures for a longer period. This decision was based on the worries of any reversing trends in the containment process. *Source* Ali Cheshmehzangi

Fig. 5.78 Maintenance of those declining or closed businesses became the added burden to those responsibilities for their management. Here, we see the change from dead plants to empty scenes of the frontages of retail units and restaurants in a shopping mall. *Source* Ali Cheshmehzangi

Fig. 5.79 As a secondary amenity, public toilets have to be remain closed during the outbreak progression. Once reopened, it is ideal to have contactless hand cleaning and hand drying methods. Also relevant information on hand washing procedures should be provided visibly to ensure everyone can comply with main precautionary measures. *Source* Ali Cheshmehzangi

Fig. 5.80 Regular fitness activities are essential. If possible, it is important to use outdoor spaces and parks for various activities such as walking, jogging, and playing sports. It is important to only do such activities when possible. These should be done either individually or in smaller groups to avoid any cluster activities of such kind. It is also important to maintain dispersed activities in large public areas, open spaces, and parks. *Source* Ali Cheshmehzangi

Fig. 5.81 In the recovery phase, construction workers are allowed in and dismissed in certain designated times. They are only allowed in and out in smaller groups of less than 10 people at a time, and in intervals with careful monitory and supervision in multiple nodes throughout the whole process. *Source* Ali Cheshmehzangi

Fig. 5.82 Utilising all display boards to showcase procedures, guidelines, and information about the disease and raising awareness to overcome the issues. These boards are placed in transit and access nodes, such as lobbies, public areas and are visible to users and passer-bys. Here are examples of display boards at the university campus. *Source* Ali Cheshmehzangi

Fig. 5.83 The newly added signs on shops and public buildings/premises included new information such as the emphasis on "code scanning access", as well as reminders of the primary prevention and safety measures, such as "wash hands", "wear a facial mask", "avoid gathering", and "keep 1-meter distance". The information for disinfection procedures was also indicated and displayed to assure the users/public of their cleaner environments. Such measures, if they last for a longer time, would have a positive impact on the hygiene improvement of the general public as well as public buildings, retail units, restaurants, markets, etc. *Source* Ali Cheshmehzangi

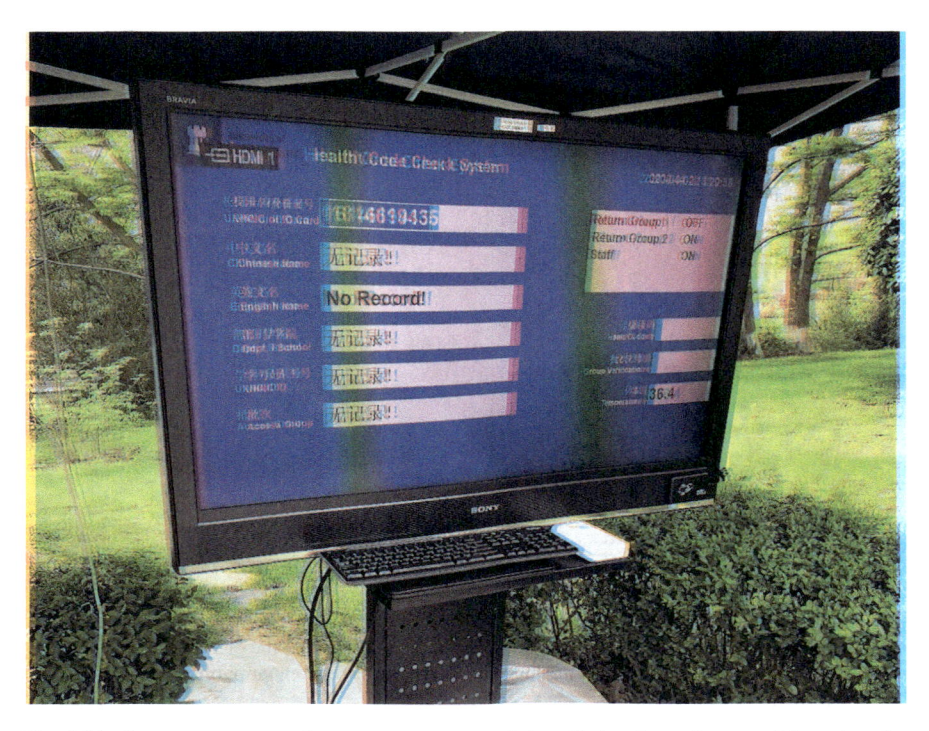

Fig. 5.84 A new access control management system is installed at the main gate of the university campus at the start of the post-recovery phase. *Source* Ali Cheshmehzangi

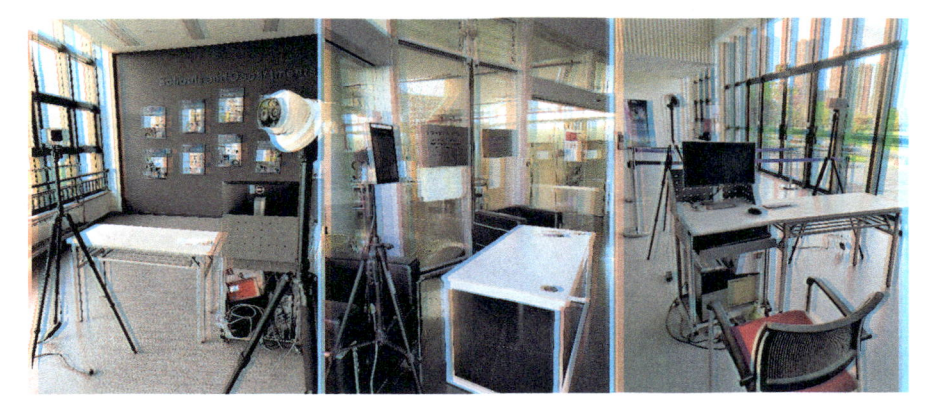

Fig. 5.85 Placement of new gadgets for temperature checks and recording of entries and exits to teaching areas/buildings. *Source* Ali Cheshmehzangi

Fig. 5.86 Allocation of security at the entry of main public places at the city level is not an easy task to manage the populated flows in and out of communities and popular public areas. *Source* Ali Cheshmehzangi

Fig. 5.87 If possible, micro-scale farming would be helpful to reduce the pressure on the local food supply chain. *Source* Ali Cheshmehzangi

Fig. 5.88 The use of public buses is subject to health check measures. The person without a facial mask was given one by the driver. *Source* Ali Cheshmehzangi

Fig. 5.89 Some restaurants may choose to alter their mode of serving, but these should be coordinated with relevant authorities in advance. This method requires health checks, hygiene checks, and food safety checks. *Source* Ali Cheshmehzangi

Fig. 5.90 In this incident, the security forces had to interfere and stop the operation of this non-coordinated street food sale. A total number of eight security members were called to the place to stop the operations with immediate effect. *Source* Ali Cheshmehzangi

Fig. 5.91 At the time, the communities were confident of the recovery stage the flow of help and support changed its direction. Many enterprises/companies/communities started to send medical supplies, sanitisation products, and medicine. Here, we see 61 kg of medical products ready to dispatch to the UK (from Ningbo) with the wording "*Keep Calm and Carry On*" taken on 23 March 2020. *Source* Ali Cheshmehzangi

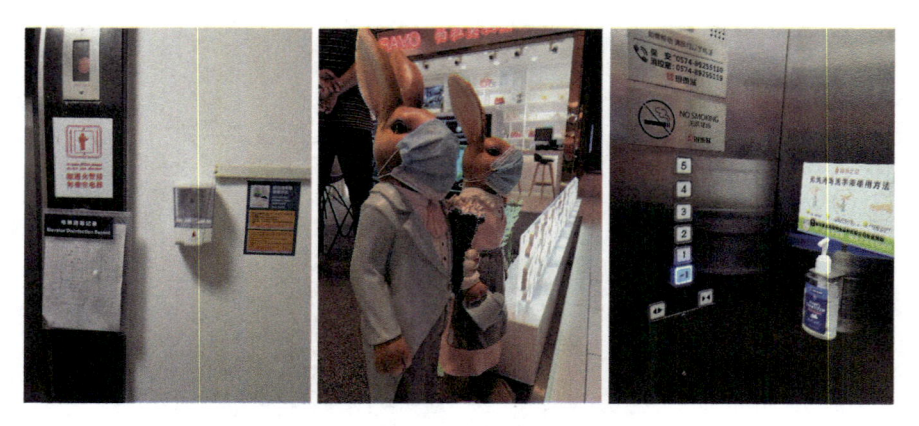

Fig. 5.92 The new additions to our main public areas aimed to promote hygiene improvement, especially that the education of such matter require time and a lot of public engagement. The new information displays and taking into consideration the essential procedures were amongst the many changes. If people do not forget fast, as they normally do, such measures could help to improve our personal and public hygiene. But, it is very likely people would forget recklessly once this pandemic is over. *Source* Ali Cheshmehzangi

Fig. 5.93 One example of extended safety measures at the front of a community supermarket during the post-recovery phase, with the continuing requirements of check and register at the entry point. *Source* Ali Cheshmehzangi

Fig. 5.94 After the gradual opening of schools, the students' temperature and health code were checked and registered before going to school and upon arrival at the school premises. *Source* Ali Cheshmehzangi

Fig. 5.95 Continued safety checks were remained in place during the post-recovery phase to ensure all risks are detected, assessed, and managed—and through prevention and safety procedures. *Source* Ali Cheshmehzangi

Fig. 5.96 Rehearsal sessions were conducted prior to opening of the education sector, including safety check procedures for all staff and students. This was then followed by several guidelines, including emergency response for fever and suspected COVID-19 case. *Source* Ali Cheshmehzangi

Fig. 5.97 Creating one-way path systems were amongst some of those early spatial arrangements (Cheshmehzangi 2020c) to have a better monitory of public places and populated areas. *Source* Ali Cheshmehzangi

Fig. 5.98 Some measures were put in place by people at the individual level. Here, we see a plastic curtain that is placed by the taxi driver to separate himself from the clients. Also, cash payments were minimal and were not advised even at the post-recovery phase. *Source* Ali Cheshmehzangi

Fig. 5.99 Adaptive measures for spatial arrangements (Cheshmehzangi, 2020c) were effective in making new temporary working spaces, separated units, and new designated areas to avoid gathering. *Source* Ali Cheshmehzangi

Fig. 5.100 This pandemic has created a unique opportunity to speed up the use of facial rendition devices, and some may cost us our privacy in the near future. *Source* Ali Cheshmehzangi

Fig. 5.101 The major impact was seen in the sector that did not or could not offer online services. Also, this affected those populated areas and secondary public facilities, particularly in city areas. *Source* Ali Cheshmehzangi

Fig. 5.102 One of the tangible impacts is the increasing poverty widespread amongst the lower social classes, particularly those that rely on their daily incomes or hourly pas jobs. Here, we see the impact on street-sellers with a significant decrease in their daily customers. In most cases, there is little financial support or legislation available to help those most vulnerable groups of society. *Source* Ali Cheshmehzangi

Fig. 5.103 The tourism industry and all industries associated with it are likely to struggle to recover in a short period. Here, we see an example of empty places in one of the most populated areas in the City of Ningbo. Despite the good weather conditions, most people avoid populated areas, especially in the city areas. The situation for the outer-city leisure areas is a different story as most people assume the green areas are safer than the urban areas. The same situation was seen in other countries where, after the lockdown, people avoided populated urban areas/destinations. *Source* Ali Cheshmehzangi

5.4   A Reflection: What Normally Occurs? and How the City Reacts?

Coinciding partly with the Chinese New Year holiday period, the situation was at first commonly perceived as an extended holiday. The prolonged phase of response and containment changes general perceptions. The seriousness of the situation was evident from the earlier days of the outbreak as most people reflected on their previous experiences of such events (i.e. mainly the outbreak situation of SARS in 2002–2003). By the time the administrative response was in place, this emergency situation already instigated major shockwaves across all sectors. In fact, it permitted very little time for considerable reactions as circumstances were unceasingly changing and uncertainties were unrestrained. The international media coverage was expectedly—but unfortunately—not very accommodating and nor they were any close to realities that could help the society. Time became scarcer and made decision making processes more subtle. In order to maintain a sensible society, it was simply vital to have the right reactions against what actually occurs on the ground. Governance includes both society and government (Fig. 5.104). Hence, public relations became ever important and how the city could or should react was of major concern to many. In the following four sub-sections, we provide an overview of those reactions that happened in the city in the recent event of the disruptive disease outbreak.

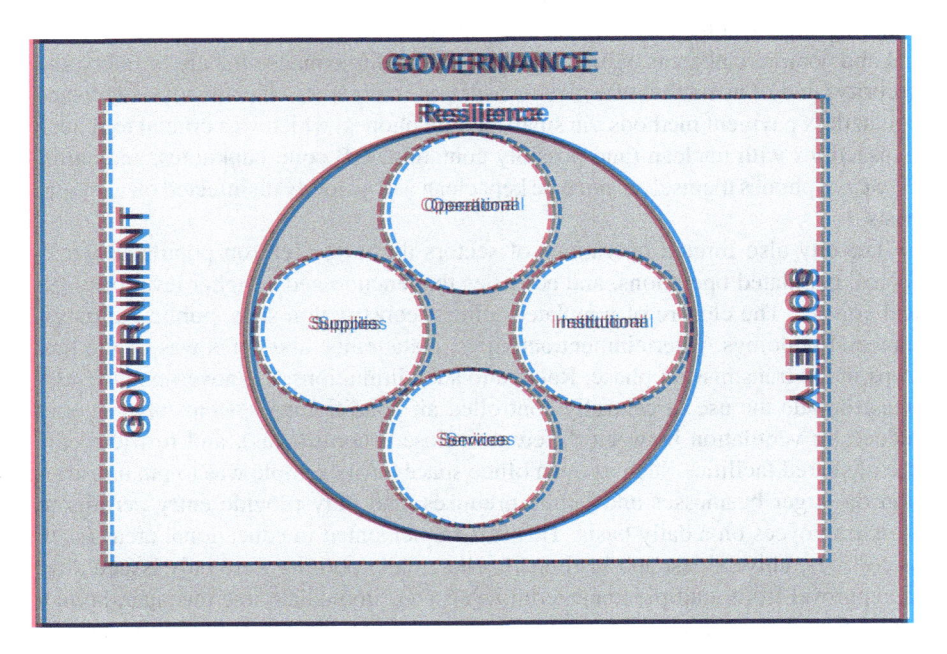

Fig. 5.104 Summary of reflections from resilience dimensions to society and government enhancement and the ultimate governance. *Source* The Author's own

5.4.1   Operational

The city simply assessed what needs to operate and when they should go into partial or full operation. An all-inclusive assessment is always a difficult task, and in this situation, it had to be procedural and with adequate planning. With immediate responses, the city had to adapt and localise the key national guidelines for quarantine and self-isolation measures. These happened in the peak of the transmission phase, first from the gradual lockdown of communities, and soon later to the nearly-full lockdown of the city. Main reflective arrangements, such as temporary closure of highways, railways, and other main modes of transportation (with restricted air travel) between provinces, were effective not only to control community transmission but also to control import and export of the disease. In this regard, we can highlight the effectiveness of city boundaries and monitory checks in the main parts of the city. Some examples are prohibited entries of non-locals to rural communities of the city as well as vulnerable groups (such as elderly care facilities), temporary isolation of residential compounds with higher risks, and limitations for mobility per household. The latter was officially announced to permit only one family member to go out of their household every two days, which was only put in place during the peak time of the transmission phase. These restricted measures were strengthened also by the control of distinct gated community layout in Chinese cities, which are generally planned with large compounds/urban blocks (Cheshmehzangi 2018). This particular planning layout helped to have a better monitory over entry and exit to residential and commercial areas, which ultimately helped to enhance the city's safety and security (ibid). On another effective measure, we can refer to the already established contactless payment methods via smart mobile phones, which was crucial to reduce transactions with unclean (and possibly contaminated) coin, banknotes, and cards. However, phones themselves must be kept clean and properly disinfected on a regular basis.

The city also limited operations of sectors that often rely on populated workforces, populated operations, and activities that encouraged a higher level of physical contact. The closure of populated offices, construction sites, public buildings, financial buildings, entertainment activities, restaurants, and cafés was of the first steps in the transmission phase. Related to such limitations, the government guidelines forbade the use of centrally controlled air conditioning systems/units (i.e. to reduce air ventilation between different enclosed rooms/areas), and restricted the use of shared facilities, such as open office spaces. An example was to put a restriction on larger businesses and public premises, and only provide entry permits to their employees on a daily basis. This was implemented in educational areas (such as colleges, universities, and institutes), where daily permits were only issued after the approval from multiple managerial levels (e.g. immediate line manager, human resources, and the central management office). Furthermore, the management of services, and in particular healthcare services, were of the highest priority. Through resource management, the city responded to the needs of infected healthcare units and facilities, to ensure the situation is controlled through the highest security and safety measures.

5.4.2   *Institutional*

The maintenance of healthcare services and facilities, as a major institutional support to outbreak control, was certainly a major urgency for the city. This developed as the main approach to keep healthcare centers/facilities as the primary safe hubs of the city. At a relatively lower risk compared to other cities, the temporary closure of small-scaled clinics and health care units was imposed to overcome any deficiencies. This strategic move was implemented due to a lack of safety measures and workforces in small-scaled healthcare units across the city, which was also decided as a reflection on recorded cases in infected community-level clinics. It was also recognised as a preventive approach against further community spread from sources that are less likely to be regularly and efficiently monitored. Hence, community-level measures were only limited to some other aspects. Also, the resourcefulness of all main healthcare units/services should have been carefully managed as part of this major institutional support. A larger stepping stone towards enhanced control measures was the compulsory use of the newly developed and immediately implemented 'Ningbo Health Code', which was mandatory for all people who had to access public places or use any of the public services. This information-based and data-driven approach was developed in the first stages of the transition phase, and soon became a city-wide instrument. It then enabled the city to have a better overview of patterns and possible situations that required immediate attention. This perhaps resembles an example of large scale surveillance, but is proven to be effective in the time of need.

Of the major concerns on institutional arrangements, we can refer to the role of media coverage and the adverse issues it may have on society. Hence, the media was regulated with the right information, updates, relevant training, and announcement of precautionary measures. The local media immediately turned into a knowledge-share platform between the professionals/experts and the general public, allowing for the right flow of information, which should be accessible, supportive, and positive. A large scale propagation of information also has significant impacts on socio-behaviour patterns, as the impacts can only be in place if the majority would comply with those regulated measures.

Moreover, in parallel to national-level updates on the outbreak progression, the local media took their position firmly in providing transparent and regular updates. Hence, the media was formalised with continuous updates and information. From the early stages of the transition phase, the media coverage also included positive broadcasting of success and updates on progression, which were meant to be reassuring measures to the general public and all relevant stakeholders.

Considering the large scale economic impacts of outbreak events, the reactions to economic management are the ones that we expect to be developed in a longer process, but including some immediate and short-term action plans. These are the actions that may require higher-level legislation, such as issuing tax exemption solutions, sick pay support, business loss insurance, recovery contingencies, adjustments in bank and mortgage interests, provision of low-interest loans, etc. As part of the immediate action plan, the step-by-step process for the closure and reopening of businesses was taken into consideration. This can be recognised as part of the early strategic plans for economic assessment and its management. Later on, this process was followed by some flexibility measures through adaptive planning approaches, allowing for alternative operations, a re-adjusted schedule of operations, and gradual support. These approaches then led to gradual normalisation steps, which certainly come with longer plans for economic revitalisation.

5.4.3 Services

As broad as they sound, urban services are extremely important for the necessary provisions of the city. With multiple sectors of services, there come also multiple demands/needs, different rearrangements, and different responses. To start with, the safety of transportation services was more of a priority than their regular operations. The early restrictions were crucial to reduce the chance of community transmission, which also restricted people's mobility with a high level of control. This specific measure only lasted for several days, but was effective in a long time. It then led to gradual operations of transportation services with extended safety and security measures. As mentioned earlier, measures such as the early restrictions on residential compounds, and restricting access points to only one or two entry and exit nodes, represent a great starting point to reduce the city-wide mobility. These measures helped to minimise unnecessary travel within the city, between different districts, and in and out of the municipality, which also provided a breathing space for other services to operate more smoothly. The same approach of restricted measures was later adapted for communal and open places, reducing the number of access points with higher monitory levels. Later on, this was also adapted for the reopening of main shopping malls/centers and markets, which were closed in the early days or early phases of outbreak progression. Hence, for the safety and security services, the whole situation became a chance to reinforce their workforces and to identify the less apparent deficiencies of their regular services. In return, the safety and security services had a chance to increase their preparedness and support other services of the city, particularly medical units, emergency medical services, and micro-level security of compounds and public services. To further support these services, the city's plan for the utilisation of military forces were to boost the provision of public services, and those that are important for the control and monitory of the outbreak.

In addition, the maintenance of social services could only be temporarily affected, hence the city had to develop contingency plans across all primary services/sectors. This also became a good opportunity to increase their support at the community level, particularly the ones that seem more vulnerable than the others. In doing so, social services of public means (or public services) had to be prioritised and not get overlooked in multiple phases of the outbreak progression. Of major achievements of the city is the quick establishment of voluntary workforces, some representing their communities and some supporting their local public services. This was beneficial from a threefold, which: a) increased the number of people operated on the ground as safety and security volunteers, b) addressed the community-level issues, needs, and/or emergencies; and c) facilitated the opportunity to employ those temporary redundant workforces, whom could then help their immediate communities. The other achievement, which is often hard to control but is very essential, was the careful monitory of price control and hoarding activities, specifically related to primary products, food, medical supplies, etc. In doing so, the city maintained healthy private-public partnerships and developed a robust monitory plan to overcome such difficulties. The involvement of the private sector and the general public in the form of community representatives were effective tools to help with the overall maintenance of the primary social services.

5.4.4 Supplies

The city relies very much on the supplies of various types. For example, food supply (intentionally distinguished based on the explanations of Chap. 3) should reflect on several factors of safety, production, availability, and distribution. Hence, the city structure like Ningbo, which includes both urban and rural areas, is advantageous to maintaining food production and delivery—i.e. in a form of the enhanced domestic food supply chain. This was carefully monitored as part of restricted access in and out of rural communities, to ensure the rural communities, especially with high food production, are kept safe and secure throughout the outbreak progression. At the same time, while the city started closing the shopping malls/centers across all districts, the operation of primary supplies such as supermarkets remained uninterrupted. This particular decision maintained the opportunity for the continuous availability of food supply and accessibility of the general public to those main amenities. Some food shortages were only experienced for a very short period, and less than two weeks. This mainly occurs as necessary readjustments are needed for the food supply chain. Moreover, different methods of food delivery and food shopping delivery were introduced and maintained throughout. The only perceptible impact was the eventual reduction in a variety of online retail companies. An example was the main supplier named DingDong, which had a temporary closure of their operations for about three weeks. Once they returned into full operations, they sent reassuring messages to their users regarding their high level of food stocks and the availability of primary products. The city also encouraged contactless online shopping by keeping the local food delivery operational. Through public-private partnerships on primary operations, they also provided mechanisms for safe delivery of necessary fresh products, which were affordable and were then delivered in only two to three days after the submission of online requests (see Fig. 5.73).

Based on our categorisation of amenities supply, the operation of primary amenities—even if partial—were kept as the priority for the provision of the city's services and supplies. This meant unnoticeable disruptions in the provision of those amenities, with an exception on goods collection points and maintenance units. These two primary amenities were mostly affected due to limited human resources, which were gradually resolved once the higher level restrictions were lifted. Both the secondary and mobile amenities were the ones affected the most. Yet, the city was able to maintain partial operations of selective and more necessary secondary services such as postal services, fuel services, and bank services, even though there were inevitable delays in their operations (particularly in the first few weeks that was also affected by the Chinese New Year holiday). Some secondary amenities also provided more opportunities for online and telephone-based services, supporting those supplies that could maintain their operations through alternative methods.

Of major concern to the general public was the smooth operations of education. Hence, the local authorities worked closely with the educational services of all levels, ensuring that they are safe at first, and they can operate through alternative methods. Therefore, with only one or two weeks of delay, most of the city's education was delivered through online teaching. With the use of relevant digital technologies and available software, the education was able to start operations with some minor disruptions. This was mostly developed and implemented in the form of self-learning, experiential teaching, and subject-based virtual teaching, and with the extensive use of online educational materials. Therefore, the outbreak was also recognised as an opportunity to empower virtual amenities and utilise them in the time of need. This was something, which was tested before but was never implemented at this scale and pace.

5.5 Capturing the Realities: The Reactions to Disruptive Time

This section captures the realities by highlighting some of the reactions at multiple levels, in China and other countries/locales. In doing so, the intention is to cover the realities of the impacts and how they were perceived, addressed, and developed. The highs and lows of such an outbreak event was simply a matter beyond just the city or the country. It only took a few weeks until it became a global health emergency. The realities are fascinating from the various perspectives of politicians, governments, policymakers, economists, scientists, academics, health organisations, businesses, and the general public. This disruptive event truly represented many examples of international relations, strategic planning, decision making, responsive methods, political determinations, public policy, predictions, preventions, community resilience enhancement, etc. From the many examples, only a handful of reactions are selected, which are specifically most relevant to our discussions here. All these perspectives from multiple stakeholders are recorded in this particular section, to reflect appropriately on the reactions to this recent (and partly unprecedented) disruptive time (Boxes 5.7 to 5.27).

BOX 5.7 "Coronavirus disease (COVID-19) advice for the public" (This extended public guideline was developed by the World Health Organization (WHO), providing: (i) Basic protective measures (with instructive videos), (ii) Protection measures for people whom are in or recently visited infected areas, (iii) Graphical guide for individual protection, (iv) Graphical guide to cope with stress, (v) Graphical guide for food safety, (vi) Graphical guide for shopping and working in wet markets in relevant contexts, and (vii) an Extended graphical guide for travellers)

Under (i)

"Basic protective measures against the new coronavirus"

This was provided in a list, explanatory guide, and educational videos to inform the general public of the basic protective measures. Key factors were highlighted, explaining not only the measures but the reasons for why they are important in this particular event. These factors include:

- Wash your hands frequently;
- Maintain social distancing;
- Avoid touching eyes, nose, and mouth;
- Practice respiratory hygiene;
- If you have fever, cough, and difficulty breathing, seek medical care early;
- Stay informed and follow advice given by your healthcare provider.

Under (ii)

"Protection measures for persons who are in or have recently visited (past 14 days) areas where COVID-19 is spreading"

- Follow the guidance outlined above;
- Stay at home if you begin to feel unwell, even with mild symptoms such as headache and slight runny nose, until you recover. Why? Avoiding contact with others and visits to medical facilities will allow these facilities to operate more effectively and help protect you and others from possible COVID-19 and other viruses;
- If you develop fever, cough and difficulty breathing, seek medical advice promptly as this may be due to a respiratory infection or other serious condition. Call in advance and tell your provider of any recent travel or contact with travellers. Why? Calling in advance will allow your health care provider to quickly direct you to the right health facility. This will also help to prevent possible spread of COVID-19 and other viruses.

Under (iii)

"Protect yourself and others from getting sick"

This is provided in four pictorial documents, which are just summarised in their categories in here:

- Wash your hands;
- Protect yourself and others from getting sick;
- Protect others from getting sick (1);
- Protect others from getting sick (2).

Under (iv)

"How to cope with stress during 2019-nCoV outbreak"
This is provided in two very detailed pictorial documents, which are just summarised in their categories in here:

- Coping with stress during the 2019-nCoV outbreak

(Including a set of six guidelines on emotional matters, healthy lifestyle, guide for health support, getting the facts, limiting worries, and development of new skills);

- Helping children cope with stress during the 2019-nCoV outbreak

(Including a set of five guidelines on children reactions, children care, guideline for support in case of separation incidents, keeping regular routines, and providing them with facts).

Under (v)

"Practice food safety"
This is provided in three pictorial documents, which cover guidelines for dealing with food (raw meat and cooked food), hygiene matters, non-consumption guide for sick animals, and meat preparation.

Under (vi)

"Shopping/Working in wet markets in China and Southeast Asia"
This is provided in three detailed pictorial documents, which cover guidelines for a variety of factors covering guidelines on washing hands, avoid contacts with animals, safety measures during and after visiting such places, and a detailed guideline for disinfection measures and practices.

Under (vii)

"Stay healthy while travelling"
This is provided in five detailed pictorial documents, which cover guidelines for a variety of factors such as, (1) non-travel guides in case of having any symptoms, (2) certain safety measures during travelling; (3) guide for dealing with precautionary measures and medical products, (4) guide for development

of symptoms while travelling; and (5) General guide for things that should be avoided while traveling.

Original source https://www.who.int/emergencies/diseases/novel-corona virus-2019/advice-for-public.

BOX 5.8 "Safety measures for local online transportation systems"

By following all protocols and new regulations, all businesses and enterprises had to provide a reflective guide and report, to maintain the safety of both their employees and customers. DiDi, as the most popular online transportation system in China, provided a range of reassuring guide as part of their reaction to the outbreak. The below examples are mobile phone adverts from DiDi Company.

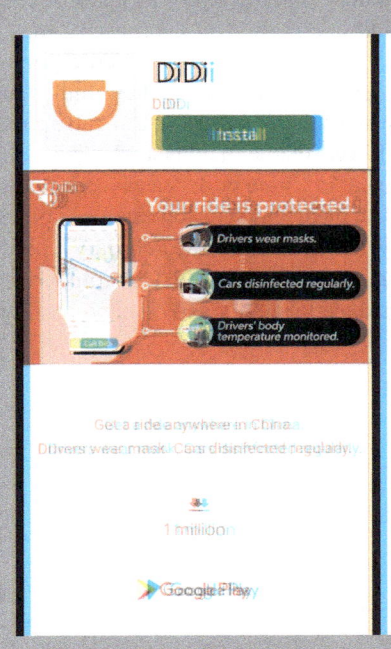 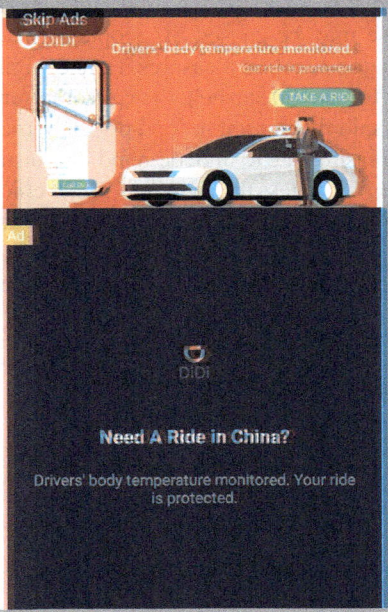

In such particular events, all companies have to be extra cautious as any reported infected cases in their premises/facilities or through their services/operations could cause adverse impacts on the sustainability of their operations. This may also cause longer economic impacts on their businesses. They also have to be cautious in case of any potential malicious acts from unpleasant customers or their rancorous competitors/rivals. It is important that in all cases, businesses keep transparency in their operations and must keep their workforces as safe as possible. The wellbeing of their employees should remain a priority. They also have to be as transparent as they can in order to release information on infected cases and provide their reflective response to those incidents. This approach helps them tremendously to tackle any false information/news from unofficial media and social media. Hence, risk assessment is generally high for businesses and enterprises.

BOX 5.9 "Issues of price control and stock checks on primary products"

The outbreak causes disruptions on primary products, especially those that are identified as useful products against the disease spread, such as disinfection products (shown below), medical supplies (such as masks), medicine, and fresh produce. There are inevitable impacts on unavailability and shortages of such products, lower stocks, higher prices, and counterfeit productions. Some of these may be caused by people's panic buying, hoarding of specific goods, and uncontrolled measures. Hence, high-level regulations and their immediate reinforcement can help to avoid or minimise such impacts. There are also cases of rapid fluctuation in food prices, some lowering due to high stocks and some increasing due to high demand. For instance, in China pork prices were soared by 116% in less than a month, while chicken meat was at its lowest due to piling up stocks in particular parts of the country. But eventually, the February records indicated a 5.4% spike in China's consumer price index (CPI), in comparison to last year's figures. Therefore, the imbalance in supply and demand requires immediate attention.

Elsewhere, shortages and rapid increase in prices were even worse. In the UK, for instance, hand sanitisers were sold 25–40 times of the original price (shown above). To overcome some of the issues, the governments responded by reinforcing their law and ensuring they are followed by all involved stakeholders, such as:

> In accordance with law and principles, the government will sternly deal with acts that interfere with quarantine efforts, illegal hoarding of medical goods and acts that spark uneasiness through massive rallies.—Quoted from South Korea's Prime Minister, Chung Sye-Kyun, on 22 February 2020

Shortages in essential food and household items could also become a serious concern for the government. After the early rounds of the supermarket rationing act in the UK (in early March 2020), the government had to step into meet with the main supermarkets of the country and discuss their responses. This meeting was arranged on the 9th of March 2020. This also was to ensure the supply chain and delivery of essential goods are well-maintained.

With temporary shortages in the first two weeks, China's Ministry of Commerce (MoC) had to step into boost food production and operations, on 6 February 2020. A response to increasing demand should be met and those who illegally increase the prices would be fined.

BOX 5.10 "Society's Response and Coping with the outbreak disruptions"

People's reactions are generally dependant on the level of seriousness taken by their governments; i.e. one reflecting precisely on the other. The media coverage is usually a big player, and society reacts to any disseminated news and information. Regardless of their false or correct information, the media coverage is effective as they can feed into the outbreak progression in various ways.

Generally, people's perception of the outbreak changes quite drastically during the whole event. It also differs from culture to culture, as we witnessed through the examples of how the same outbreak was handled differently in various parts of the world. In China, the society went through the cancellation of the New Year celebrations in a bid to contain the outbreak. Amid all uncertainties, people took the issues less seriously at first, but there were signs of growing anxiety. The immediate comparisons to the 2002–03 SARS outbreak were widespread and the lockdown of Hubei Provinces was a major turning point in people's perception of the disease. The local media coverage of the hospital construction site started from 24 January 2020, then intensified the seriousness of the outbreak. State-owned news outlet Changjiang Daily highlighted: *"The project will solve the shortages of existing medical resources and would be built fast and not cost much*". By the day after, 12 provinces in China had growing numbers, which was then doubled by the 26th of January 2020. With more than 50 cases outside mainland China by 28 January 2020, we witnessed diverse responses at the global level. On the last day of January 2020, Zhejiang province (where Ningbo is located) became the first province outside the epicenter with more than 500 confirmed cases. By then comparisons with previous cases, such as the H1N1 outbreak in 2009, were mostly to reduce the initial shocks. By then, there were growing concerns in various parts of the world. But it was only by the end of February 2020, when the situation was worsened with fewer more countries labelled "high-risk countries", and the spread took a much faster pace outside the boundaries of China.

But how serious is the outbreak?

We witnessed general global neglect on the seriousness of the outbreak. The persisting dismissive attitude from meme-making and jokes about the disease was not necessarily aimed to ease the anxiety, but they were simply adding to the informality of the situation. In general, such an approach gave a false perception of the seriousness of the situation. This impacted the general public as they often only comply with general requirements when the situation is worsened and in extreme conditions. This is unfortunate, but we have witnessed many examples of such incidents on social media, particularly in the areas where the number of cases was gradually mounting. This can be considered perhaps as a natural reaction of self-comfort, but we have to also understand the possible adverse impacts of being dismissive on something that is quite serious, disruptive, and impactful. It is a good reminder that positive thinking is different from dismissive thinking. Amongst these though, there were some comments that had dual tones, such as: the question "*why the disease outbreak became so intense in South Korea?*" And the response was, "*because South Korea didn't learn from North Korea*".

On the other side of society and from those at top of the pyramid, the issues were not any better. Some thought just because they are above the law, they would not get infected or pass the disease. This reflects on the amount of care some of those individuals may have for the general public. Unfortunately, we have seen dismissive responses from those that people are devoted to the most, from famous figures, social media influencers, sportsmen, royal family members, politicians, etc. Undoubtedly, such ignorant behaviour has a pervasive ripple effect on society.

Started with spreading the video and images of 'bat soup' in social media, the society's response was relatively dismissive at first. It took society some time to realise the complexity of the situation and have a more serious understanding of the consequences of the outbreak. As the outbreak started in different locations at different times, and its severity was also different, the responses were quite different, too. In general, however, the seriousness of the situation and how it is handled by the society is dependent on how it is driven or handled by the government or the authorities. As proven, if handled less seriously through the power of governance, then it is likely it is also followed less seriously by the general public. This was evidenced through the issues associated with gatherings, people's mobility/movement, people not following precautionary measures, people not following the procedures, and even in some cases ended as breaking the lockdown rules imposed by local governments. In some countries, the local governments also imposed fines against those who acted against the lockdown measures. For instance, there were more fines and restrictions related to issues of hoarding (in the form of stock piling up), unofficial increased prices, and unapproved opening of businesses. Moreover, there were common societal issues with mobility matters as many methods of movement control order were not taken seriously. In some cases, these involved the military or defence forces for the better control of the situation. Other examples include the restrictions or measures imposed for certain procedures to control the transmission and even delay or reduce the impacts. In some countries, we realised the use of medical masks were only practiced by certain groups of people. Also, there were official announcements in those countries to only use such products in case of having any related symptoms. This was somehow perceived as an immediate action against racism or hatred incidences that were growing rapidly.

More focused on general health concerns, the immediate economic impacts were also of major concern to the majority. This, in particular, turned into much bigger adversity for the vulnerable groups of poor and low-income people. The struggle of such vulnerable groups reflected on the knowns and unknowns of the society, the deficiencies of social services, and the ineffectiveness of governmental actions.

For the first time perhaps, the general public noted how a simple act of 'washing hands' is done wrongly on a large scale. Then, there were many educational guidelines from both official and social media to cover such assumingly simple matters. Moreover, the later natural reactions to survival, such as panic buying, were amongst those that represent our greedy nature. On this matter, the way how people reacted and how some responded were noteworthy. The situation made us aware of how vulnerable our societies have become. One anonymous response stated: *"vultures emptied my local supermarket of all meat, frozen food, tins, bread, pasta, milk, toilet roll and even washing tablets. But they left all the high-end pricey stuff. Bless them. It's hard to root for mankind sometimes"*. Another response was on issues of greediness, which argued that *"hopefully when the greed runs out, the rest of us can go and do a normal shop without this aggravation. Sadly, only when the stock piling has filled their fridges, freezers, garages, garden sheds, and the lofts"*. Also blended with some blames against the political system, there were disappointing feelings against how people are not considerate against the vulnerable groups and elderly: "*be sensible, take care and look out for the vulnerable*". Moreover, they were unprecedented scenes of fights and arguments in supermarkets around the globe, which was then reminded us the famous statement by Margaret Thatcher (quote from 1987) that says, "*There is no such thing as society: there are individual men and women, and there are families*". There were also some valid comparisons of these generations with those who fought for their countries, who saved lives, and were remembered to be more sympathetic to each other.

Under the growing global concerns and wide-range disruptions, society's inactions were also seen as a major challenge. The ripple effect from informal approaches of the government to lack of compliances by the general public was seen as a major matter that showed the level of collective thinking in our contemporary societies. There were growing worries against certain technologies, some treatment measures, and many examples of anxiety in society. The elderly assumed the disease was an attempt to get rid of their generation, while the youngsters were still in denial about the impacts of the disease on their assumed-stronger immunities. There is no wonder why some countries starting to blame the others as early as they could, and later asked for unrealistic compensations.

With the many hypothetical arguments, non-evidenced knowledge shares, biased discourses, blame games, and deteriorating power relations, there were no surprises that we witnessed random responses from the society, too. We see the growing impacts of such informal approaches on how society perceives the realities and then copes with the adversities. The impacts of such thinking always become widespread and affect society the most. Moreover, in this new chapter in our contemporary age, a novel disease made so much impact on our everyday life and is expected to make more changes in the years to come. Although this event enabled us to realise more about the realities, it is expected the society forgets faster than the normalisation of our operations.

While some countries started to invest in their healthcare systems and strengthen their resilience measures/plans, other countries were threatening to cut their budgets to global health organisations. While some were progressing faster through the development of medical sciences and adaptation of new technologies, some were stuck with their divine narrative. While some countries were accepting patients from their neighbouring countries, some were developing fear and anxiety against the idea of 'contaminated' foreign nationals. While we witnessed much act of love and care, many opportunists went against the flow of humanity and chose to go with the economic values and benefits. While there were growing inequalities in every corner, we had concerns about who were the real heroes of society.

During this event, the need for global governance was ever important, for cross-countries and sub-regional measures to help enhance our unity rather than towards individualism and fragmented decision-making procedures. And also for national guidelines to help regional/provincial strategies to become stronger for the support of better implementation in the network of cities, the city and its environments, and the network of communities. While the larger scaler strategies are not realistic, they should be passed on to smaller scales that are more feasible for better implementation and more effective results. In countries where health checks and quarantine (not lockdown) measures were in place most rigidly, those people under quarantine were under full monitory and regular checks.

With 14 days of strict quarantine measures, those people had to undergo a significant number of health cases before they could be provided with a "notice on the removal of medical quarantine observation". Hence, more than quarantine itself, it was the observation measures that were important throughout the process. One of these letters indicated: "*Everyone is the first person responsible for their own health. These actions are surely in favour of the health of yourselves, family members, and the public to taking good personal hygiene and living habits, as well as to further improving the awareness and health accomplishment for the epidemic prevention and treatment. It also helps the prevention and control of the novel coronavirus pneumonia epidemic. May we work together to overcome it*" (extracted from the notice provided by the Chinese Prevention and Control Teams). This example, along with many others, shows the seriousness of this event, as well as the collective measures that are needed to overcome the issues.

BOX 5.11 "Political and International Responses"

Here are extracts from a famous British sitcom TV series from 1986 to 1988, "Yes, Prime Minister" (a sequel from the earlier "Yes Minister"), written by Antony Jay and Jonathan Lynn, in its episode on "A Victory for Democracy" (1986), which suggest how the standard foreign office response may be in a time of crisis. This is the communication between two main characters of this episode, Sir Richard Wharton (A), and Sir Humphrey Appleby (B):

(A): In Stage one we say nothing is going to happen. (B): Stage two, we say something may be about to happen, but we should do nothing about it. (A): In stage three, we say that maybe we should do something about it, but there's nothing we *can* do. (B): Stage four, we say maybe there was something we could have done, but it's too late now.

All of a sudden, the above communication was viral on social media.

It is indeed unfortunate to see public health become political. Nevertheless, the outbreak event with its broad international angle, adds to the complexity of politics around public health. The mental wars of one country against another (or one group against another) play a big part in the political issues of international relations. The fact that at first COVID-19 was globally symbolised as 'Wuhan Disease', 'Wuhan Coronavirus', or 'Wuhan Virus' showed how one-sided this mental war could be in one of the most difficult times of the country. In the early days of the outbreak, China was under international criticism, but was later appraised as its efforts turned into eventual success. At first, China's lockdown approach was condemned by the global powers as it was deemed to have issues against its effectivity from various perspectives. A few weeks later, the approach was not appraised but was branded as a role model for those high-risk countries. Soon after, several countries implemented the same measures. It was on 10 March 2020 that Italy, as the whole country, was under full lockdown.

The US-China relations are a good example of political relations in this particular event. The United States was the first country to impose bans for entries from China, and their government representative referred to the situation as an opportunity for the US economy. The outbreak also exacerbated the already heightening trade wars between the two countries. The situation was escalated so rapidly that the two countries' presidents had to hold their first outbreak-related phone meeting on 7 February 2020. The concluding points of their meeting were on better communication, better coordination, and joint efforts towards the containment of the disease. The meeting then helped to partly ease the already escalated situation. But the political stances may not be overlooked so easily.

Countries that lack preparedness also suffer at the time of response. The delayed announcements or action plans could only worsen the situation. For example, the UK's slow pace in preparation was distressing as the government's 4-stage battle plan seemed to lack adequate resilience thinking. Politically, it is correct that any country should do its best to reduce the negative impacts. But it is utterly unacceptable to utilise this as an approach to reduce awareness and then dwindle the societal preparedness. Moreover, those countries with no reports of infected cases, if not genuine, face three scenarios: (1) they may not know if they have cases for whatever reason, (2) they do not want to know as they may seem to be unprepared, or (3) they do not want others to know for continuity of their international trade, tourism, etc. In the latter instance, the game played on is to make a bigger problem as small as possible. Also as per records, the poorer nations and conflict zones either claimed they had no cases or they simply delayed announcing it as their economies could not cope with disruptions.

BOX 5.12 "Government Responses"

Government responses are generally different depending on the level of risk, the needs, and the preparedness they may have to deal with the outbreak. Some of these were strong and reassuring, while the other came with delays and more uncertainties. Some call for solidarity while the others are passively observing the progress. In all cases, the provision of guidelines is essential to be a completely top-down approach. Any informality of responsiveness would have negative impacts on how the plans are taken forward or delivered at later stages. The (national) guidelines come in various forms and represent an array of responses. The Chinese president first announced, *"I have, at all times, monitored the spread of the epidemic and the progress in prevention and control work, and continue to give oral orders and instructions"*. This reflects on the provision of national guidelines to those lower governmental levels for the implementation plan. This led to lockdown strategies. After his first visit to Wuhan during the outbreak on 10 March 2020, the Chinese President acknowledged, *"The situation of coronavirus prevention in Hubei and Wuhan has shown positive changes and achieved phased results, initially realizing goals to stabilize and turn around the situation"*. This was a clear message to reflect on successful progress.

With growing criticism on the general public, the Italian Prime Minister requested people to follow the public health emergency that was imposed on 21 February 2020. He urged people that *"The Future depends on us and everyone must do their part"*. Soon after, on 10 March 2020, the whole country was under full lockdown. Meanwhile, the UK government suggested a 4-stage plan (as shown below) in early March 2020, announced by the UK Prime Minister: *"The plan has four strands. Containing the virus, delaying its spread, researching its origins and cure, and finally mitigating the impact should the virus become more widespread. That is, contain, delay, research, mitigate"*. The research stage was later described as a continuous and prolonged stage.

STAGE 1: CONTAIN	STAGE 2: DELAY	STAGE 3: RESEARCH	STAGE 4: MITIGATE
a) Detect early cases, isolate them to prevent clusters; b) Follow up close contacts of patients; c) Screen travelers to and from high risk areas.	a) Launch public information campaign on washing hands; b) Allow pupils to do schoolwork from home; c) Relax sick pay rules as one in five could be forced to stay home; d) Ban large-scale gatherings.	a) Pump money into researching illness; b) Ignore low-level crime if lots of police officers test positive; c) Prepare for multiple waves of virus	a) Cancel non-urgent operations to prioritise vulnerable; b) Deploy troops on the streets; c) NHS staff called out of retirement.

There were examples of responses that reflected on dismissive thinking against the handling of the situation. For instance, as the Grand Princess cruise liner was stuck in limbo off the Californian coast with 21 confirmed infected cases on board (at the time), the US president responded "*I like the numbers being where they are. I don't need to have the numbers double because of one ship that wasn't our fault*". Perhaps this was an inconsiderate reflection on the earlier situation experienced by Japan. This issue also came at the time the stock market had its biggest hit since the 2008 economic crisis and oil prices had the highest drop in value in a day since 1991. This was 9 March 2020, when the US president informally announced: "*So last year 37,000 Americans died from the common flue. It averages between 27,000 and 70,000 per year. Nothing is shut down, life & economy go on. At this moment there are 546 confirmed cases of coronavirus, with 22 deaths. Think about that!*".

BOX 5.13 "Health System Responses"

According to the Network for Public Health Law (2020, online source), the health system response includes a wide variety of actors and requires to be developed as part of policy development procedures. These actors include: "*public health agencies, health care workers, emergency managers, and policymakers are grappling with core legal preparedness and response efforts, particularly related to emergency powers*".

Also shown here is an explanation of legal and policy decisions during emergencies:

Public health officials may face many critical legal and policy decisions during public health emergencies, including:

- *Inter-jurisdictional legal coordination of federal, tribal, state and local actors in real-time emergencies under changing legal norms;*
- *The ability to issue isolation or quarantine orders, or other social distancing methods, to control public health threats;*
- *Whether to close or dismiss schools, or other public assemblies, temporarily or for prolonged periods to prevent the spread of communicable diseases;*
- *The authority to mandate vaccinations for minors or autonomous adults, including health care workers;*
- *Licensing, credentialing and privileging out-of-state health practitioners;*
- *Inter-jurisdictional management of scarce resources including personnel, vaccines, shelter and sustenance.*
- *Omnipresent concerns over liability of public health practitioners during emergencies*

From travel advisory plans as shown in (a) the representation of government action plan (example from Japan), (b) health system responses are vital to the management of the outbreak (example from Australia). The health system responses should be widespread (example from the UK) as shown in (c), or location-based and immediate to any arising matters (example from the US) as shown in (d).

(a) Under the measures of "Coronavirus (COVID-19) Advisory information" the Government of Japan (2020) proposed measures for travel, safety tips, information for main official resources, information on 24/3 medical needs (multilingual support), insurance, news, and temporary closures of places of interest/attractions. A detailed health advisory is helpful to provide necessary and right information to the general public. It is important these are regularly updated and everyone has the right to access the resources. *Source* https://www.japan.travel/en/coronavirus/.

(b) Alongside many health advisory guidelines, the Department of Health under the Australian Government released action plan information in the forms of "government response to the outbreak" (2020a) and "emergency response plan" (2020b), both accessible by the general public. In the former action plan, the information includes:

> "what we aim to do", "who managed the response", "our response plan (also the latter source)", "what we are doing now", and "what you can do".

This is an informative source that covers issues related to overall aims, economic management plans, health matters, reviews and responses, guides, approaches and steps. In the latter source (2020b), the plan is aimed to "guide the response of Government health agencies", "make sure the actions are well coordinated" and "outline four main stages of the response— initial action, targeted action, stand-down, preparedness". This document is intended for the general public, and includes an overview of the national approach, and the government's operational plan.

Source for 2020a: https://www.health.gov.au/news/health-alerts/novel-cor onavirus-2019-ncov-health-alert/government-response-to-the-covid-19- outbreak.
Source for 2020b: https://www.health.gov.au/resources/publications/aus tralian-health-sector-emergency-response-plan-for-novel-coronavirus-cov id-19.

(c) The UK's National Health Service (NHS) is the country's main publicly funded healthcare system. At first, they contacted people in early February via text message notifications:

> If you have recently returned from mainland China, or have been in contact with some who has, and are experiencing flu like symptoms then we advise that stay at home and call 1111. Please do not attend the surgery.

Almost five weeks after in early March 2020, NHS disabled their online booking system due to the uncertainty around the disease. By then, they developed an online general advice (NHS, 2020).

Source https://111.nhs.uk/covid-19.

(d) Reported Coronavirus Case (an email source)

> We have been notified today a resident of [xxxxx] that they have tested positive for
> the novel coronavirus (COVID-19). The individual and roommates are currently
> self-quarantining inside their apartment.

The notification message then continues with safety and cleaning procedures, as well as health and well-being considerations. It then provides a
detailed information about required actions, including seven precautionary
measures recommended by public health officials in the New York City, the
US. The guidelines are then provided and communicated with those directly
and indirectly affected.

BOX 5.14 "Responses and Predictions from Experts"

Amid all disruptions, some issues never change and some governments do not
reflect considerately on the situation. In the middle of this crisis-like situation,
The US successfully tested its first supersonic missile, claiming that it could
target any point around a globe in less than an hour. Apart from these threatening
messages, North Korea kept testing its missiles into the sea at the same time.
And of course, our biased media only covers the latter. This is a reminder that
some issues truly stay the same or even worsen in any situation. However, the
pandemic outbreak was more than just one or two military businesses. It was
seen a game-changer event that could have longer impacts around the world.
For instance, failures in the European Union (EU) management and control
of borders, angered many who thought the restrictions and bans came later
than expected. Some countries planned to print out money and some even
suggested to print money and pass their financial support packages to people's
doorsteps. Soon after, those who advanced in the name of human rights or
becoming united, only looked after their (own) priorities and imposed their
(own) country-level strategies; so much for the spirit of unity, that has a range
of different benefits for different parties.

In a race to showcase economic resilience, technology, and transparency,
there are always many examples of governmental resolutions and actions to
combat the disease outbreak. It is a major time to showcase the capacity and
capability of resources, military forces, economic supremacy, political influence, technologies, scientific advancement, etc. For instance, we could see
many countries announced their success in vaccine development as early as
they could. But this time, the disease had more to change that just some policies,
relations, and economic situations. Under the climate conditions of calamities,
we cannot only anticipate selective impacts, but should foresee those impacts

that can make changes in different ways. In an extended article by Politico Magazine on the 19th of March 2020, COVID-19 was identified as a large scale crisis, which can reorder society in many different and dramatic ways. They also highlighted several US-based experts' predictions, which are extracted and summarised here:

Under 'Community' subject:

(1) **'The personal becomes dangerous'**—*by Deborah Tannen, Professor of Linguistics*

This is regarded as a new order, which comes with the loss of innocence or complacency. It also highlights the paradox of distance communication and how our greater comfort and safety may just be achieved through the absence of others.

(2) **'A new kind of patriotism'**—*by Mark Lawrence Schrad, Associate Professor in Political Science*

This is narrated as an approach to salute those services and their workforces who sacrificed their lives and times to combat the disease. This was suggested as a potential de-militarisation of patriotism, which could represent patriotic figures more than just the military forces or military veterans.

(3) **'A decline in polarisation'**—*By Peter T. Coleman, Professor of Psychology*

This is interestingly discussed as an opportunity to reduce the escalating political and cultural polarisation. This is argued based on two reasons of "common enemy scenario" and "political shockwave scenario", which could help to enhance more constructive patterns in the contemporary cultural and political discourse. However, this only depends on the political climatic conditions of the country and how it has developed in recent years.

(4) **'A return to faith in serious experts'**—*By Tom Nichols, Professor of National Security Affairs*

The discussions are narrated through worries on how unserious things have become in recent years and how affluence and high levels of consumer technology have changed our attitudes. The situation in return forced people to accept facts and recommendations given by the experts, in an attempt to possibly bring more seriousness to the mentality and behaviours of the society.

(5) **'Less individualism'**—*By Eric Klinenberg, Professor of Sociology*

This is narrated as a turning point toward authoritarianism, a chance for the reorientation of politics and possible recoveries of our primary systems and services, such as public health, healthcare system, and public services. This is recognised as an opportunity to reconsider our values and enhance what may be lacking in our economic and social orders.

(6) **'Religious worship will look different'**—*By Amy Sullivan, Director of strategy*

The arguments are compared with the conditions and challenges of war or diaspora or precaution, and how they may impact the daily operations of religious activities. Evaluated through cultural and social differences, the discussions were focused on our common good merely to revitalise our interconnected humanity.

(7) **'New forms of reform'**—*By Jonathan Rauch, Senior Fellow and Activist*

The pandemic event is identified as a transformational time, similar to those that are experienced by certain communities or groups of people. The impacts are predicted to be on our healthcare system, as well as how the government can become more conscious of interdependency and communities. These would then lead to potential landmark reforms.

Under 'Technology' subject:

(8) **'Regulatory barriers to online tools will fall'**—*By Katherine Mangu-Ward, Editor-in-chief of Reason Magazine*

This is narrated as an opportunity to fast-track online tools and virtual platforms. Examples of such transformative opportunities are seen across multiple sectors or systems, from education to simple tasks of medicare billing. These changes will provide patterns that become norms in how we practice our work routines, meetings, appointments, classes, etc. Through remote working opportunities, we expect to see innovative or simplified methods of doing the same tasks.

(9) **'A healthier digital lifestyle'**—*By Sherry Turkle, Professor of Social Studies*

This was suggested as an approach to use our devices more thoughtfully and more creative, and to make them work better for better uses than just social media. In order to do so, we have to apply our human instincts to our devices and make progress in how digital lifestyles can become more useful and supportive.

(10) **'A boon to virtual reality'**—*By Elizabeth Bradley, Scholar of Global Health*

This is a chance for virtual reality (VR) to be more innovative and even provide a programme to help with matters of socialisation and mental health. Interventions of such kind are expected to become more popular and applicable.

Under 'Health and Science' subject:

(11) **'The rise of telemedicine'**—*By Ezekiel J. Emanuel, Chair in medical ethics and health policy*

The pandemic is identified as a shift in the paradigm for healthcare delivery. The ultimate impacts of such a paradigm shift would be to offer more vide call medical checks, staying home and keeping out of the transit system. This is likely to reduce the pressure on our healthcare systems and also overpopulated facilities/services.

(12) **'An opening for stronger family care'**—*By Ai-Jen Poo, Executive Director of the National Domestic Workers Alliance and Caring Across Generations*

The pandemic event has highlighted deficiencies in our care infrastructure. With larger impacts on the health and financial conditions of families, it is important to seek additional political support for family care and overcome the challenges that most people face in the changing conditions of the outbreak. This event also highlighted the growing ageing population, and how their issues should be addressed, through a better sense of care that is shared responsibility.

(13) **'Government becomes Big Pharma'**—*By Steph Stirling, Vice President of advocacy and policy*

This is discussed mainly from the perspective that highlights flaws in our health systems, often due to high costs, inefficient operations, and a market-based system for medicine development and production (including vaccines). In return, this should help to make resilient government approaches and promote the public sector to take direct responsibility for the development and production of medicines.

(14) **'Science reigns again'**—*By Sonja Trauss, President and Founder of Yes In May Back Yard*

This is narrated as an opportunity to shed light on truth and give more credit to scientific findings. This event should also become an opportunity to enlighten science doubters, and increase the pubic respect for expertise in specific areas of public health and epidemics.

Under 'Government' subject:

(15) **'Congress can finally go virtual'**—*By Ethan Zuckerman, Associate Professor of the Practice in Media Arts and Sciences*

The pandemic event provides an opportunity for many governmental activities and institutions to operate virtually. The continuity of governmental operations is as important as dealing with governmental reapportionment and expansion. The use of technologies at such a level could not only virtualise the institutions but also help to expand the size of the government.

(16) **'Big government makes a comeback'**—*By Margaret O'Mara, Professor of History*

This pandemic event provided an opportunity to enhance the visibility of the government at multiple levels. This is also a chance to highlight the role of "big government" and its effectiveness in society. This is aimed to also support the public infrastructure, health, and wellbeing of the communities, and include economic support when needed.

(17) **'Government service regains its cachet'**—*By Lilliana Mason, Associate Professor in Government and politics*

This event is identified as a chance for the functionality of government, and a chance to prove responsive and responsible support to the society.

(18) **'A new civic federalism'**—*By Archon Fung, Professor of Citizenship and self-government*

This means providing the opportunity for local governments to become the main centers of justice, solidarity, and problem-solving. The event also helps to harness what needs to happen at the local level to reduce inequalities and provide transformative measures for economic inclusion.

(19) **'The rules we've live by won't all apply'**—*By Astra Taylor, Film maker*

The pandemic event is described as a rule-changing opportunity to ensure we reflect on the realities and address the needs of vulnerable groups. These are specifically related to those rules that made society more delicate and unequal.

(20) **'Revived trust in institutions'**—*By Michiko Kakutani, Author and critic*

This is narrated as an opportunity to regain public trust which is crucial to governance. This is required to happen through transparency, highlighting what John M. Barry mentioned in his 2004 book on '*The Great Influenza*': "*...those in authority must retain the public's trust...[and]...the way to do that is to distort nothing, to put the best face on nothing, to try to manipulate no one*".

(21) **'Expect a political uprising'**—*By Cathy O'Neil, founder and CEO of the algorithmic auditing company*

The prediction here is a new political uprising in the aftermath of the pandemic event, and perhaps a movement towards "Occupy Wall Street 2.0". This should lead to drastic changes in the political system, and those associated with our economic systems and addressing the needs of the society.

(22) **'Electronic voting goes mainstream'**—*By Joe Brotherton, Chairman of Democracy Live*

With proven technologies, there are now more possibilities for more mobile and at-home voting. These types of voting are already practiced worldwide, but it is yet to become the new normal. In addition, such technology use could help to enhance security, transparency, and cost-effectiveness of voting procedures. By adopting more advanced technologies, we could be steps closer to making electronic voting to become permanent.

(23) **'Election day will become election month'**—*By Lee Drutman, Senior Fellow*

The pandemic outbreak could simply change the normal election procedures. One of these potential changes would be to have early voting procedures and provide a fair, equal, and transparent experience to all communities. While there are flaws related to issues of fraud and inconsistency in the voting process, this idea could possibly become a new experience to citizens.

(24) **'Voting by mail will become the norm'**—*By Kevin R. Kosar, Vice President of Research Partnerships; also Dale Ho, Director of voting Rights Project*

The pandemic poses a threat to people who normally vote in person. Therefore, it provides a chance to encourage and promote alternative voting systems, one of which is voting by mail. There are certain requirements for some adjustments such as issues of language barriers, disabilities, limited postal access, or late arrival of the mails. Nevertheless, voting by mail can reduce the risks and help to promote the safety of individuals and communities.

Under 'the Global Economy' subject:

(25) **'More restrains on mass consumption'**—*By Sonia Shah, Author*

The pandemic event is recognised as a force on society to reduce consumptions, or to accept limitations on high-level consumer culture. This is likely to change some of the increasing outsizing culture or those already-developed habits of mass consumption. This is a chance to reduce our industrial footprint.

On the contrary, *"the end of mass quarantine will unleash pent-up demand for intimacy and a mini baby-boom. The hype around online education will be abandoned, as a generation of young people forced into seclusion will reshape the culture around a contrarian appreciation for communal life"*.

(26) **'Stronger domestic supply chains'**—*By Todd N. Tucker, Director of Governance Studies;* also *Dambisa Moyo, Economist and author*

The pandemic event highlighted some of the main issues around our supply chains. It was also proven that how one country's disrupted economy could impact others. Hence, the obvious opportunity to switch to:

> …a more robust domestic supply chain would reduce dependence on an increasingly fractured global supply system. But while this would better ensure that people get the goods they need, this shift would likely also increase costs to corporations and consumers.

These measures are to be taken into future considerations for better supply chain systems, and better management of product supply and shortages.

(27) **'The inequality gap will widen'**—*By Theda Skocpol, Professor of Government and Sociology*

There are significant issues around inequalities. The adversities would have more impact on not only the vulnerable groups, but also the less affluent groups. There are issues regarding less ability to work from home, or certain jobs that may put people in danger of coming into contact with the disease. There are also concerns about lack of home-schooling or education for children of those who cannot handle such methods due to any reason, such as lack of resources, lack of access to high-speed internet, lack of time, lack of abilities, etc.

Under 'Lifestyle' subject:

(28) **'A hunger for diversion'**—*By Mary Frances Berry, Professor of American Social Thought, History and Africana Studies*

It is anticipated that the current pandemic outbreak would probably accelerate some trends already underway, mostly related to security technologies and control measures that were practiced during the event. After the event is over, we may witness how people will find their relief and support in their communities.

(29) **'Less communal dining—but maybe more cooking'**—*By Paul Freedman, Professor of History*

As our eating behaviours have already changed significantly during the outbreak, it is likely we require less demand for sit-down restaurants. It is expected that in a collective way, we will be temporarily less communal than before.

(30) **'A revival of parks'**—*By Alexandra Lange, Architecture critic*

It is anticipated that after the pandemic event, people would value their parks (or other big spaces) more than before. People may start using parks for other reasons than just for selective or seasonal activities. This may result in more public investment in places that are open, accessible, and are suitable for all seasons—i.e. parks.

(31) **'A change in our understanding of change'**—*By Matthew Continetti, Resident Fellow*

We must anticipate some paradigm shifts and those that may change our collective notions. This will eventually force us to revise our conception of "change". This event may simply change the issues related to societal frivolity and ceaseless activities.

(32) **'The tyranny of habit no more'**—*By Virginia Heffernan, Author*

This pandemic event was a turning point against the "*fantasy of optimising a life*". This is likely to lead to major changes in our regular and established behaviours, such as those that are phrased as "*our favourites: driving cars, eating meat, burning electricity*". This event may potentially recharge our commitment to raise new concerns about our surroundings and happenings. Perhaps this can lead to a better understanding of "*living peacefully and meaningfully together is going to take much more than bed-making and canny investments. The Power of No Habits*".

Although the opinions and predictions of these experts are focused on the context of the US, most of them are applicable and relevant to other contexts, too. After the recovery from this pandemic outbreak event, we anticipate having its impacts lasting for long. The aftermath of this pandemic event is likely to include changes that will be widespread, and perhaps more impactful than we can predict now.

Original source: extracted from Politico Magazine (article on 19 March 2020), online source available from: https://www.politico.com/news/magazine/2020/03/19/coronavirus-effect-economy-life-society-analysis-covid-135579#superComments.

BOX 5.15 "Lockdown, Limitations, and Bans"

During the outbreak, and following from diverse government responses, there was also a range of measures addressing different methods of lockdown, limitations, and bans. There were different responses, again depending on the level of risks, the situation of each region (or sub-region), and some also included geographical situations, such as the vulnerability of small island countries/areas in populated areas like Singapore, or countries that are transit points between major hubs like Italy and Iran, or countries that faced issues associated with cruise arrival to their ports or tourism. Higher risks such as high level of mobility both in China at first, and later in the US, South Korea, and Italy indicated faster progress of community transmission in the regions with higher mobility. Here are selective measures:

China's lockdown was at both city-level and provincial-level. By the time the situation was worsening in other countries, and to avoid reversing the progress internally, China imposed new restrictions for people traveling back to the country. There were new and different quarantine measures in different categories. This was first imposed by Beijing on 4 March 2020, and soon after was followed by larger and secondary cities across the countries. In this approach, the cities localised the national-level guidelines with a new categorised quarantine policy including four distinct categories of (1) high-level input countries, (2) relatively-high risk input countries, (3) middle risk input countries/regions, and (4) Other countries with low risk. In each category, the quarantine measures were different:

1. *High-Risk Input Countries—People travelling back from these countries within 14 days (back-tracking from 23 February 2020) need to perform quarantine for 14 days in government appointed hotels.*
2. *Relatively-High Risk Input Countries—People travelling from these countries within 14 days (back-tracking from 23 February 2020) need to perform quarantine for 14 days in government appointed hotels if they done have residence here. If they have residence here, the quarantine can be in their residence.* (This category was later modified to: *people in this category may also now have to undergo enhanced medical observation in a government designated hotel for 14 days*).
3. *Middle Risk Input Countries/Regions—Same as category 2, without the later modification.*
4. *Other Countries: Low Risk—People traveling from these countries need to complete the information register and report health condition periodically fi they do not show health symptoms entering the city.*

Note Later on, from 24 March 2020, the quarantine measures applied to all people returning to China.

The detailed guide and restriction measures were announced alongside detailed quarantine requirements for any food delivery, courier, and garbage/rubbish. In the guidelines, it was also recommended to include medical observation checks, including initial registration and final cancellation of the observation process. One of the main reasons behind these new measures, as explained earlier, was to avoid reversing the outbreak progress. In the earlier case of eight infected foreigners traveling from Italy to Huzhou in Zhejiang Province, East China was an example that made faster changes to the situation's guidelines. While globally, the new measures by China were not welcomed, it was received very importantly internally for a country that just suffered in the first wave of outbreak. On 11 March 2020, the 14-day quarantine rule applied to all foreigners arriving in Beijing. After the outbreak was declared a pandemic on the same day, we witnessed a new round and higher level of travel restrictions and bans. This started with the US immediately banning any travels between Mainland Europe and the US for 30 days. As the decision deemed to be political and the number of cases was escalating in the UK, similar measures were imposed on other parts of Europe after two days. Similar measures were already developing in smaller nations, but now it was widespread.

The early school bans in Iran, particularly before the start of the Persian New Year holidays, was perceived less seriously by the general public. In return, the closure announcement led to higher intra-provincial mobility. The significant impact was seen in Northern provinces from a twofold: (A) popularity for local holiday destinations, and (B) clustered layout of cities in a large corridor around coastal areas of the Caspian Sea. This led to extensive bans in many provinces, the lockdown of selective cities, and higher-level bans in highly infected districts in the capital city of Teheran. Similar patterns of mobility were experienced in other places/countries, soon after the closure of schools. In the UK, it was recorded that only one day after the school closures people were clustered in beeches, parks, and popular attractions.

With larger fears over major religious activities and events, Saudi Arabia had to cancel some of their major annual religious activities, and put more restrictions on travel bans for people exiting the country. Eventually, the Qatif region as one of the main industrial regions for oil production was under full lockdown. The restrictions went beyond just the physical lockdown (similar to those in the region), and move into the next level of travel bans: *"The new travel restrictions are far-reaching, and will impact citizens, residents, and those who regularly commute to Saudi Arabia for work"*. Later on, all foreign entries were banned.

The heightened restrictions were not only limited to the following bans from entries from high-risk countries/regions, or bans for Saudi Arabia's own citizens and residents. By 3 March 2020, the bans also affected citizens and residents of several other countries in the region, partly the smaller Arab states in the southern region of the Persian Gulf, including the UAE, Bahrain, Kuwait, Oman, and Qatar.

From the early days of the outbreak, there were growing strikes in Hong Kong Special Administrative Region (HK SAR) demanding immediate border closure. With an intentional fire incident in one of the treatment centers, the situation led to the closure of border crossing and extended limitations on air travel to and from Mainland China. This also coincided with the first recorded death case outside Mainland China. The early updates on flights to/from Mainland China indicated: *"In light of the Government Response Plan to combat the coronavirus infection, we will be reducing the capacity of our flights to and from Mainland China by 50% or more, from 30 January to the end of March 2020"*. Effective after, from 2 February 2020, local airlines stopped their operations to selective destinations.

Globally, most regional flights either reduced or suspended their operations to and from China. Amid growing fears about the disease spread, the same measures were replicated for other countries/regions. The continuous and larger scale travel bans brought major disruption to global travel.

Countries like Mongolia and Russia closed their border. And North Korea banned any mobility between the two sides. North Korea also implemented high-level measures and warned against any negligence.

Closure of border and suspension of flights by the 29th of January 2020, was soon followed by the official governmental announcements by the UK and France. As recorded by the UK's Foreign and Commonwealth Office (FCO) on 4 February 2020, they provided:

> …advise against all travel to Hubei Province due to the ongoing novel coronavirus outbreak. The FCO advise against all but essential travel to the rest of mainland China (not including Hong Kong and Macao). The British Consulates-General in Wuhan and Chongqing are currently closed. If you're in China and able to leave, you should do so. The elderly and those with pre-existing medical conditions may be at heightened risk. Later on, the same measures were applied to other high-risk countries.

The US travel ban on foreign nationals who have been in China was the first international measure, suggesting for *"temporarily suspending the entry into the United States of foreign nationals who pose a risk of transmitting the 2019 novel coronavirus"*. This was effective as early as 2 February 2020, as announced by the Health and Human Services Secretary.

Singapore was also one of the first countries, second in the row, to put travel bans in place on 3 February 2020. Their approach was taken forward from the assessment conducted by a multi-ministry task force dealing with the virus, announcing *"Our approach is to…[look]…at the evidence, at the source of where the virus is, how widely it is spreading, and which are the at-risk sources"*. Their approach was also very reflective of the outbreak progression as the fluidity of the situation remains unpredictable. They then added: *"We will not rule out more action on travel restrictions but it has to be based not just on geographical proximity or distance but based on evidence on the nature of the virus [and] how widely it is spreading"*.

A similar ban was also imposed by countries like Malaysia, Brunei, etc. These were implemented firstly due to high mobility between China and Southeast Asian countries, and then later due to a similar flow of travels between the Middle East and Southeast Asia.

On other examples of high-level measures, Israel imposed 14 days of quarantine for anyone entering from abroad, a high-level ban which was partly followed by Jordan and Saudi Arabia amid the growing fear and escalating cases in the region. This was initially imposed for 14 days.

Some smaller countries had more strict restrictions, such as the Kingdom of Bhutan, who banned all foreign tourists from entering the country from the early days of the outbreak. Another example is the Maldives, which temporarily forbidden entry and exit to selective resort islands and those inhabited by locals. Also, higher restrictions were imposed from earlier stages by smaller Pacific island states, such as Kiribati, Marshall Islands, and Samoa.

As Italy entered a critical stage by 10 March 2020, also regarded as its "*darkest hour*", its Prime Minister confirmed: "There won't be just a red zone…There will be Italy". The impacts of this message were heard in other parts of the European continent. The UK Prime Minister then later announced on the same day, that "…*containing is extremely unlikely to work on its own*".

The list goes on, and each case of isolation/quarantine and bans/restrictions were different from one another. They also developed and updated regularly, depending on multiple factors associated with the outbreak progression. Decisions were made differently by different authorities, even at multiple governmental levels. For some countries, this could also vary between different cities. In most cases, the measures were continuously changing depending on the outbreak progression in a specific region. Relevant authorities and airlines continued with regular updates on their temporary or longer-term adjustments. While this book is completing, the situation continues globally. Hence, we can only capture some of those cases that represent a variety of responses and measures. This is just part of the realities during this outbreak.

Note 14 days is seen several times across many examples of travel bans and restrictions. This was mainly due to the information on the disease incubation period in an infected person's body as a carrier. However, for nearly all cases the 14 days restrictions were extended and/or intensified.

BOX 5.16 "From Travel Bans to Global Disruptions"

It did not take long for many countries/regions to escalate their measures on travel bans and entry restrictions. For some, this was from the inception, and for few, it came after ineffectiveness of the first 14 days bans and restrictions. For instance, Israel started to ban all foreign nationals' entries from abroad (with some exceptions for self-isolation for 14 days) and began digital tracking of confirmed cases.

The same or even more restrict travel bans were imposed (before or after Israel) by many others, such as Argentina, Australia, Chile, Costa Rica, Cuba, Ecuador, El Salvador, Ethiopia, Ghana, Guatemala, India, Jordan, Kenya, Malaysia (with very limited exceptions), New Zealand, Nigeria, Peru, Russia, Saudi Arabia, Syria, the Philippines, Ukraine, and many more. Some of these travel bans were the highest measures with the highest hit on the airline industry worldwide. Some countries of Latin America (including South and Central Americas), Asia, Oceania, and Africa had their bans and restrictions in place prior to high alert situations or just at the beginning of their upsurge in infected cases. This perhaps resembles a lesson learned from the case of the European Union countries, and not to repeat their mistakes.

On 17 March 2020, the European Union closed all borders of 26 countries to all visitors for at least 30 days. Soon after, the European Central Bank (ECB) launched an emergency package worth US$820 billion to ease the impact across the Eurozone, by buying government and company debt. One day after, the borders between Canada and the US were closed by mutual decision and only opened for trade purposes and some cases of exceptions (such as for work purposes). Some bans also appeared to be apparent reactions against one another, almost a retaliation act that worsened some of the international relations. The examples were bans between Colombia and Venezuela, China and South Korea, the EU and the US, etc. On 15 March 2020, South Africa announced a national state of disaster. Others declared a state of emergency, and some regarded it as a wartime crisis.

The impacts on countries under sanctions or in war zones were worrying and catastrophic, reminding us of the fact that the so-called international community should consider the matters more thoughtfully at the time of need. The inhumane acts could only worsen the conditions of those countries/regions that already suffer from other existing or ongoing adversities. With poorer infrastructures and widespread shortages, we could face a disastrous situation and no longer just cases of crisis or emergency.

On the day all the states of the US had confirmed cases, the oil price dropped again to its lowest in 17 years, at US$25 a barrel. The economic impacts were already global and affected many. On 18 March 2020, China had no new domestic daily cases for the first time from the start of the outbreak. This was recorded by China's National Health Commission (NHS) who reported no domestically transmitted cases in China for the first time. This was a positive record that brought signs of hope for a country that faced the disease in a shock, with little time for preparedness and responsiveness. However, this came under the threat of the second wave of imported cases, particularly for China, Japan, Singapore, and South Korea. By then, all four countries have been successful in their robust and effective control and monitory measures. In Japan, as the situation was changing by lifting its state of emergency. Their focus was meant to be more on recovery and sustaining social and economic activities. This came at the time that other Asian and European governments were begging their people to stay indoors and follow the precautionary measures. A remarkable example is the announcement of Malaysia's Director-General of Health on the day the country started its semi-lockdown (18 March 2020): *"We have a slim chance to break the chain of COVID-19 infections…*[and]*…Failure is not an option here. If not, we may face a third wave of this virus, which would be greater than a tsunami, if we maintain a 'so what' attitude"*.

Also recognised as an unprecedented challenge, UNESCO (2020) confirmed that education disruption affected more than half of the students around the world (by mid-March 2020). This resulted from the nationwide education closures in at least 107 countries (till 18 March 2020) and more to come (12 proposed them soon including Bhutan, India, Myanmar, Russia, UK, US, Uruguay, Vietnam, etc.). Till that date, these closures impacted 861.7 million students at various levels and it was expected that they continue with further disruptions. UNESCO also reminded us of the importance of scientific cooperation and education and knowledge sharing during this time. This also comes with recommendations to provide immediate support to those in need and with minimised education infrastructures and facilities. As the disruptions created outrageous situations around the globe, the impacts on 'health and economy' were phrased alongside each other. On many occasions, the economic impacts were perceived or even fabricated (to some extent) to be more than the health impacts. On the contrary, there are many debates on prioritising public health and allowing for the economy to recover in a longer period. Nevertheless, it soon became a question 'which one is the priority?' Health or Economy? And then, should they be regarded as important as the other? Clearly, 'health' and 'economy' are two different things, but they have significant impacts on one another. An unhealthy society would mean an unhealthy economy, and vice versa.

Started as a health emergency, the outbreak transformed into a much larger event for many countries. This came at the time when China was recovering gradually with high concerns about any reversing trends. The event turned into a case of crisis or a wartime situation for some countries. Despite the time, there was for early preparation and prevention, it was as if the world was just watching in the first few weeks. The blame game was not yet finished, but was just more severe and became a tool of retaliation between different countries. As the blames against China were increasing, there were signs of more negative reports or opinions from various places. The continuous efforts to politicise the situation were unfortunate but somehow expected. For example, on 15 March 2020, the Jerusalem Post had an article that not only attacked China's role in the containment of the disease, but also allegedly and offensively undermined China's role model figure in anything else. The wrong accusations, however, were not deflecting or reducing any of the issues faced by the outbreak. Unfortunately, this is just an example and there were many more of this kind, or even worse ones than had signs of racism attacks, too. The many unthoughtful examples of accusations and mental wars are normally aimed to demonise and demoralise nations, governments, and groups of people. And unfortunately, the game of blames continues as always and our absolute unity in any adversity is not yet achieved.

In the midst of all these mischievous behaviours, there were signs of humanity, too. Those countries that helped China at the beginning of the outbreak later received help and support from China. Cities that were under full lockdown, were receiving external support, even from the military forces. Communities that suffered, received support from their neighbouring communities as well as their authorities who could assist. There were many volunteers on the ground at the community-level who helped to sustain the control and monitory measures. People had to help one another, and families had to become more resilient. These happened individually and collectively, too. Sooner or later, inactions had to get replaced by more robust reactions. Those that were yet to go on the list of highly-infected countries had to reflect on the dynamism of the situation. They had to learn and avoid repeating the same mistakes as others. They had to prepare and respond effectively. Many alerting reminders were given to the developing and low-to-mid-income countries of Eastern Europe, South East Asia, Africa, Central America, and the Middle East. Meanwhile, the possibility of the disease being a bioterrorism act was still on the table.

BOX 5.17 "Economic Impacts, Business, and International Trade"

With high-level restrictions and significant impacts on travel and mobility, there were also many adverse impacts on other main sectors, of particular economic impacts, businesses, and international trade. As expressed earlier, we anticipate these impacts would last longer than anything else.

> Downside risks to our outlook mainly come from a longer outbreak, more severe disruption. If the outbreak outside Hubei persists beyond the first quarter, the disruption could be more severe than we assume.—Quoted from Jian Chang, Chief China Economist for Barclays in Hong Kong.

Monday 9 March 2020 was marked as a day of major economic crisis when global stock markets crashed significantly with £130 billion losses in the FTSE 100, and Brent crude oil was as low as US$30 a barrel. There should be no further surprises if these sudden losses and price drops happen again.

> Oil firms, and US shale oil companies too, will face serious pressures with oil now trading around $30 for WTI (West Texas Intermediate) and $33 for Brent. Such pressures will include funding problems, spreading through the broader economy and ramping up expectations more rate cuts, and potentially monetary easing.—Quoted from Chris Beauchamp, Chief Market Analyst at IG Group, UK, 10 March 2020

While keeping the economy as safe as possible remains the priority of many governments, the eventual impacts of the outbreak are somewhat uncontrollable. The impacts were so significant that within the first six weeks of the outbreak, there were reports of shortages of many commodities globally, less oil demand (and a later significant drop in prices), reduced sale from smartphone to passenger cars, reduced sale of luxury bags and products, reduced tourism (and travel in general), etc. In China alone, in January and February 2020, there were significant impacts on consumer-behaviour metrics, such as $60 billion reductions of consumer spending on food and drinks (Jan and Feb), 80% reduction in hotel occupancy (in 2nd half of Jan, and 1st half of Feb), 92% fall in retails sales of passenger cars (1st half of Feb), and 37% drop in smartphone sales (in Jan) (McKinsey Analysis 2020). The figures are expected to continue to either fall or create instability in the market.

With the growing number of consumer-driven economies relying heavily on high production and consumption and international trade, an outbreak of such scale could cause a serious economic crisis, and even a step closer to a probable recession. Similarly, McKinsey analysis (from McKinsey & Company[1]) developed three scenarios of *"(1) Quick recovery, (2) Global slowdown, and (3) Global pandemic and recession"*. While each scenario is expected to have different outcomes both for China and the rest of the world, their assumptions were based on four key areas of "public health response", "seasonality", "fatality ratio", "change in behaviours". Of the major economic issues, supply chain challenges were highlighted as the most worrying ones:

Perhaps the biggest uncertainty for supply-chain managers and production heads is customer demand. Customers that have prebooked logistics capacity may not use it; customers may compete for prioritization in receiving a factory's output; and the unpredictability of the timing and extent of demand rebound will mean confusing signals for several weeks.—Quoted from McKinsey Analysis, 2020

[1]McKinsey Report on COVID-19 implications for business, extracted from original source available from: https://www.mckinsey.com/business-functions/risk/our-insights/covid-19-implications-for-business

BOX 5.18 "Policy, Regulations, and Public Health"

During the outbreak, we should anticipate developing policies, regulations, and regular updates on public health. Here are also a selection of examples that how diverse but how important these factors can be. An example of which is the early development of *"Twelve Opinions on Stabilising Labour Relations and Supporting Enterprises' Resumption of Work and Production"*, released by China's Ministry of Human Resources and Social Security on 7 February 2020 (Yang 2020). This indicates the importance of adaptive measures that should be taken into full consideration as early as possible. The same flow that was seen in the US and the UK, and surely in other affected countries too.

With 117 territories having confirmed infected cases by 11 March 2020, new policies and regulations on public health were inevitable. On this day, the outbreak was declared 'pandemic'. In the US, the Centers for Disease Control and Prevention (CDC) (2020b) continuously worked on matters of 'risk assessment and management' against the disease, particularly for the provision of guidelines for actions by public health authorities of those communities that are not experiencing sustained community transmission. In the UK, similar measures were developing in the form of guidelines (Public Health Matters 2020) with early precautionary measures and warning for the future stages. By highlighting in particular the societal impacts including probable impacts on lifestyles and possible measures, the UK's Public Health England (PHE) urged the general public to "*plan ahead …*[and]…*consider how you or your family would manage if you had to self-isolate for a couple of weeks*". This message and many similar ones are preparatory measures for what may be regulated in the early future. This was followed by what the UK Prime Minister confirmed in a news conference: "*We are preparing various actions to slow the spread of this disease in order to reduce the strain it places on the NHS*".

Singapore's Ministry of Health (MoH) also provided extended guideline that was also regularly updated (MoH 2020). These updates were meant to clarify any remaining public misinformation and how the country's 'Disease Outbreak Response System Condition (DORSCON)' was responding to the progress. As informative as such reports or sources could get, they are meant to provide precautionary measures, health advisory, travellers' guide, and cover any areas under overarching matters of public health.

Issued under 'Report of the WHO-China Joint Mission on Coronavirus Disease 2019 (COVID-19)' (WHO-China Joint Mission, 16–24 February (2020)), this internationally developed comprehensive document provided clarity on six areas: the virus, the outbreak, transmission dynamics, disease progression and severity, the China response, and knowledge gaps. The information is provided through an established organisational structure and response mechanism comprised of nine working groups to coordinate the response:

> Each working group has a ministerial level leader. Emergency response laws and regulations for the emergency response to public health emergencies, prevention, and control of infectious diseases have been developed or updated to guide the response (ibid, p. 27).

These are just a handful of examples from countries at high risk and those that are close to high risk. The general guides can often be followed by the other countries, but they should be adapted from a reflective approach considering the conditions of the society, addressing what regulations may be missing, and what policies may be required. It is important that these get updated on a regular basis and in a reflective approach.

BOX 5.19 "Media Coverage"

International media covers anything negative against their rivals of any kind (e.g. political, economic, trades, geographical, historical, etc.). Like many other occasions, politicising the outbreak was perceptible since the inception. As expressed earlier, this should be avoided as much as possible. The impacts from the growing international negativity could be reduced by making the national media and multi-level resources, efficient, transparent, informative, and reliable. The official records should be available and acceptable to all by various means.

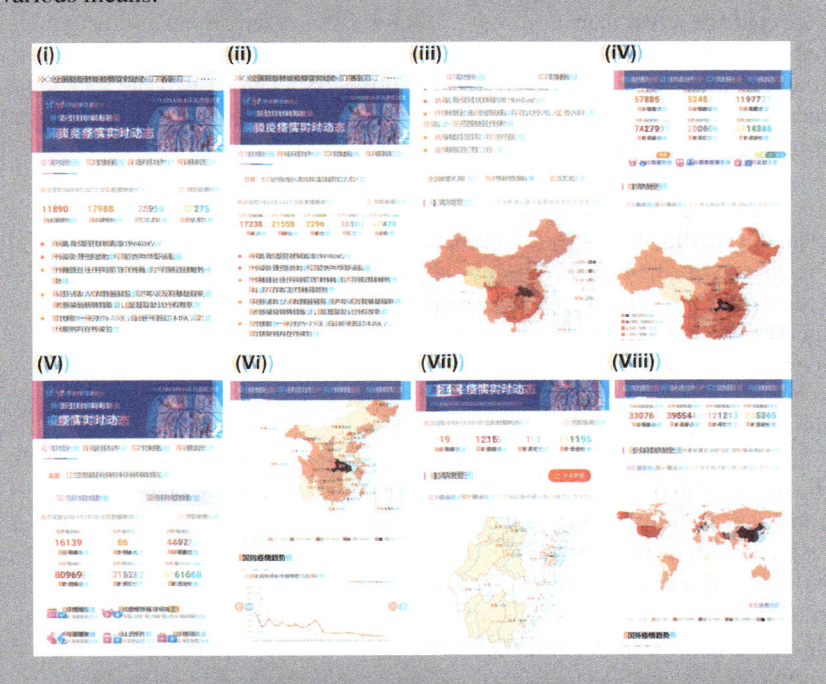

One of the best examples of this outbreak's media coverage, amid all the unofficial misinformation and international criticism, was the development of a multi-functional digital platform on social media. This official platform (as shown above in eight segments) was continuously upgraded during the outbreak. It started with very basic information and precautionary measures (i) and indicated the progress (ii). While the interactive map was utilised from the beginning (iii), more interactive tools were added at a later stage (iv). In the later stages, there more tools integrated into this platform (v), while all data were recorded with graphs and statistical reports from inception (vi). There was also available information for the smaller scale of provincial and municipalities (vii), which were later developed further to cover the global situation (viii).

Perhaps, this is a good lesson to take on board; i.e. to make media helpful rather than harmful.

As it was described by many who faced lockdown measures for their first time, media was seen in the state of meltdown. Social media played a major part in developing manmade and manipulated scenarios, those that came with disbelief and with a lack of scientific grounds. While the so-called draconian social distancing measures believed to save lives during the pandemic event, some still phrased it as the situation of "house arrest". On the contrary, some believed in the effectiveness of the lockdown measures but also argued that the cure could be worse than the disease. Despite the many creative activities people put on social media, the ones under more pressure of facing more adversities found it unbearable and unrealistic. Nevertheless, even though the lockdown measurers were implemented late in some countries, the results then changed the situation by reducing the disastrous impacts.

The media coverage should avoid any retaliation reactions. While they may assume they could raise awareness, they could also increase the growing criticism of specific sectors. In doing so, people who are more dependent on social media may grow in disbelief and may pass their disbelief and agony to the others. This may also increase the disparities between people and their governments. An anonymous example was *"The UK will likely be forced to an even worse lockdown, yet the underground tubes keep running in the capital, London is not isolated from the rest of the UK, and our borders are wide open. I am no longer taking the government responses seriously"*.

Some also regarded the event as a manipulating act of the government and the media who provide wrong figures, some requested more clarity, and some could not see transparency in the actual happenings. With so much of disbelief also comes narrow-minded disbelief against the actual threat of the disease, its severity, and its impact on a larger scale. Echoing some politicians, an anonymous comment was *"Seasonal flu deaths far exceed COVID-19 deaths. The mass panic and deliberate power grab justified by the disease is likely to unravel with huge backlash when the public realises the size of the lie"*.

Despite the larger scale acceptance of science and scientific reports, some argued that the situation is manipulated by scientists and people were spoon-fed. The attacks were often on virologists and epidemiologists that relate to the nature of the disease. These so-called manipulation scenarios were two-sided, one that argued the government and scientists are hiding the truth and the situation is much worse, and the other assumed the situation is fine and manageable and the relevant sectors are overreacting. Both scenarios represent the group of people who could not trust the official reports and are likely to be the ones who do not comply with the official measure either.

With changes in records and methods of recording, the media could add to the uncertainties of the situation. As such, people argue about *"a mockery of all graphs that appear in publications as those graphs are not and cannot be adapted to illustrate those changes*". This came at the time, when the media was highlighting a social media material as positive news, an incident in the northern town of Wales highlighting "*A herd of goats invade a village*". We argue that the media should rather take a much stronger position, avoid the false spread of news, provide training, positive thoughts, and reactions, and help to raise awareness, and educate the general public.

Amid all concerns of the lockdown and multiple measures, the situation of the outbreak is always a chance for politicised discussions between countries and between different parties/groups of a country. This is specifically more visible in those nations facing the round the corner elections. The media often takes a biased position, which becomes an effective tool to manipulate society.

BOX 5.20 "Micro Management of Outbreak Control and Protection"

At the time when facial masks became part of our daily clothing attire, society experienced more fear than ever. While toilet paper shortage became a phenomenal matter, all countries already experienced disruptions of some sort. With continuous falls in the global stock market, crumbling airline industries, and drastic actions in newly-worried countries, it was clear that nearly all countries were affected by travel bans and restrictions. The pace outside China seemed much faster: the headline of every news, regular updates, upsurge in numbers, shortages, new bans almost every day, new lockdowns every few days, etc. In all these pandemonia, a simple cough was an alarming sound, and social distancing became a norm. In many places, social contacts were minimised, sudden regulations (often related to bans and restrictions) were imposed, and health systems were worryingly readjusting. For the first time, the total number of infected cases outside China overtook those in the country (as per records of WHO on 16 March 2020). In China, the push was towards full recovery, which was promised to happen by the end of March 2020. The provincial race to reach zero cases was realised as a big push for cities to accelerate treatments. Cities entering the recovery phase had to intensify their control and protection of their residents or those that were coming to the city. The extended measures were in place to take up a gradual pace of the recovery phase. Amid all these, we can refer to the importance of micro management, which is identified as the main ground of implementing the measures.

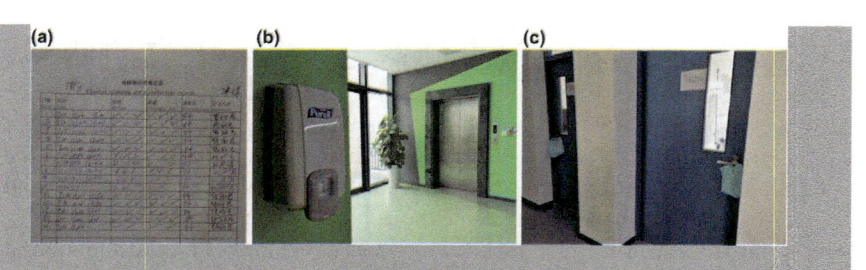

An example of this micro management is the implementation of guidelines (inclusive of control and protection) at the university campus, where we highlight some of the main measures. Apart from the earlier asset management and operation action plans (with contingency plans), there were plans for the gradual reopening of the premises and in line with the governmental announcements and planning. The facilities update that included the use of indoor sports facilities, teaching area facilities, and amenities of living areas were the highlights of gradual progress in the recovery phase. The enhanced measures also included: regular and detailed entry checks, restrictions to external non-registered members, limited working hours of amenities, controlled sports facilities access (i.e. a maximum of seven personnel at a time), monitory of indoor environments, etc. Alongside the regular disinfection procedures of premises, more attention was given to details—e.g. for as lifts with three times disinfection procedures a day (a). Moreover, personal disinfection products were provided near any major transit points (b), with a regular check on supply. More importantly, protection measures and guidelines were provided in various forms and in various places. These included protection and prevention notes on doors (c) or in other areas. The necessary supplies (such as facial masks, disinfection sprays, and gels) were also provided to all members of staff (c), and those entered the university premises for the first time after the outbreak.

BOX 5.21 "City-level Recovery under the Shadow of Global Emergency"

With polarised decision-making processes in fighting the pandemic, we could argue one model as a more gradual 'Western Model' and the other a more controlled 'Eastern Model'. There were also signs of hybrid models, including what South Korea did with large scale lockdowns and testing. Some countries then ended up replicating the models of lockdown in their own contextualised way, which appeared to be more effective if they happened earlier. Also, in other regions, like in Eastern Europe, North Africa, Central Asia, Middle East, East Africa, there were signs of early lockdowns and early restrictions on mobility and transportation in and out of countries. This cautious reaction, while overshadowed by vast scale disease spread in specific regions, seemed to be a sensible move towards the Eastern Model.

The widespread reductions in connectivity and mobility helped cities to have better control over the community transmission and import/export of the disease. The applicability of city borders and city boundaries proved to be a useful way to help the enhancement of city-level control measures. While cities like Ningbo were experiencing the recovery phase, risk management was highly visible across all sectors. The high level of monitory and control were intensified to avoid reversing the progression of the outbreak containment. This was an experience of city-level recovery under the shadow of a global emergency. Some more common examples of these measures were: (a) continuous health checks and temperature checks via the 'Ningbo Health Code' as the city's self-reporting health information system, (b) official records of temperature checks for people who pack and deliver food, and (c) protection guidelines provided by companies/businesses/employers to their employees/customers.

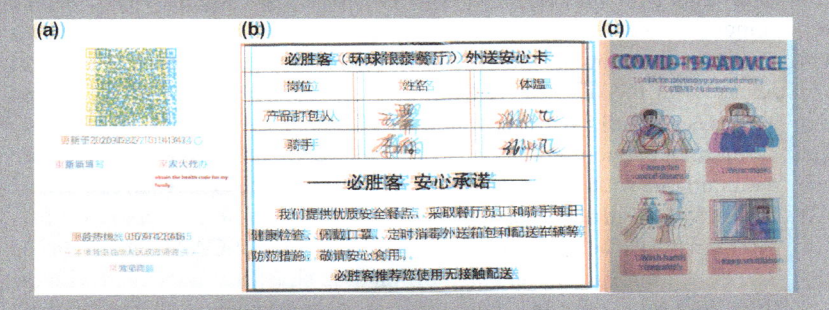

At the same people, people could not wait any longer to regain their social life and experience a life beyond their living compounds/areas. In the Chinese language, this situation is phrased as 'Wéichéng' (in Chinese 围城), which is translated as 'siege'. But in here, it actually means people inside a place would like to go outside (often due to dissatisfaction), and those who are outside would like to come inside instead. The situation in the recovery phase resembled a moment of wéichéng. With restrictions easing gradually, inner-city mobility was once again high. Soon, people found signs of hope and gradual stability.

BOX 5.22 "Rapid Changes under Continuous Alert"

Under the continuous alert situation, many changes happened so rapidly at the global level. Even though most Chinese cities were in their recovery phase (or close to recovery), they were under full alert for any escalated matters, or possible second wave of the disease outbreak. Under the shadow of global emergency and some very serious situations, China also rapidly tightened some of the regulations, restrictions, and control procedures. One of these was the immediate reduction of international passenger flights, announced on the 26th of March 2020. This came at the time when the number of cases was increasing rapidly in the US, Russia stopped all its flights in/out of the country (except for Russian citizens stranded outside or returning to Russia), and there was a state of alert and emergency in many regions.

Notice on Further Reducing International Passenger Flights during the Epidemic Prevention and Control Period 26/03/2020 (Notice on Further Reducing International Passenger Flights during the Epidemic Prevention and Control Period)

To All Transport Airlines: In order to resolutely contain the increasing risks of imported COVID-19 cases, and in accordance with the requirements of the State Council for joint prevention and control of the epidemic, it is decided to further reduce the number of international passenger flights. Details of the requirement are as follows:

(1) On the basis of the Information on International Flight Plans (Phase Five) released on the official website of CAAC on March 12, each Chinese airline is only allowed to maintain one route to any specific country with no more than one flight per week; each foreign airline is only allowed to maintain one route to China with no more than one weekly flight.

(2) Airlines shall, in accordance with the requirements above, apply for Pre-Flight plans to the Operation Supervisory Center of CAAC in advance.

(3) The Operating Permits and take-off/landing slots, etc. associated with the flights cut by airlines in accordance with requirements of this Notice will be retained.

(4) Airlines shall strictly implement the latest edition of Preventing Spread of Coronavirus Disease 2019 (COVID-19) Guideline for Airlines issued by the Office of China Civil Aviation Prevention and Control COVID-19 Leading Group, take stringent prevention and control measures on the flights to/from China and ensure passenger load factor no higher than 75%.

(5) In accordance with the need for epidemic containment, CAAC may issue policy to further reduce the total number of international passenger flights. Airlines therefore are required to closely follow information released, analyse the situation and make contingency plans in advance, and handle in a proper way the issues such as extension and refund of sold air tickets, etc.

(6) Airlines may operate all-cargo flights with passenger aircraft, which will not be counted against the total number of passenger flights operated.

(7) Flight plans adjusted by airlines in accordance with paragraph 1 of this Notice shall be implemented as of March 29, 2020.

(8) This Notice shall take effect on the date of its issuance, and the expiration date will be notified separately. As of the date of taking effect of this

Notice, the Notice MHF [2020] No. 11 issued by CAAC shall become invalid.

Civil Aviation Administration of China (Extracted directly from: http://www. caac.gov.cn/en/XWZX/202003/t20200326_201748.html?from=groupmess age&isappinstalled=0).

Only hours after the earlier announcement on 'international passenger flight restrictions' (shown in the previous page), a much higher announcement was made by two relevant officials in China, namely the Ministry of Foreign Affairs and the National Immigration Administration. Under the growing uncertainties, global emergency, and concerns about the new waves of the outbreak, the new restriction was to temporarily ban the entry of foreign nationals to China. This was aimed to build protections against any possible blowbacks and a chance to research and maintain the full recovery at the national level; the full recovery that was showing lights of hope by the reduction of restrictions in Wuhan, the original epicenter of the outbreak. This sudden and heightened measure was seen as a reflection on other countries, and those success examples that introduced and implemented early and/or restricted travel control measures. Something that was criticised at first, but was recognised as a reflective response to the failures of the others. This can be regarded as what Fukuyama (2014) narrated as political decay in the ostensible 'democracies'. The way that ends up with weaker governance as it gets distracted by special interests to the detriment of the public good (ibid). Somehow, this can be seen as the failures of the new capitalism. It is important to note, this announcement came at the time that there were growing neglect by the US administration, while the US was already a new epicenter of the disease outbreak. A call that made it look like a retribution act at first, but it was a sensible move to keep out of the growing chaotic situations elsewhere. A call that depends on the other countries' success in flattening their curves of the outbreak progression (i.e. based on both infected cases and mortality rates). Also partly, it was also meant to avoid and reduce the growing domestic perceptions of contaminated/infected foreigners. This came after the earlier concerns of Premier Li Keqiang (CGTN News, 2020), the worries of the second wave of the outbreak, and the growing number of non-domestic cases (i.e. mainly those people travelling back to China).

This joint announcement is summarised here:

Ministry of Foreign Affairs of the People's Republic of China National Immigration Administration Announcement on the Temporary Suspension of Entry by Foreign Nationals Holding Valid Chinese Visas or Residence Permits, March 26, 2020

In view of the rapid spread of COVID-19 across the world, China has decided to temporarily suspend the entry into China by foreign nationals holding visas or residence permits still valid to the time of this announcement, effective from 0:00 a.m., 28 March 2020. Entry by foreign nationals with APEC Business Travel Cards will be suspended as well. Policies including port visas, 24/72/144-hour visa-free transit policy, Hainan 30-day visa-free policy, 15-day visa-free policy specified for foreign cruise-group-tour through Shanghai Port, Guangdong 144-hour visa-free policy specified for foreign tour groups from Hong Kong or Macao SAR, and Guangxi 15-day visa-free policy specified for foreign tour groups of ASEAN countries will also be temporarily suspended. Entry with diplomatic, service, courtesy or C visas will not be affected. Foreign nationals coming to China for necessary economic, trade, scientific or technological activities or out of emergency humanitarian needs may apply for visas at Chinese embassies or consulates. Entry by foreign nationals with visas issued after this announcement will not be affected.

The suspension is a temporary measure that China is compelled to take in light of the outbreak situation and the practices of other countries. China will stay in close touch with all sides and properly handle personnel exchanges with the rest of the world under the special circumstances. The above-mentioned measures will be calibrated in light of the evolving situation and announced accordingly.

Ministry of Foreign Affairs of the People's Republic of China and National Immigration Administration (Extracted directly from: https://www.fmprc.gov. cn/mfa_eng/wjbxw/t1761867.shtml?from=groupmessage&isappinstalled=0).

Only a day after the earlier joint announcement on the heightened travel bans for foreigners traveling to or returning China, the ripple effects were seen in new regulations at the provincial and municipality levels. Bearing in mind that quarantine restrictions were already tightening for a while before it reached the highest level of travel bans (with some exemptions), this new measure was for the cities and provinces to focus more on their local level containment plans. After all, since the only eradication plan is based on new vaccine development, it is vital to take steps seriously and allow for no failures in the containment and recovery procedures. Here, we demonstrate an example of extended measures from the city-level foreign affairs of Shenzhen, south of China. Shenzhen is a major economic and technological hub in China, bordering Hong Kong SAR (Special Administrative Region).

The new announcement includes some of the details of the national level announcement and then includes:

- *The Office of Guangdong Headquarters for Prevention and Control of COVID-19 announced that effective from 06:00 a.m., March 27, all inbound travellers (including travellers from Hong Kong SAR, Macao SAR, and Taiwan, and transit passengers) entering via ports of entry in Guangdong shall accept nucleic acid testing and undergo a 14-day collective quarantine for medical observation.*

- *The accommodation and meals during the collective quarantine for medical observation shall be at their own expense. Specific persons for ensuring normal production and livelihood in Hong Kong SAR and Macao SAR, and persons such as cross-border drivers and ship crew who transport goods urgently needed in Guangdong Province, Hong Kong SAR and Macao SAR, are not subject to collective quarantine for medical observation temporarily but shall all undergo nucleic acid testing.*
- *Their health is strictly monitored and managed in an end-to-end manner to ensure closed-loop operation.*
- *For epidemic-related consultations, please call Guangdong Service Hotline 1258088 (in Chinese, English, Japanese and Korean), Shenzhen Government Service Hotline 12345 (in Chinese and English), Service Hotline of Foreign Affairs Office of Shenzhen Municipal People's Government 0755-88121224 (in Chinese and English), Public Hotline of Shenzhen Center for Disease Control and Prevention 88812320 (in Chinese) and Call Platform of Shenzhen Family Doctors 4401191160-2 (in Chinese).*

Entry policies will be adjusted in accordance with the epidemic situation. Please follow closely. With our concerted efforts, we believe the epidemic will be over. Thank you for your understanding and support.
Foreign Affairs Office of Shenzhen Municipal
People's Government
Health Commission of Shenzhen Municipality
March 27, 2020

(Extracted directly from: The homepage of Shenzhen Foreign Affairs, China).
Under the fear of imported cases from outside China and the intensified travel bans and restrictions, there were signs of change at the city level. These were seen as examples of no-longer-gradual change, but those that were rapidly in place on 28 March 2020. For instance, "more than 30 commercial complexes in Ningbo have cancelled temperature measurement and health code display before customers enter, and the entrances and exits of shopping malls have been fully restored". While previously major supermarkets were conducting synchronous temperature measurements and health code checks, from this day, they were all cancelled. This was seen as the first stage towards gradual restoration, and bringing normal operations back to the city. With the return to normal opening hours, it was the start of the post-recovery phase under the shadow of outer-China growing cases, emergencies, and crises. This also came at the time that the original epicenter of the outbreak, the City of Wuhan, gradually started its operation. Also, Wuhan started operations in shopping malls, some public transportation systems, and inbound transport from across the country. All these occurred under the alert of a potential second wave of the outbreak. Hence, preparations were made for the gradual changes and eventual

travel plans of people in Wuhan and other locations (i.e. cities, towns, villages) of the Hubei Province.

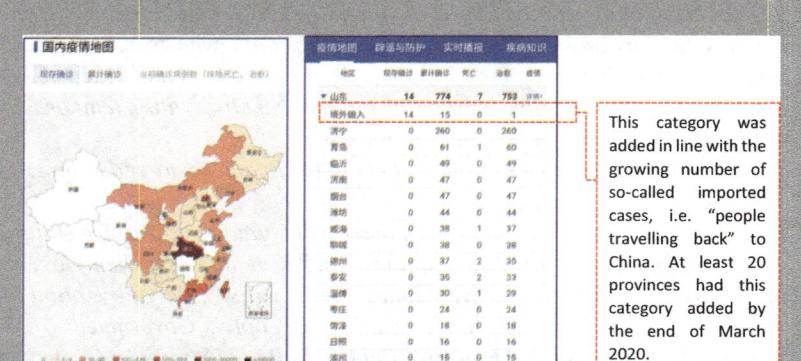

This category was added in line with the growing number of so-called imported cases, i.e. "people travelling back" to China. At least 20 provinces had this category added by the end of March 2020.

At this time, at least 20 Chinese provinces were reporting a growing number of imported cases of those travellers coming back to China (both locals and foreigners). Under the fear of retaliation acts of racism and stigmatisation, the imported cases were also seen as a possible backfire and towards having another wave of the outbreak. As shown above, the allocation of a new category stating "people coming back to China" was a sign of growing imported cases, particularly for those cities and provinces that previously managed to successfully bring their inflected numbers to zero. This may also reflect negatively on the international media as well as social media, and from various perspectives.

In line with the early signs of the post-recovery phase, the need to revitalise the main urban systems of the city was essential. This requires sector-based guidelines to first restart the operations, and then help to gradually revitalise sectors, businesses, enterprises, amenities, and services. This requires time, and it is recommended to take gradual steps. One of the early examples of change was to revitalise the public transportation system, which was under pressure for a long time. With a period of the temporary closure of operations, the later limited operations, and finally with a very low number of customers, public transportation was in need to regain its face and operations.

Hence, the plan of the local government was also to support this major urban system as early as 23 March 2020 and then 28 March 2020. Here is the summary of the action plan and proposed support to revitalise this primary system:

Starting from March 23, the discount program for Ningbo metro e-tickets will last until September 30, with up to 50% discount for passengers using Ningbo metro APP.

Details:

Within one discount period (30 days), Ningbo Rail Transit will give free e-tickets of different times (without mileage limitation) to passengers according to their use of e-tickets (including the ride code, the same below). Specific as follows:

- **Item A**: Ningbo Rail Transit will give free e-tickets for 20 trips without mileage limitation if passengers use the e-tickets for 25 or more subway rides with each ticket price no less than RMB6 within one discount period.
- **Item B**: Ningbo Rail Transit will give free e-tickets for 15 trips without mileage limitation if passengers use the e-tickets for 30 or more subway rides with each ticket price no less than RMB4 within one discount period.
- **Item C**: Ningbo Rail Transit will give free e-tickets for 10 trips without mileage limitation if passengers use the e-tickets for 35 or more subway rides with each ticket price no less than RMB3 within one discount period.

Rules:

- A 30-day discount period starts from the date the passenger first uses the e-ticket to travel.
- The start time of the next cycle is the date when the passenger uses the e-ticket again after the previous cycle.
- In the same discount period, when the passenger uses the e-ticket to take the subway, and the single ticket price meets the above multiple discount conditions, the times of use can be counted simultaneously.
- If the passenger's fare and times meets one or more conditions within the same discount period, the passenger can choose the time of collection and the number of complimentary tickets. But only one collection is allowed within the same period.
- Within the same discount period, the rides times after the passenger receives complimentary tickets can no longer be accumulated.
- After receiving complimentary tickets, passengers are free to use it within the same period.
- At the end of one period, all the cumulative times will be cleared, and the unused and uncollected complimentary tickets will become invalid.

(Extracted from: Ningbo Subway, and provided by Ningbo Focus, news article on 28 March 2020).

The more people stayed indoors, the bigger the perception that the outdoors are dangerous. The change of perceptual requires to happen at a gradual pace, as the normalisation also occurs in several steps. Under the growing fear of contaminated everything, the city has to propose innovative ideas for the use of public places, public services, public transportation, and support the local economies. A major move was to expand on the earlier changes to public transportation use, which was announced in early April 2020:

From now on, free rail and bus travel during off-peak hours will be imple-mented in Ningbo, and the end date will be announced later. Passengers will be charged as usual from 7:00 to 9:00 and from 16:30 to 18:30. Free rides are available except these two periods. The starting and ending time of the free ride is subject to the boarding time of the passenger. During the free period, passengers do not need to pay for metro tickets or scan QR codes at the ticket barriers, and they do not need to pay cash or swipe cards when taking the bus. But passengers must pay for a ride during non-free hours. And passengers who enter the subway station by swiping cards should be sure to swipe cards again when exiting the station in order to avoid the delay of charging.

(Extracted directly from Ningbo Focus News, on "free Bus and Subway Rides during Specific Periods").

At the same time, public amenities and attractions were reopened with free admissions, mainly to scenic spots last up to 2 months. The support that was needed to revitalise the local economies were essential in the post-recovery phase. From the beginning of April 2020, these were proposed at the district level:

For Beilun District

*From April 1 to May 31, **Beilun Nine Peaks Mountain Tourist Area, Meishan Bay Beach Park and China Port Museum** will offer free admission. Besides, cultural, tourism and sports gift packages will be released, including general consumption coupons with a total value of 10 million Yuan and corporate vouchers. There will also be various culinary perks. During this period, various discount coupons with a total value of 10 million Yuan will also be issued in Beilun. With the discount of up to 50% off, people can enjoy the delicacies in Beilun.*

For JianBei District

*In addition, Jiangbei has issued package tickets worth nearly 20 million Yuan, which covers six scenic spots in Jiangbei District, including **Cicheng ancient town, Baoguo Temple, Daren Village, Lvye Daren Valley, Ali's Farm and Train Nice Theme Park**. A single package is worth 475 Yuan, and there are 40,000 packages in total. Before May 31, tourists can get a package ticket worth 475 Yuan on UnionPay to visit six scenic spots for free if their consumption reaches the standard when they use Union Pay or the Apps of major banks to book accommodation or shop in the business area.*

For Yinzhou District

*From now to April 6, **the pear garden of about 100 mu (6.67 hectares) in Tiangong Fazenda will be open to tourists for free for the first time**. Visitors can raise a pear tree by paid adoption and come to pick the pears after they are ripe. And the yield of a pear tree can reach 10 to 25 kg.*

(Extracted directly from Ningbo Focus News, on "Free Admissions to Scenic Spots Last up to 2 Months").

BOX 5.23 "Beyond the Adversities"

Global attention was at its peak at the time when the global economy was completely disrupted, more countries were getting into the list of countries with confirmed cases, and borders were closing every day. In most Chinese cities, people were already at the recovery stage while there were still signs of new cases coming back to the country. People resumed their daily routines with major changes in their behaviours and lifestyles. Many things seem to be settled as the new normal: temperature checks, health monitory, wearing masks, social distancing, online working, online shopping, etc. One of the common topics of conversation in the recovery phase was to compare the total weight gained during the lockdown period. The other communications were about limited exercises, new outdoor experiences, changes in daily routines, and new snack consumptions. For many, the combined effect of China's recovery stage and global emergency issues meant a prolonged recovery stage and not reaching post-recovery early.

Globally, there were many cases of recorded theft or piling up certain primary products. There were some arrests of those who stole products, who piled up products (e.g. mostly to sell them later at a higher price), who fraudulently scammed people in various ways, who made and provided fake products and necessary supplies, who violated the quarantine/lockdown measures, who rejected to self-isolate, etc. For instance, criminal scams can happen in various ways of formal and informal set-ups, from officially-looking letters and notifications to a simple knock on the door. These incidents involved tradesmen knocking on doors of the elderly, people who acted as officials, and many other cases that were targeting certain groups of people—often the vulnerable. Also, in the early days of the outbreak event, the World Health Organisation (WHO) noted many scams that were happening in their names. They then had to make an official announcement to raise the awareness of such cases and to ensure people get the right messages through the right channels. They also warned against those opportunist criminals who try to steal money or sensitive information, mainly seen through emails and online methods (UN News, 2020).

Moreover, people were yet to know how to adapt to new changes and measures, learn how to comply with precautionary measures, have collective methods to societal wellbeing, and understand the difference between lockdown, quarantine, and self-isolation. Many aspects depend on the level of trust needed between people themselves, between people and their communities, between communities and their governments, and between local governments and their upper-level governments.

All these had to happen so fast and responses had to be more humane than ever. All these happened at the time when the UN agency was continuously asking people to comply with the basic precautionary measures:

- *Frequently wash your hands with an alcohol-based hand rub or warm water and soap;*
- *Cover your mouth and nose with a flexed elbow or tissue when sneezing or coughing;*
- *Avoid close contact with anyone who has a fever or cough;*
- *Seek early medical help if you have a fever, cough, and difficulty breathing, and share your travel history with healthcare providers.*

These actions we previously thought to be simple but then appeared not so simple after knowing the fact that many people are yet to learn the basics. All these happened at the time when vulnerable groups were even more vulnerable, and many global health organisations were warning against the inactions of governments and people. These came at the time when governments were worried about the lack of people's compliances, and people were worried about the uncertainties of the event.

Several weeks in the recovery phase, we are still asked where we are going when leaving the compound areas. Our temperature is checked before leaving, upon entry to any premises, and once we return home. Notifications are displayed everywhere reminding people about safety measures, procedures, recommendations, and requirements. Our world is changing with this invisible enemy. From 'another hoax', the disease outbreak turned into a crisis-making machine. People report "things are changing every day not only in terms of the COVID-19 situation but also the responses of different levels of government". As it has been warned by many, "things will get worse before they get better". As we expect the situation to last longer and numbers of infected cases and mortality are increasing rapidly, we expect harder measures and stricter restriction methods. Not all can manage with a strong central government, and that is only part of the problems. We also have to bear in mind that lockdown is not the same for everyone, and the economic hit will be more significant than initially calculated. It would make the vulnerable groups more vulnerable, and it will enlarge the size of our vulnerable groups.

While the UK was entering a new phase and reaching 20,000 infected cases, its government was talking about preparedness. While Iran was reaching 30,000 infected cases, cities were just moving towards the less-appreciated lockdown measures across the country. While the number of infected cases went over 100,000 in the US, they then talked about cooperation. There were many more country-level examples that represent late actions, which seem to be the after-effect reactions to a period of initial inactions. By entering their third week of lockdown, the UK government proposed to write to all 30 million households across the country, including details about the 'national emergency' and a leaflet about official advice. This is a rather late and unnecessary burden, at a minimum cost of £5.8 million (equivalent to US$7.2 million at the proposed time). Then we question what is the use of our digital technologies? Can we not innovate in any better ways at this so-called technologically-advanced age? It is important to remind those deceiving actors that their business-as-usual actions and scenarios are not fit the conditions of the outbreak. At the same time, the UK's motto was 'Stay Home, Protect the NHS, Save Lives', something that could have been done several weeks earlier and could have saved the country more effectively. To sum briefly, we confirm the measures have to be implemented at the right time. During the outbreak, governments should be more realistic and responsive. Mobilising support is not an easy task but is doable. For instance, on the 29th of March 2020, the Australian government made a significant change to its foreign investment framework and reduced it to $0. This type of adaptive measure is important for governments to prioritise the right priorities and at the right time—not any time later, and not with any hesitation or neglect. On the contrary, we anticipate significant reduction or changes to foreign investment plans and procedures. This is at the time that many countries were admitting the success of China in the handling of the outbreak; bearing in mind that China was the original epicenter that was affected with a sudden wave of shocks and not with a long period of disbelief, neglect, and inactions.

As the international health organisations urge for more 'testing, testing, and testing', we may not have the right record of cases and mortality rates. As many cannot cope with such adversities, we see a growing number of deaths at home and non-recorded cases, those that were unattended for several days. At the time, Latin America is on the verge to become the next epicenter and African countries are closing their borders, a new chapter of this pandemic event is about to start.

For years, many unwanted wars and adversities were forced on people and their homelands leaving them in despair while leaving many of us unprepared in our safe zones. Now, the invisible enemy seems invincible and is right at our doorsteps. And for many of us, we have struggled to manage and recover.

We noted the treatment of infected cases became more difficult in compared to earlier stages of the disease outbreak. Hence, a hypothesis was developed utilising the data up to the end of March 2020.

When did SARS-CoV-2 virus mutate? And did this impact treatment procedures?

As the infected cases of the COVID-19 disease outbreak are increasing exponentially, there is evidence that the virus is already mutated. As coronaviruses often mutate, it is important to know when and where this mutation has occurred in this pandemic event. Until the end of March 2020, this pandemic outbreak has developed in three major phases: (1) the original epicenter in China and later in the region of East and Southeast Asia, (2) the second epicenters in Iran and Italy, both then affecting their regions (mostly Europe) and beyond, and (3) the rapidly-growing epicenter in the US. This is at the time that Latin America may potentially become the next epicenter of this pandemic event. Early signs of the virus mutation are tracked back in the early-to-mid stages of the second phase.

As the official records indicate, the severity of the virus is strengthened as early as the first week of March 2020. This is based on the accuracy of country-level data, but also on several factors that are very important to this pandemic progression. Started as a small outbreak then an immediate pandemic, this disease outbreak event was only become pandemic after about 50 days from its official announcement. There is a period of uncertainty at the first stages that require further investigation, which may then result in better clarity on the source and starting point of the outbreak, including more information on the first index case. Nevertheless, four factors could potentially have significant impacts on methods of recording and the development of the outbreak:

- *Our knowledge of the disease progressed and developed over time*

It is natural when this happens. It is unfair to argue that our knowledge of the disease should have been the same from the inception. This is unlikely to happen in any case of the outbreak, or else they may not turn into a case of an outbreak at all. As the outbreak started and based on the fact it was a novel disease, our knowledge was relatively little. It also remained like that for a long time, even at the time when the number of cases was increasing rapidly elsewhere. It took some time for us to know how the disease may transmit between people, its level of severity, and its impacts on our bodies. It took almost three months for the disease to show its first sign of transmission to other species, which could have resulted from the possible mutation of the virus.

- *Early measures and early responsiveness worked effectively but it was only done by selective countries/regions*

The measures in different parts of the world were not consistent. And not all of them happened at the right time. Some failed to implement early measures, and some had many inactions that caused later effects on the increase of infected cases and fatality rates. Some sacrificed their long term societal health and wellbeing with their short-term economic and political goals. Those were the ones hit the most later on. Examples of these were the hesitation to put adequate lockdown measures, prevention and control plans, travel bans, health checks, etc. For instance, it was recorded that London's Heathrow Airport had minimal health checks even days before the country's Prime Minister was hospitalised because of the disease. Countries that put early restricted measures, such as smaller countries in Oceania, were hit the least. Also, we see much better progress in countries like China, Singapore, South Korea, and Japan, where high-level measures were in place. We have to bear in mind this event was announced as a global health emergency only 10 days after the official announcement of the outbreak.

- ***Forms of measurement changed during the outbreak progression***

Testing was only increased after the event was declared pandemic. Hence, many cases could have remained hidden or unrecorded. Apart from this, as our knowledge of the disease developed, our forms of measurement also developed. It is, however, important to strictly adhere to systems of measurement to the values and records of the monitory measures that are in place for the outbreak progression.

- ***Potential (early) mutation of the disease based on multiple factors***

There are several aspects related to this. First, the information about early transmissions requires further investigation to shed light on the mutation pattern of the disease. As respiratory diseases often mutate, this has likely occurred at earlier stages. Also there are factors, such as geographical or genetically-associated issues that need to be studied further. For instance, there is a hypothesis that the virus became stronger or even more aggressive in colder climates.

Furthermore, it is evidenced that since early March 2020, the numbers of recovered patients are dropping significantly, particularly for those affected in the second phase and carried on to other regions. This is evidenced by the early treatment cases in countries like the UK, when the patients were able to get recovered under just 10 days. The period of treatment is now extended at 2–3 weeks, which remarks on the gradual strengthening of the virus. Apart from the longer period of treatment compared to earlier days, the resilience of the virus is also evidenced-based on the available data of fatality and recovered cases in multiple regions and at different times.

This fact is also evidenced through the apparent changes in the fatality rate. In phase 1, the fatality rate was assumed below 2%, but this increased to a much higher rate now. With the exception of China's fatality rate at 4.05% (to 31/03/2020), the highly-affected countries of phase 1 have managed to maintain their low fatality rates to date; such as Singapore at 0.32%, Thailand at 0.61%, South Korea at 1.66%, and Japan at 2.87%. However, currently, we see the fatality rate increasing to a much higher rate of 6–9% with the exception of Italy, which also represents inconsistency in the fatality rate in different locations. Based on the country-level figures, we confirm the fatality rate of those affected in the second phase (until 31/03/2020) are: Iran at 6.50%, France at 6.79%, the UK at 7.19%, the Netherlands at 8.25%, Spain at 8.76%, and Italy at 11.75%. These numbers may slightly decrease later on. At the same time, the main exceptional cases are Germany (at 1.04%) and the US (at 2.05%), both increased by a minimum of 0.1% to 0.15% from the day before.

They both managed to keep their country-level fatality rates close to the initial 2%. These numbers may increase later on. Nevertheless, in Germany, we see a major difference in highly-affected regions, as the country's three most infected regions of Bavaria, Baden-Württemberg, and North Rhine-Westphalia all have 0% recovered cases. As the number of cases in the US increases drastically in phase 3, we also identify a tangible change in recovered cases. Hence, we also see much lower recovered cases in highly affected states of the US, such as Florida at 0%, Illinois at 0%, Michigan at 0%, New Jersey at 0%, Washington at 0%, as well as two other most affected states of California at 0.08%, and New York at 0.14%. These figures are likely to increase as the outbreak progresses.

Furthermore, the longer recovery period of infected cases from the second wave of the outbreak in China, and those returned to the country, indicate an extension of at least 1–2 weeks for treatment procedures. This was despite the easing pressures on the healthcare systems. Hence, for later assessments, the correlation between time, fatality rate, and recovered rate should be taken into consideration. Also, it is important that newly infected cases are monitored more carefully and treatment measures should be enhanced reflectively in the case of a mutated virus. This is required to be studied in a procedural way, one that could also assess the other aspects that may influence the treatment procedures. As the numbers of infected cases surpass 1 million, treatment becomes ever crucial.

BOX 5.24 "Reflecting on the Time of Disruptions"

It is unfortunate that we witnessed many countries playing chase-up games. Something that should be avoided in such serious health emergency situations that not only affects the city and its communities, but also the health of people, the health of the economy, health of urban systems, the health of industries, the health of governance, etc. In order to make effective progress in containment and recovery, we cannot remain with the absence of actions, and we cannot delay decision-making procedures. These decisions need to be effective and at the right time. Hence, either actions should be gradual and procedural from a reflective approach, or they should be immediate and at a high level at the early stage of outbreak progression. The examples of less-connected island countries in Oceania and their earlier high-level measures prove this point. The combination of high-level restrictions and a later response is not only not effective, but it also can be quite dangerous.

In facing the widespread adversities, indifferent attitudes should not be tolerated. When governments are seen pleading their people, we know this is not a common emergency. Also, when governments are seen relaxing their measures in the midst of the outbreak progression, we know their priorities are not right.

And when people are seen not complying with the measures and regulations, we know what type of society has developed over the years. Unlike the common perceptions about fragile social and political systems of developing countries, we witnessed an array of failures in developed countries, and some of those assumingly-advanced and socio-politically stable countries of the global North. We saw some serious inactions of those affluent countries. Their societies would later ask, where do their taxes go if their governments cannot protect them in the time of need? Thus, the dysfunctionality of measures and regulations is a two-sided matter that involves both people and their respective governments. Some may be associated with the decline of democracies (Fukuyama 2014), some are simply based on our decaying social and political structures, some are based on our weakening institutions and institutional operationalisation, and some are truly based on our flawed methods of the global economy.

While it was assumed there are at least some sort of positive news on environmental revitalisation due to the outbreak, the US government initiated to ease their environmental regulations on businesses and productions, such as for the oil industry. This came at the time when the US just reached the top of the chart with the highest number of infected cases. Also, with the highest unemployment upsurge in recent decades, the US government had to tighten the measures, increase cooperation, and provide more support. At the same time, the US Senate passed a US$2 trillion coronavirus bill. But there were doubts about tightening the measures and saving the economy.

The earlier environmental benefits include: general reduction in productions, general reduction in oil production and consumption, less travel and transportation, less air traffic, etc. The later studies showed signs of improvement in cleaner water around coastal areas, sudden—but temporary—drops of airborne Nitrogen Dioxide (NASA Earth Observatory 2020) and Carbon Dioxide (CarbonBrief 2020). At first, such studies analysed China, and later other affected areas and regions. Some of these environmental benefits of the outbreak were regarded as "*environmental enthusiasm*" by Gomes (2020), who then argued "*the silver lining is that a large portion of those infected are recovering. After all the hardship and tragedy from COVID-19, we can expect some clean and pollution-free air to breathe*". Apart from temporary reductions in some of our pollutions and emissions, the outbreak also reminded us about methods of reuse and recycling, more considerate consumption, less demand (only to some extent), less unnecessary travels, more efficient ways of operations, etc.

CarbonBrief (2020) summarised some of China's earlier reductions in electrical demand and industrial output, reported by 4th of March 2020, including:

- *Coal consumption at power plants was down 36%;*
- *Operating rates for main steel products were down by more than 15%, while crude steel production was almost unchanged;*

- *Coal throughput at the largest coal port fell 29%;*
- *Coking plant utilization fell 23%;*
- *Satellite-based NO2 levels were 37% lower;*
- *Utilization of oil refining capacity was lowered by 34%;*
- *At their peak, flight cancellations were reducing global passenger aviation volumes by 10%, but the sector appears to be recovering, with global capacity down 5% on year in February as a whole.*

Nevertheless, there were also adverse environmental impacts, such as imbalance in food supply chain, upsurge in production and consumption of certain products (Cheshmehzangi 2020a), as well as unhealthy disposal of medical products (Mukhopadhyay 2020). In the latter example, there are warnings regarding environmental impact and waste management failures (WHO and UNICEF 2015; WHO 2018). Hence, the deficiency of waste management systems and wrong methods of disposal would lead to adverse and unexpected environmental impacts.

Furthermore, in many countries, the so-called 'coronavirus scare' eventually turned into reality. With a non-symptomatic disease and a high level of person-to-person transmission, people had to cope with lockdown measures, quarantine, social distancing, etc. These types of disruptive times require time, financial support, responsiveness, and cooperation. Holding just about 80% of the global economy, G20 members announced they would inject US$5 trillion into the global economy (on 27 March 2020). A decision that is likely to continue and get extended. A decision that requires regional and national attention, especially to those more vulnerable countries already facing other structural issues. In this situation, everybody is vulnerable, but there are some that appeared to be more vulnerable than the others. Eventually, from a country shutdown, the situation turned into the global shutdown.

Later on, as all countries and regions were struggling with the adversities of the outbreak, either directly or indirectly, the United Nations (UN) requested for an articulated or coordinated response to having more collective measures towards containment, treatment, and the eventual eradication of the disease.

The suggestions by the UN also includes cooperation and strengthening of the institutions. Hence, as threats are becoming more global, it was suggested to strengthen multilateral institutions and overcome the issues of dysfunctional international relations. This was narrated from the perspective that power relations are unclear, the global supply chain is under pressure, and county-level fights or attempts are not necessarily successful in the long run. It was urged that we have a choice between "chaos or united actions" against such disruptions. This reminds us of the notes on those humanitarian support that said "one world, one heart". It also reflects on the poem that was added at the beginning of this book. Finally, the UN responds to the issues of the trade-off between the global economy and fighting the disease. And in achieving this, we have to consider a sustainable and inclusive approach to recovery. The main message from the UN was to be responsible, be smart, and have solidarity to tackle this challenge. After all, this is not the first time we deal with such adversities, and surely, this will not be the last time either.

From governmental denial to individual disbelief, we see major disruptions caused by both. Since the inception, we were advocating the use of facial masks for enhanced prevention and control measures. Nonetheless, many countries advised against the use of such medical products, perhaps mainly because of limited stocks at first. But then this required much more thoughtful attention from the early stages, which could have reduced the scale of the spread of the disease. It was only after the publication of a scientific paper that some countries started to revise their recommendation on facial mask use. While many Asian countries already practiced this main prevention measure, some countries also made it compulsory, such as China and South Korea. At the time the number of global infected cases surpassed 1 million, some countries made the use of facial masks compulsory, such as Austria, the Czech Republic, and Slovakia. The ripple effect was also seen in parts of Germany, the US, and other countries.

It took less than three months for the globe to reach 1 million cases and 50,000 deaths. The numbers almost doubled in just a week, as the spread of disease continued rapidly in Europe and North America. By then, only a handful of countries and regions were left without any confirmed cases. With some earlier inactions, new epicenters emerged and the speed of transmission in some countries was drastic and horrific. This primarily sparkled by the act of waiting than the more effective acts of preparing and preventing. While some governments were in denial, they raised the level of disbelief in their societies. Once the situation turned into a crisis inside their boundaries, this large scale disbelief turned into a shock and then a mistrust of the government. While there were worries about more societal disruptions, people started to worry about the effectiveness of the measures. An anonymous comment was *"I am giving the lockdown another 9 days, and then I am back to normal life with no isolation for me. It is obvious the lockdown is not working. You have to look at Spain and Italy. I won't be locked up like a caged animal for much longer"*. There were also voices of doubts and escalating mistrust: *"The virus is obviously slowing down as the news is drying up and just being repeated now and that emergency hospital was supposed to be operational yesterday. But there have been no more stories on that for a few days"*. Some also argued against the supremacy of the governmental guidelines: *"Remember, these are guidelines and not law. Not saying they should not be followed where possible, however, the police and other people who take it upon themselves to behave like they are in charge of the public, should remember that and stop trying to enforce guidelines, which are simply that"*. While the others, under the assumption of short-term lockdown optimism, argued that: *"Under the quarantine, no-one should go outside even for exercise in public places. Staying at home for 14 days is not that hard*". This came at the time when governments were warning about longer unimaginable lockdown measures of 3 to 6 months, and even a full year.

These were all at the time that a range of advice was given, but enforcement was still lacking. Health checks and testing were still not done as fast as they should, which indicate a lack of preparedness and wrong delays in prevention and control procedures (Cheshmehzangi 2020b). Some countries had to start reflecting on the realities more seriously, and some actions had to be passed to those who are more effective in the adaptation and implementation of the guidelines and measures, i.e. the regional and local governments (ibid). The disruptions continue as the outbreak grows more and more. At this time, many investigations are ongoing to shed light on the origin of the disease, questioning if the disease started much earlier or from other locations. At this time, the world is at a halt. People are stressed and economies are falling. With the increase of unemployment, we foresee further unwanted disruptions, some that could be crime-related. Furthermore, one small virus has done many things that we failed to do on our own and collectively as the so-called global citizens.

BOX 5.25 "Normalisation Versus the New Normal(s)"

Under the worries this outbreak may be a prolonged pandemic, it is expected for longer-term adversities and impacts. From a warlike time, there emerge new opportunities some in the form of innovations, and some development to the new normal(s) of what we may do afterward. This is evident in previous experiences and even major wars that helped to boost our technologies and techniques. In a sense, the normalisation in the post-recovery phase is not the step back to the business-as-usual situations or doings. There are expected to be tangible changes and some that will remain as the new normals.

For instance, this outbreak could potentially enhance societal resilience, or can instead worsen it through poorer relations or more vulnerabilities. The latter is highly likely based on our non-collective attitudes. Hence, we hope this does not mean an opportunity for the expansion of our growing individualism. This outbreak should be studied reflectively for us to stop overlooking our continuing social issues and political misconceptions. In other examples, there is a likelihood for the enhanced safety measures, not only those that include the safety of our supplies, but of regular operations, distributions, workplaces, and workforces. For instance, supermarkets released information on several matters that were not as important as they were before, including:

(1) 'Food for all' and especially addressing the needs of those vulnerable groups and those in need for extra help and support,

(2) 'Safety for everyone', which is already part of the daily operations but are expected to be enhanced with further monitory measures, and
(3) 'Supporting workforces' that will include the multiplicity of safety measures, enhanced training, welfare and wellbeing, and support.

This outbreak event should give us a chance to boost some of our neglected institutions, those that are essential to our societal wellbeing, safety, and security. It should provide a chance to enhance our primary/basic social services and help to address some of the deficiencies in our institutions. Nevertheless, these all depend on how countries or cities would reflect on the afterward situations, and how they would react to nurturing some of the forgotten institutions and services. This should provide a chance to boost those unattended factors through better financial support and timely management of their integration into the enhancement of cities and communities (Cheshmehzangi 2016). This outbreak would possibly change our trade matters, and should help countries and cities/regions to look into local productions and self-sufficient approaches to enhance our supply chains. The global trade would then require to be assessed and revised, to have a better multilateralism approach and to be further developed to become all-inclusive and effective.

The most perceptible of all would be to boost our digital technologies. The possible increase in the use of smart technologies is also very likely. With the boost in our digital platforms, this outbreak opens up a unique opportunity for many digitally-advanced innovations such as: data-based and information-based operations, enhanced monitory through data-driven methods, distance learning and online education (or e-learning), cashless operations, integrated mobile applications, e-commerce, digital product deliveries, online operations, etc. As the digital platforms were amongst those that benefitted the most from the outbreak are expected to develop opportunities for the new normals. These may appear and develop differently in different locations, depending on multiple factors of societal trust, cultural factors, social matters, and political systems.

There are also many other possible new normals, and we highlight only a few here.

Increased poverty—There is no hiding from the economic impacts, particularly on the most vulnerable groups. This event will increase poverty in those severely-hit countries. The economic crisis will last for a while before it could save some of those poorer nations or those highly-affected ones.

Cybercrime and cybersecurity—We will see an increase in both. Cybercriminals find this event as a great opportunity to innovate methods of cyberattacks and cybercrime. In return, governments and companies will be forced to spend more cybersecurity measures.

Facial masks—The use of masks may become more common than ever. While it is wrongly advertised about the ineffectiveness of masks, it is important to note that they can be used to avoid passing the disease to others, similar to what the Japanese culture indicates. In the future, we are likely to see more use of facial masks when people get ill, or they are in busy and/or contaminated places, etc.

Security and safety services—We anticipate more power to be given to those security and safety services. This likely happens in those declining economies or those with poorer institutions. The empowerment of such services would possibly take a different turn in autonomous governments or those of totalitarian structures. Thus, such empowerment should be considered with added values or else it becomes a tool to further policing the society. It is important to put wellbeing central to such enhancement opportunities and to allow for a revitalised or strengthened institutions.

Postal and delivery operations—We anticipate seeing an adjustment in operations, from safety measures to issues of workload, and distribution methods. The postal and delivery services are likely to change, allowing for more contact-less opportunities, and reduction of paper-based deliveries for letters/mail. The use of digital replacements will take over much faster.

Changes in consumption patterns—We expect to see changes at two sides of the consumption spectrum after the outbreak. Some changes would likely be temporary and some will remain for a longer period for a group of people. It is likely a larger group, affluent or not, will consume more in the after the event period. This is due to their less consumption period, which will be assumed as a time of deprivation. For those who are more reflective or become more vulnerable, then this becomes a lesson learnt that indicates a possible chance to reduce some of the habits of high consumption. This will be taken positively particularly if this can happen at a larger scale. However, in the longer term, this may not last so long.

Hoarding behaviour—Similar to the impact wars can have on people's behaviour, this outbreak could also increase the chance of hoarding behaviour. From panic-buying incidents to hoarding activities, this event would reflect negatively on some people's behaviours. The hoarding behaviours are expected to continue and this may also impact some of our consumption behaviours.

Online shopping—With the success of e-commerce and online services, we anticipate the opportunity for the enhancement of online shopping services. In countries, where online shopping services were already common, we would see the upsurge in the use of what we called the virtual amenities. In countries, were online shopping services were not as common, we would see the emergence of such services and later decline of physical shopping. This will be a turning point for them and will mark as an opportunity for online systems.

Hygiene improvement—Globally, we may see signs of hygiene improvement. This may support new regulations or develop in the form of sector-based guidelines. Also, there is a likelihood that individual hygiene and the use of cleansing products will stay more popular than before. Some companies/enterprises/institutions would adopt enhanced cleaning procedures in their daily practices. People are likely to become more aware of asymptomatic incidences.

Social distancing—The longer effect of social distancing (or physical distancing) would be the reduction of unnecessary physical meetings, events, and gatherings. They would either be reduced or they would be replaced by alternative online meetings, webinars, and online forums/groups.

The emergence of high-level access management and control systems—As it is happening already, we would see more similar tools emerging for various reasons, such as security, anti-terrorism, etc.

More Caution before travelling—It is likely people would check more information about their travel plans and travel destinations. People would potentially learn from this event, knowing that outbreaks could happen anytime and anywhere. People may also be more aware that smaller outbreaks are common.

Demand for faster emergency construction—This will include faster methods for the assembly of temporary hospitals, shelters, etc., which will not only be useful for future disease outbreaks but also for war zones, natural disaster events, other crises, and emergencies, etc. We should also see more innovative design and planning strategies for transformative/flexible spaces and building typologies.

Airport health check procedures—Some measures will continue to remain in place. In some countries and regions, particularly in East Asia, temperature checks were already part of the procedural checks. This is expected to become a common practice in more countries (if not all), and hopefully more sensible than those already-practiced global security measures. Such health checks may remain in other dense transportation hubs, such as the main railway stations, ferry stations, bus stations, etc.

Regional measures and regional planning—This is a good opportunity to empower regional level strategies. National-level planning alone may not be effective and will require regional-level support. If not regional, local governments should be empowered as the main actors of the implementation process.

Surveillance will increase—This will happen both by people and the machines. People will monitor and record any misbehaviours more than before. This will be a chance for facial recognition devices for multiple uses, starting with a more acceptable temperature detection and then into security measures.

Artificial intelligence (AI) and detection measures—This event will boost the use of AI in our detection and prediction methods and will become widely-used for various reasons.

The earlier mishandling of the outbreak in multiple regions caused many changes, and some of those changes would turn into our new normals. Nevertheless, normalisation would not mean 'going back to the previously normal'. There will be several major changes, and some new practices to increase global awareness and preparedness in the case of future outbreak events. We also have to note, with some caution, that in the name of security, this experience may just increase our societal surveillance through the so-call smart strategies. This can be perceived positively and negatively, depending on the actual intentions behind the use of such techniques or platforms. As it always has been, governance could become stronger by utilising the tool of societal fear. More than before, we may end up using robots, artificial intelligence, facial recognition devices, tracking devices, drones, etc.

This outbreak will open a new chapter.

BOX 5.26 "Tools, Tools, and more New Tools"
Knowing that we have a growing number of redundant positions, and many opportunists took advantage of the vulnerable situations, there were signs of more effective and supportive actions, too. Under the growing demand for digital tools and many new integrated systems that were proposed during the outbreak, we can see a positive take in how they played part in promoting education, training, awareness, medical support, charity fund, remote working solutions, real-time reporting, medical and health checks, etc. Here, we showcase some of these examples that were utilised during the outbreak.

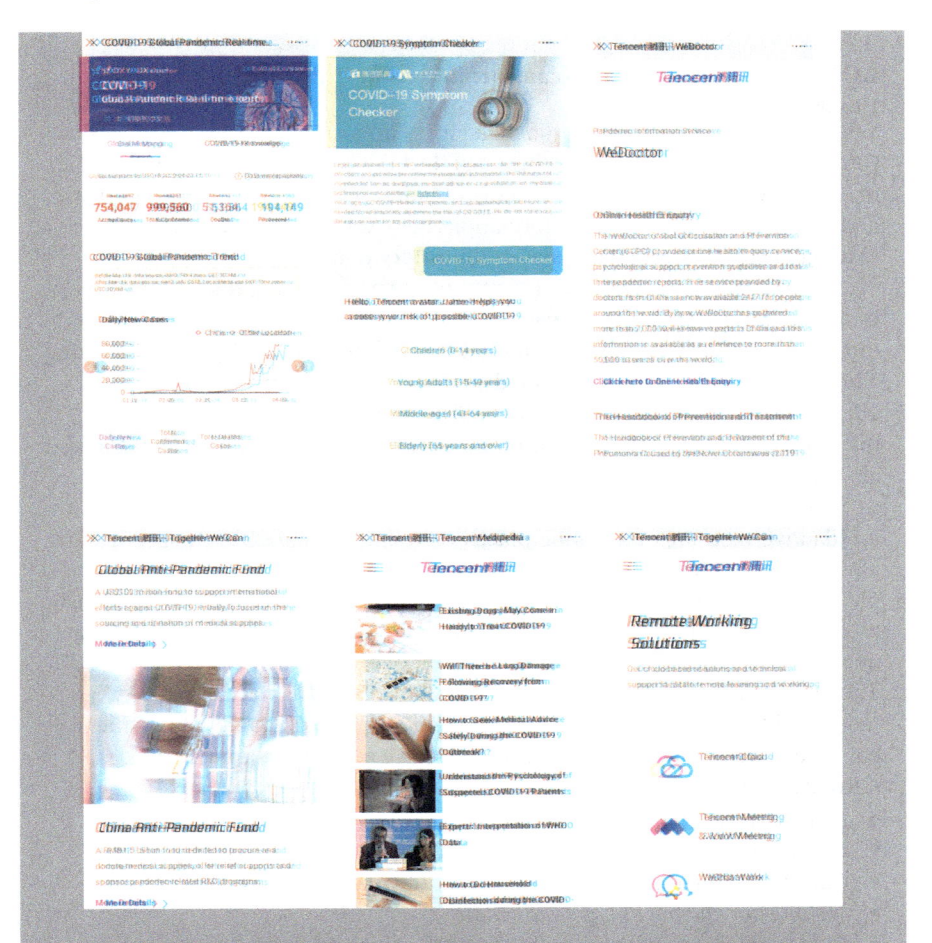

Here, we elaborate on some of these examples: Tencent Mediapedia *"collaborates with medical experts to provide users with pandemic-related articles and videos that include tips on disease prevention, self-quarantining at home, and other related information"*. The WeDoctor, is a platform for *"Global Consultation and Prevention Center (GCPC), and provides online health enquiry service, psychological support, prevention guidelines, and real-time pandemic report in China"*.

These are all Chinese or China-based digital platforms. We are not doubting or neglecting the fact there many similar examples developed and implemented in the countries. This is just to show some of the examples we experienced during this outbreak, as they are more, and there will be more on the way.

(Extracted from: Tencent at: http://www.tencent.com/en-us/responsibility/combat-covid-19.html).

BOX 5.27 "Transitions vs. Transformations in Cities and the Built Environments"

While we see there are transitional changes, we also anticipate several transformations that may affect our cities and the built environments. These are the ones that we believe may lead to new policies or regulations, new practices, and new paradigms. Such transformations are likely to differ from one context to another, and some of them may be transitory than long-term changes.

Now, as time passes, we see many studies emerging that reflect on various transformations, mostly associated with business and economic impacts and those that are affecting the policy and polity matters. Some of these are addressed based on three identified periods of "pre-COVID-19", "COVID-19, and "post-COVID-19", which itself is quite interesting for us to identify the scale of this pandemic event on every sector. Some of these impacts are also seen as transitions and transformations in the built environment sectors.

In a recently published article, some of these transformations are identified as adaptive measures for public places (Cheshmehzangi 2020c), specifically aimed to help cities and city managers to face the COVID-19 impacts in public places of cities. In this brief article, two main questions are raised: (1) what measures could we apply to safeguard our public places?; and (2) to what extent, does the public place play a part in making our city and society safe for all? The two characteristics of contemporary public places, namely 'flexibility' and 'adaptability' are highlighted as methods of adaptive planning and resilience enhancement. In this study, 10 specific adaptive measures for public places are introduced and briefly discussed. These adaptive measures include:

- *Limited access nodes for better management of Public Places*
- *Establish a one-way mobility circulation*
- *Checkpoint allocation for monitory and recording opportunities*
- *Making social spaces safe and viable*
- *Public places as informative nodes in the city*
- *Flexible public places to support essential sectors*
- *Restrictions on shared facilities in public places*
- *Closure of secondary public places*
- *Provision of community-level designated zones as key public places*
- *Temporary regulations to monitor and restrict certain activities in public places*

However, these measures are the ones specifically suggested for cities and communities that are facing the outbreak. These provided guidelines, from the perspective of adaptive planning, suggest methods of spatial management and spatial arrangements (or rearrangements) that could help the city to safeguard its populated urban hubs and public realms. Some of these ideas could also develop in the way we could innovate new design and planning strategies, which could reflect on some of the deficiencies or flaws of our current planning thinking or design methods. In particular, we could address some of the main design principles; for instance, how adaptability could be enhanced or how public realms could be better articulated in the practice of urban design, landscape architecture, and public place design.

To summarise, there certain reflective approaches that could be utilised for future design and planning approaches. Those are the ones that could be developed further as potential design/planning innovation, new transformative strategies, and enhanced design guidelines. In doing so, we are able to provide a new range of design and planning paradigms that may help us to better contain the disease.

In at least two following short communication from the earlier work, the author has highlighted progressive thinking in adaptive planning for future development and resilience enhancement approaches. These are two brief articles currently under review. The first is a reflection on early lessons for urban resilience enhancement during the COVID-19 (Cheshmehzangi 2020d), and the second article is a brief comment on possible development changes in the construction sector, and in particular, the build environment sector due to this current pandemic outbreak (Cheshmehzangi 2020e).

In the first article (Cheshmehzangi 2020d), a set of reflective viewpoints is provided mainly from the perspective of the current happening (the first six months of the COVID-19 outbreak). The state of the art in this brief study is to highlight the role of urban resilience in practice, which provides a reflective narrative on early lessons that could help us make our cities and communities more resilient. They are identified as early lessons that successfully worked in cities and regions that are more confident in relaxing their high-level measures at a faster pace. These suggestions are narrated in a way that could be adapted for any context. These are suggested as eight reflective viewpoints as shown below:

- Consider a holistic spatial management (similar to Cheshmehzangi 2020c);
- Pay attention to details in a reflective manner;
- Early asset management and prioritisation plan;
- Regional level and local level approach for implementation;
- Early high-level measures are more effective than later restrictions;
- Include community representatives;
- Make good use of the right platforms;
- Regular safety checks, hygiene, and prevention.

In the second article (Cheshmehzangi 2020e), the narrative changes to develop several predictions that would potentially change the way we build our cities, communities, and buildings, the three main spatial levels of the built environments. In this study, the development changes are defined based on five key areas of: (1) resilience, (2) health, (3) policy, (4) practices, and (5) perspectives. These are then assessed in terms of the level of importance or potential changes in six primary areas in the civil engineering sector (i.e. construction engineering, construction technology, structural engineering, transport engineering, infrastructure engineering, and building/construction management), and six primary areas in the built environment sector (i.e. building engineering, architecture, urban design, urban planning, landscape architecture, and interior design). These development changes are in 10 categories as shown below:

- Decline in car-based transportation infrastructure;
- A push for information-based construction management methods;
- Increase in off-site construction and engineering;
- Increase in lightweight structural systems;
- Opportunities for new materials for performative insulation;
- Revisions in Density and Compact Design;
- Spatial planning considerations;
- Smaller and individual internal layouts;
- A push for meso-scale strategies at the neighbourhood/community level;
- New opportunities for non-centralized building systems.

To summarise briefly, we cannot neglect the potential changes as this pandemic has been a big lesson learned for all. Nevertheless, the major changes in our primary systems are yet to be seen and this may just be the beginning of a new chapter. As the pandemic comes to an end, and if everything is not.

5.6   A Summary: Facing Disruptive Disease Outbreaks

This chapter provides a holistic reflection on disruptions caused by the recent disease outbreak. This is a representation of many (similar) outbreak events, not necessarily as widespread as the one experienced, but from the central perspective of disruptions. This event shares many similarities to those regular outbreaks that are less impactful. With the example of the COVID-19 outbreak, we looked into a variety of aspects, from city-level local responses to global reactions. In all levels, we hope public health is not turned into a political tool and is rather supported by the enhanced urban resilience and adequate city management. As shown in Chap. 3, preparedness is essential through urban resilience, which should be comprehensive and practical by all means. Then in Chap. 4, we highlighted the role of responsiveness through all-inclusive city management and the many considerations we should have to overcome the adversities of different phases of the outbreak progression. In this chapter, we summed up these aspects through a reflective viewpoint, to manage the city in need and ultimately save it, too. This sort of thinking is vital to cover all aspects, from an individual's health to the large-scale impacts on international trade.

In this chapter, we highlighted two very important factors that could simply equip the city in the case of an outbreak event. One is 'society' and the other is 'government'. In the former, we narrated the society's experience, which in fact demonstrates the realities of the outbreak. In the latter, we covered an array of responses to show how the city could react in the case of a health emergency. Both together form a better approach to high-quality governance (see Fig. 5.77). What we have not fully covered is the other side of the coin, where society could suffer the most and the government may fail to react and respond properly and timely. Nevertheless, we note some of the main concerns addressed in both society and government—and then on what represents the role of all-inclusive governance. For instance, public disorders would still exist during the outbreak. In fact, there is a high chance that they turn into added disruptions. In such cases, the individuals who often ignore the regulations and procedures may take the city's vulnerability as an opportunity, knowing that there is little monitory on their misdemeanours and misconducts. Therefore, it is likely to see an upsurge in some of those disorders both related and unrelated to issues of the outbreak. These include—but not limited to—a range of public and social disorders, such as non-compliance with the guidelines, hoarding of primary products for resale and benefit, misuse of social services, the spread of false information/news, misguiding the public, jeopardising the society, riots, and unrests, exploitation of medical and emergency services, public fights, domestic violence, unsafe driving, unsafe littering, etc. There is also a possible increase in crime, such as theft, looting, loan sharking, arson attacks, cyber-attacks/cybercrime, physical and verbal assaults, aggressive behaviour, racism, stigmatisation behaviours, scams/fraud, false claims/accusations, prison break/unrest, and even intentional transmission of disease to others. Believe it or not, all these issues happened during this recent outbreak event; but providentially, not clustered in one location. In reality, to respond to these issues, the city needs to be fully prepared and the society needs to have the right mindset. Without any of the two, the struggle would be prevalent.

In conjunction with some of the highlights of the chapter, we referred to the significance of already emerged virtual and digital platforms, from e-commerce to education delivery, and from infected patient detection to data-based approaches that enhance control and monitory measures of the city. We did not necessarily explore associated issues related to the lack of inclusiveness of smart technologies and platforms, but we highlighted their values and support in the time of need. This was highlighted as an opportunity for innovations and smart-resilient approaches. We also highlighted some values of certain spatial and urban planning, which help to speed up some control measures. We looked into a variety of implementation factors and what they mean to various sectors and stakeholders. We also found the necessity to be resilient and less political in facing the disruptions of outbreak events. We also end this chapter with two sides of the spectrum, society and government, which in reality are complementary to one another. The combined forces would help the city in need; hence, could it be the governance of the society? Or the society that drives healthy governance? Or simply both? In sum, the government does not need to be political, all they have to do is to reflect and respond appropriately. And lastly, people do not need to be selfless, all they have to do is not be selfish in such events. Finally, if any of the two (i.e. society and government) becomes a burden to another, then it may just worsen the situation. The ultimate goals should be to reflect promptly on potential disruptions, and ultimately enhance the governance for the city in need.

References

Al-Nasrawi, S., El-Zaart, A., & Adams, C. (2017). The anatomy of smartness of smart sustainable cities: An inclusive approach. *proceedings for 2017 International Conference on Computer and Applications (ICCA)* (pp. 348–353). Qatar: Doha.

Alterman, R. (1988). Adaptive Planning. *Cognitive Science, 12,* 393–421.

Ambat, A. S., Zubair, S. M., Prasad, N., Pundir, P., Rajwar, E., Patil, D. S., et al. (2019). Nipah virus: A review on epidemiological characteristics and outbreaks to inform public health decision making. *Journal of Infection and Public Health, 12*(5), 634–639.

Ascher, K. (2007). *The words: anatomy of a city.* Middlesex: Penguin Books.

Australian Government Department of Health (2020a). Government response to the COVID-19 outbreak. Retrieved March 10, 2020, from https://www.health.gov.au/news/health-alerts/novel-coronavirus-2019-ncov-health-alert/government-response-to-the-covid-19-outbreak.

Australian Government Department of Health (2020b). *Australian Health Sector Emergency Response Plan for Novel Coronavirus (COVID-19).* Retrieved March 10, 2020, from https://www.health.gov.au/resources/publications/australian-health-sector-emergency-response-plan-for-novel-coronavirus-covid-19.

Baumol, W. J. (1967). Macroeconomics of unbalanced growth: The anatomy of urban crisis. *The American Economic Review, 57*(3), 415–426.

Benson, F., Musekiwa, A., Blumberg, L., & Rispel, L. C. (2016). Is the South African notifiable diseases surveillance system effective in preventing outbreaks? Perceptions of key stakeholders, *International Journal of Infectious Diseases, 45,* Supplement 1, poster presentation, P. 293.

Bibri, S. E. (2019). The anatomy of the data-driven smart sustainable city: instrumentation, datafication, computerization and related applications, *Journal of Big Data, 6,* article number: 59.

CarbonBrief. (2020, February 19). *Analysis: Coronavirus has temporarily reduced China's CO_2 emissions by a quarter.* Retrieved March 10, 2020, from https://www.carbonbrief.org/analysis-coronavirus-has-temporarily-reduced-chinas-co2-emissions-by-a-quarter.

Centers for Disease Control and Prevention (CDC). (2020a). *Coronavirus Disease 2019 (COVID-19).* Retrieved March 5, 2020, from https://www.cdc.gov/coronavirus.

Centre for Disease Control and Prevention (2020b). *Risk Assessment and Management,* on Coronavirus Disease 2019 (COVID-19). Retrieved March 10, 2020, from https://www.cdc.gov/coronavirus/2019-ncov/php/risk-assessment.html.

CGTN News. (2020, March 9). Premier Li Keqiang sounds alert over imported COVID-19 cases amid rapid spread overseas. Retrieved March 11, 2020, from https://news.cgtn.com/news/2020-03-09/Li-warns-on-imported-cases-amid-rapid-spread-of-COVID-19-overseas-OIXj18mtzy/index.html.

Cheshmehzangi, A. (2016). City enhancement beyond the notion of "Sustainable City": Introduction to integrated assessment for city enhancement (iACE) toolkit. *Energy Procedia, 104,* 153–158.

Cheshmehzangi, A. (2017). Sustainable development directions in China: Smart, Eco, or both? In *The 2017 International Symposium on Sustainable Smart Eco-city Planning and Development,* 11–13 October 2017, Cardiff, Wales.

Cheshmehzangi, A. (2018). The changing urban landscape of Chinese cities: Positive and negative impacts of urban design controls on contemporary urban housing. *Sustainability, 10*(8), 2839. https://doi.org/10.3390/su10082839.

Cheshmehzangi, A. (2020a). *Comprehensive Urban Resilience for the City of Ningbo*. Ningbo, China. (in Chinese: 宁波市城市综合抗灾弹性框架), Report submitted to local government units in February 2020.

Cheshmehzangi, A. (2020b). *Resilience Enhancement for Cities and Community during the Outbreak*, at the first round of EU-China knowledge-share events on COVID-19, organised by SPI Porto, Portugal, on 3 April 2020.

Cheshmehzangi, A. (2020c). *10 Adaptive Measures for Public Places to face the COVID 19 Pandemic Outbreak*. City & Society, special issue, pp. 1–10, https://doi.org/10.1111/ciso.12282.

Cheshmehzangi, A. (2020d). *Reflection on early lessons for urban resilience enhancement during the COVID-19*, currently under review, 13 pages.

Cheshmehzangi, A. (2020e). *10 Possible Development Changes in Civil Engineering and the Built Environment Sectors Due to COVID-19*, currently under review, 6 pages.

Cheshmehzangi, A., & Dawodu, A. (2018). *Sustainable Urban Development in the Age of Climate Change–People: the Cure or Curse*. Singapore: Palgrave Macmillan.

Cohendet, P., Grandadam, D., & Simon, L. (2010). The anatomy of the creative city. *Journal of Industry and Innovation, 17*(1), 91–111. Creative Jobs, Industries and Places.

Deng, W., & Cheshmehzangi, A. (2018). *Eco-development in China: Cities, Communities, and Buildings*. Singapore: Palgrave Macmillan.

Desjardins, J. (2019). *Infographic: The anatomy of a smart city*. Retrieved March 4, 2020, from https://www.visualcapitalist.com/anatomy-smart-city/.

Foreign Affairs Office (FAO) of Ningbo Municipal People's Government. (2020, January 30). A letter to all foreign friends in ningbo. Retrieved February 2, 2020, from http://english.ningbo.gov.cn/art/2020/2/1/art_931_1041824.html.

Fukuyama, F. (2014). *Political Order and Political Decay: From the Industrial Revolution to the Globalization of Democracy*. London: Profile Boos.

Gallotti, R., & Barthelemy, M. (2015). Anatomy and efficiency of urban multimodal mobility. *Scientific Reports, 4* (6911).

Goldwyn, E. (2018). Anatomy of a new dollar van route: Informal transport and planning in New York City. *Journal of Transport Geography*, in press.

Gomes, S. (2020). *Environmental Benefits of COVID-19*. Retrieved March 4, 2020, from https://www.calamitypolitics.com/2020/03/15/environmental-benefits-of-covid-19-12120/.

Hellsten, I., Jacobs, S., & Wonneberger, A. (2019). Active and passive stakeholders in issue arenas: A communication network approach to the bird flu debate on Twitter. *Public Relations Review, 45*(1), 35–48.

Jauregui-Fung, F, Kenworthy, J., Almaaroufi, S., Pulido-Castro, N., Pereira, S., & Golda-Pogratz, K. (2019). Anatomy of an informal transit city: mobility analysis of the metropolitan area of Lima, *Urban Science, 3*(3), article number: 67.

Kang, Y. (2019). Food safety governance in China: Change and continuity, *Food Control, 106*(106752).

Lamorgese, A. R., & Petrella, A. (2016). *An anatomy of Italian cities: Evidence from firm-level data* (p. 362). Paper no: Bank of Italy Occasional Paper.

McKinsey Analysts (from McKinsey & Company). (2020, March 9). *COVID-19: Briefing note*. Retrieved March 11, 2020, from https://www.mckinsey.com/business-functions/risk/our-insights/covid-19-implications-for-business.

Monge, S., García-Ortúzar, V., López Hernández, B., Lopaz Pérez, M. A., Delacour-Estrella, S., Sánchez-Seco, M. P., ..., Dengue Outbreak Investigation Team. (2020). Characterization of the first autochthonous dengue outbreak in Spain (August–September 2018). *Acta Tropica, 205*(105402).

Mukhopadhyay, S. (2020, March 20). COVID-19: Unmasking the environmental impact. Retrieved March 22, 2020, from https://earth.org/covid-19-unmasking-the-environmental-impact/.

NASA Earth Observatory. (2020). *Airborne Nitrogen Dioxide Plummets over China*, Retrieved March 9, 2020, from: https://earthobservatory.nasa.gov/images/146362/airborne-nitrogen-dioxide-plummets-over-china?utm_source=FBPAGE&utm_medium=NASA+-+National+Aeronautics+and+Space+Administration&utm_campaign=NASASocial&linkId=83339382&fbclid=IwAR0NyeTl3Z9KTgBE8_p-b3IRa6j-iwdHwUSjiAUj2u5R66HDBAOtTkrORdg.

National Health Service (NHS). (2020). About Coronavirus (COVID-19). Retrieved March 9, 2020, from https://111.nhs.uk/covid-19.

Ningbo Government Official page. (2020). The application procedure for "Ningbo Health Code". Retrieved March 1, 2020, from https://mp.weixin.qq.com/s/tMk5tfXIXyxf_uaFObSwuQ.

Ningbo Municipal Economic and Information Technology Bureau. (2020, February 13). *The provincial epidemic prevention and control supervision group visits Ningbo for supervision and inspection.* Retrieved March 3, 2020, from http://www.nbec.gov.cn/art/2020/2/13/art_1907_4048855.html.

Ningbo Science and Technology Bureau. (2020, January 29). *Notice for preventing 2019 novel coronavirus from spreading.* Retrieved February 2, 2020, from http://kjj.ningbo.gov.cn/art/2020/1/29/art_9940_4045698.html.

Politico Magazine. (2020, March 19). Coronavirus will change the world permanently. *Friday cover article.* (Materials extracted for Box 5.14). Retrieved March 20, 2020, from https://www.politico.com/news/magazine/2020/03/19/coronavirus-effect-economy-life-society-analysis-covid-135579#superComments.

Public Health Matters. (2020). Coronavirus (COVID-19)–5 things you can do to protect yourself and your community. Retrieved March 10, 2020, from https://publichealthmatters.blog.gov.uk/2020/03/04/coronavirus-covid-19-5-things-you-can-do-to-protect-yourself-and-your-community/.

Quirk, V. (2013). Anatomy of a Chinese city. *ArchDaily article.* Retrieved March 4, 2020, from https://www.archdaily.com/338302/anatomy-of-a-chinese-city.

Renner, R. (2018). *Urban Being–anatomy & Identity of the City.* Salenstein, Switzerland: Niggli.

Ritchie, B. W., Dorrell, H., Miller, D., & Miller, G. A. (2008). Crisis communication and recovery for the tourism industry. *Journal of Travel & Tourism Marketing, 15*(2–3), 199–216.

Schultz, G. W. (1968). *The Anatomy of Turkish Cities.* United States National Research Council.

Shears, P., & Garavan, C. (2020). The 2018/19 Ebola epidemic the democratic republic of the Congo (DRC): Epidemiology, outbreak control, and conflict. *Infection Prevention in Practice, 2*(1), 100038.

Stone, E. C., & Zimansky, P. (1992). Mashkan-shapir and the Anatomy of an old Babylonian City. *The Biblical Archaeologist, 55*(4), 212–218.

Tillett, P. (2018). *Shaping Portland: Anatomy of a healthy city, Series: Routledge Research in Planning and Urban Design.* Oxon: Routledge.

The Government of Japan. (2020). Coronavirus (COVID-19) advisory information. Retrieved March 10, 2020, from https://www.japan.travel/en/coronavirus/.

The Network for Public Health Law. (2020). *Emergency Legal Preparedness and Response–COVID-19.* Retrieved March 10, 2020, from https://www.networkforphl.org/resources/topics/emergency-legal-preparedness-and-response/covid-19/.

UNESCO. (2020). *COVID-19 educational disruption and response.* Retrieved March 18, 2020, from https://en.unesco.org/themes/education-emergencies/coronavirus-school-closures.

UN News. (2020, February 29). *UN Health Agency warns against coronavirus COVID-19 criminal scams.* Retrieved March 12, 2020, from https://www.news.un.org/en/story.2020/02/1058381.

WHO-China Joint Mission. (2020, February 16–24). *Report of the WHO-China Joint Mission on Coronavirus Disease 2019 (COVID-19).* Retrieved March 8, 2020, from https://www.who.int/docs/default-source/coronaviruse/who-china-joint-mission-on-covid-19-final-report.pdf.

World Health Organization (WHO). (2018). *Health-care waste..* Retrieved March 2, 2020, from https://www.who.int/news-room/fact-sheets/detail/health-care-waste.

World Health Organization (WHO). (2020, January 9). *WHO statement regarding cluster of pneumonia Cases in Wuhan, China.* Retrieved March 6, 2020, from https://www.who.int/china/news/detail/09-01-2020-who-statement-regarding-cluster-of-pneumonia-cases-in-wuhan-china.

World Health Organization (WHO). (2020*). Coronavirus disease (COVID-19) advice for public, including precautionary measures, education of right and wrong information.* Retrieved March 8, 2020, from https://www.who.int/emergencies/diseases/novel-coronavirus-2019/advice-for-public.

World Health Organization (WHO) and UNICEF. (2015). *Water, sanitation and hygiene in health care facilities: status in low-and middle-income countries.* Geneva: World Health Organization.

Yang, G. (2020, February 10). China releases twelve opinions on how companies resuming normal operations should treat their employees, breaking news material. (online document of 40 pages) Retrieved March 9, 2020, from https://www.chinalawblog.com/2020/02/breaking-news-china-releases-twelve-opinions-on-how-companies-resuming-normal-operations-should-treat-their-employees.html.

Zhao, S., & Ma, C. (2020, March 7). China doing unprecedented job in battling unknown disease: WHO expert. *News material.* Retrieved March 7, 2020, from http://www.chinadaily.com.cn/a/202003/07/WS5e62dcf8a31012821727d146.html.

Chapter 6
Recommendations for 'The City in Need'

Man needs difficulties; they are necessary for health.
—Carl Jung.

6.1 Expected Disruptions and Unexpected Impacts

As the body first deteriorates and then reaches immunity against a disease, the city also first suffers and then becomes more resilient by the end of an outbreak event. The city may not become fully immune, but will be more experienced and prepared with a much enhanced resilience for the future. The body tolerates the disease and pain as it first struggles and then grows to become sturdier; and so does the city and how it manages to pass through the multiplicity of destructive and disruptive impacts. The city will only become stronger, if not healthier. It will then regain its health and full potential in a process of responsive management, improvement, and resilience enhancement.

Based on what we have covered in the previous chapters, the only way towards progress is to overcome disruptions. If the city can assess the expected disruptions in advance, then in return it can make the pathway(s) to respond more effectively to those unexpected impacts. Some of the disruptions may not be necessarily manageable at the city level but with adequate urban resilience and city management, we can help to avoid micro-level crises or failures. With our growing international trade and economies, the issues of economic management can only be partially accomplished at the city level. Hence, the main contribution of the city would be to reduce the impacts and vulnerabilities that are likely to increase the adversities of the communities. The example of a comprehensive framework, or scenario-based assessment of the outbreak progression, enable us to better realise the expected and unexpected issues of the outbreak. Undeniably, disruptions are expected; even if some may try to delay the impacts and buy out more time to reach early containment. Hence, realising the realities is something that we have addressed, and as we have shown in the example

A. Cheshmehzangi, *The City in Need*, https://doi.org/10.1007/978-981-15-5487-2_6

of the previous chapter. In most cases, unexpected factors are the actual impacts on multiple systems and sectors and various stakeholders.

As the outbreak progressed within China, in the region, and at the global scale, the multiplicity of disruptions were just piling up. The world was no longer only at the status of a global health emergency, it was (and will be) facing a major economic crisis. By the time it was officially declared a pandemic, the impacts were already global. From shortages of many products to health care emergencies, from travels bans to cancellation of major events, and from lockdown of a city to lockdown of countries and regions. The impacts dictate: changes in lifestyles and the socio-economic values of the cities, changes in mobility and the interconnectivity of the regions, fluidity of new regulations and facing the emergencies in various countries, and the momentous impacts on international trade and relations, as well as the public health of the globe. The situation pushed the meetings of the Group of 7 (G7), an intergovernmental economic organisation, to become virtual. It overshadowed the 2020 Olympics events in Tokyo, which was eventually postponed on 24 March 2020. It put a halt on many operations and trades at multiple levels. It led to lockdown measures of many provinces. It turned vibrant cities into instantaneous ghost towns. It made atrocious impacts on highly infected communities. It crumbled the lives of many families. And it took so many lives. It simply felt like a moment of pause.

This book has so far covered a wide range of knowledge regarding the issues of outbreak events and how cities should cope in such unexpected situations. It then provided comprehensive thinking on methods of preparedness through urban resilience, as well as approaches of responsiveness through city management. This was later elaborated by realising the realities, capturing the progress, and facing the disruptions. In this concluding chapter, we first delve into matters of progress and how we should move from vulnerability to containment and then recovery. These discussions will be followed by a set of 10 recommendations for the city in need before we conclude with the final remarks of the book.

6.2 Preparing the City in the Outbreak Events

So far, we have covered what needs to happen in terms of preparedness and respon-siveness in multiple phases of the outbreak progression. Here, we address how such preparation progresses at different levels as the city becomes closer to recovery. This requires a step-by-step but progressive approach to ensure the city is prepared, it can respond effectively and reflectively, and it can ultimately manage the situation (Cheshmehzangi 2020a). To reach the recovery, the city has to be able to control the outbreak, especially if there are cases of sustained transmission or the growing number of cases in clusters and multiple locations. In doing so, such control helps to avoid a probable shift from the case of an emergency to the case of a crisis; a shift that everyone would like to avoid at all costs. To avoid any possible discordances and incongruities, the city has to respond through its capacity of resilience and its ultimate powers of management. Such responsiveness has proven to be the most effective tool in facing the outbreak events. The whole process is then narrated in three levels of: (1) from vulnerability to resilience, (2) from resilience to management, and (3) from management to containment and then recovery.

At first, it is evident that vulnerabilities are the immediate bottleneck to progress, and hence, the main objective ought to be in line with the advancements towards resilience enhancement. Nevertheless, as we learned through the earlier chapters that resilience alone is not effective. It only becomes effective with an adequate level of city management across multiple levels and multiple sectors. Thus, at the second point of progress, the integration of resilience planning in city management methods is crucial to ensure the stability of the city, operationalisation of urban systems, as well as assessment and management of risks. At last, the progress is for city management to achieve containment and then recovery.

6.2.1 *From Vulnerability to Resilience*

If the city has no preparedness, there are signs of more vulnerabilities in multiple sectors that could ultimately affect larger groups of stakeholders, and specifically the general public. There are also certain context-specific vulnerabilities that need to be identified and addressed as early as possible. Some examples of these vulnerabilities include, socio-cultural factors, education levels, institutional arrangements, geographical conditions, interconnectedness and mobility patterns, etc. Also, as there was an earlier case of transnational transmission from a conference held in Singapore, we noted the level of severity of the disease transmission. This case can be regarded as one of the early trackable cases from one location to multiple countries. Later on, the situation was reflected as a matter of urgency. It was through the highlights of the co-chair of Singapore's Task Force on the coronavirus (MoH website 2020) that countries with open economies were recognised to be at higher risk:

> We are vulnerable, but we have to everything that we can do contain that spread of the virus…We are mindful that we are indeed an open economy, we are an international travel hub…We are putting information in a very transparent manner and we continue to work with all health authorities overseas.

This multi-ministerial task force was established only two days after the official announcement of the outbreak in Wuhan. This immediate response represents an example of reflectiveness through early identification of vulnerabilities and responding to them without significant delays. This task force was established to: "*ramp up precautionary measures*", "*expand on border controls*" (including also land and sea checkpoints), "*direct a whole-government response to the outbreak*", and "*step up Singapore's overall preventive posture*" (ibid).

Apart from these top-down managerial factors and monitory measures, the vulnerabilities of critical infrastructures (Monstadt and Schmidt 2019), urban systems (Cheshmehzangi 2020a), and the society are major factors for resilience consideration, as well as for the important factor of resilience enhancement. From various viewpoints, the examples of societal vulnerabilities such as social matters (Fraser 2003), spatio-temporal dynamics of the population (Shafqat Akanda et al. 2013; Nieddu et al. 2017), socio-economic factors (Bacallao et al. 2014), and even from the consideration of gender (Rancourt 2013), and specific systems (Fraser et al. 2005), are all important to how the city can reduce its vulnerabilities. Such considerations should also be utilised for further assessment of multiple factors, inclusive of economic risk and institutional limitations (Parkins and MacKendrick 2007), those that are related to potential poorer compliance with advised health-protective behaviours and wrong perceptions (Schemann et al. 2013). Consequently, through a collection of scholarly work, Ali and Keil (2011) put together a range of methods to understand *"global cities, networks, and governance in a post-national era of public health regulations and neo-liberalisation of state services"*. Thus, the interconnectivity and networks are considered as important factors to identify vulnerabilities and then to enhance resilience. Of these factors, the roles of people and the network of people play their fundamental parts in how governance could eventually succeed. In many cases, people mistakenly find refuge in social media and other sources of misinformation, not knowing there are hidden traps to increase their anxiety and reduce their awareness. As discussed before, the lack of awareness leads to an eventual lack of compliance with precautionary measures and could turn into more disruptions in the overall management of the outbreak and its full containment. To sum, the resilience building should be heavily relied on governance, through interdisciplinary methods, integrative approaches, and participatory opportunities (Cheshmehzangi and Dawodu 2018). In doing so, the city's resilience becomes all-inclusive and reflective of the conditions of the outbreak.

6.2.2 *From Resilience to Management*

As part of the overall preparedness, the most crucial step is to integrate resilience planning into city management. In order to increase government accountability and effective responsiveness, we should utilise an adequate number of guidelines, assessment, and management considerations—in other words, resilience-based city management. In different urban studies, there are signs of resilience and management together, from the perspective of urban change (Zapata Campos and Zapata 2012), planning and climate change (Jabareen 2013), public-private-people partnership processes (Marana et al. 2018), multiple stakeholder involvement (Gimenez et al. 2017), urban eco-system services (McPhearson et al. 2014), risk assessment of various kind (Coafee 2016; Pyrko et al. 2017), urban sustainability matters (Girard 2011; Chelleri et al 2015), city infrastructures (Ng and Xu 2015; Reiner and McElvaney 2017), etc. The two complement each other, which is also evident from the studies of cities and their enhancement (Cheshmehzangi 2016). This approach of combined resilience and management is also applied to studies of flood resilience (Liao 2012; Batica et al.

2013), multi-stakeholder networks for building city resilience (Gimenez et al. 2017; Spaans and Waterhout 2017; Gimenez et al. 2018), disaster management (Brogt et al. 2015), risk city resilience trajectories (Jabareen 2015), and how they could be integrated into urban systems (Chelleri 2012; Cheshmehzangi 2020a). Such measures feed into adaptive thinking approaches, for the overarching ideas of adaptive planning (Alterman 1988; Kato and Ahern 2008; Rosenthal and Brechwald 2013; van Veelen 2016), adaptive governance (Folke et al. 2005; Brunner and Lynch 2010; Termeer et al. 2010; Bronen and Chapin 2013; Seeliger and Turok 2014; Green et al. 2016; Karpouzoglou et al. 2016; Hong and Lee 2018), or adapting/adaptive city (Verebes 2013; Errigo 2018), which have roots in social sciences or social studies (Trist 1976) and are meant to address the uncertainties in planning (Kato and Ahern 2008; Rauws and de Roo 2016). In other studies, this approach is used to guide transformations (Rauws 2017), some through integrated adaptive planning (Jim et al. 2015; Hudec 2017), institutional design (de Roo 2015), and policy development (Peck 2012).

Furthermore, the adaptive approach, particularly from a resilience-based management perspective, can lead to the development of an action plan. This is regarded as an essential factor for the integration or transference of resilience thinking to city management. In doing so, the city has to address any matters of resilience planning and conduct comprehensive sets of risk assessment and risk management. This requires adaptive measures that lead to the development of new and temporary policies, regulations, and guidelines (see Chap. 4). Some may appear simple but would be effective. For instance, on the contrary to our usual sustainable measures, it is important to allow for more private transportation use during the outbreak, and perhaps provide parking fee exemption to encourage private car use rather than public transportation. This is a simple adaptive measure but it certainly helps to reduce the transmission through our often-crowded public transportation. In our highlights on resilience thinking, we urge for the city-level outbreak action plan, which should be informative and reflective through regular updates. This is no longer just the readiness of the city against the outbreak, but should indeed represent feasible and practical planning for the purpose of multi-sectoral city management. As discussed throughout the book, this city-level approach requires to be all-inclusive in order to address any vulnerabilities, shortages, deficiencies, and possible impacts on the city and its communities. In summary, the management should be similar to a larger scale incident management, which includes preparedness, mitigation, response, and recovery (EMSA 2014; CHA 2015; Banach et al. 2017). These are recognised for the basics of comprehensive planning, in order to cover prevention measures, risk reduction plan, and the required activities and actions for the case of a health emergency. In doing so, the city can strengthen its responsiveness through resilience-based management.

6.2.3   From Management to Containment...and then Recovery

In this level, the city should prove its robustness and should indeed accentuate how it can progress successfully. This is a critical stage in which all the efforts are eventually paid off. If successful, the city would manage to reach containment before finding its path to recovery. The cities that succeed to reach early containment are the ones either well prepared for the event, or not affected significantly by the disease outbreak. In the former group, they have learnt from their experiences and have developed a robust action plan. And in the latter group, which occurs more often, the city would still require to have an action plan. Hence, the ultimate goal of any city that faces the threats of a health emergency would be to reach early containment. The earlier this occurs the lower the impacts on the communities. Indubitably, an early containment enables the city to manage the adversities that may cause significant effects on public health, socio-economic factors, and primary operations. The early containment should be regarded as a prevention stage, where later transmission and transition phases could be avoided.

Consequently, to govern the city in need, one has to understand the conditions and reflect on those promptly. This is certainly more than just the issues of social cohesion and economic factors (Keil and Ali 2007), and embraces an array of factors that require all-inclusive governance and support. The process of management to containment requires active surveillance (Nash et al. 2001) or syndromic surveillance (Heffernan et al. 2004), and a combination of safety procedures with careful control and monitory measures (Cheshmehzangi 2020a). As expressed before, the city needs to develop scenario-based approaches and regulatory approaches to management (Chamberlain et al. 2017; Weiss et al. 2017), which should be all-inclusive, reflective and adaptive. Incontrovertibly, the city has to create an emergency-based ecosystem of management. In addition, the utilisation of any possible techniques should not be circumvented. In the existing examples of a scenario-based approach to support containment and prevention strategies, we highlight only a few, such as the use of disease-spread simulation models with artificial intelligence (Tsui et al. 2013), the provision and exercise of mock models and practices (Henderson et al. 2001), the use of the random-effects model for comparisons of the disease spread (Garrett et al. 2006), a range of continuous analytical methods and preventive practices (Onowhakpor et al. 2018), the multiple methods of tracking and profiling the disease spread (Obhubunwo et al. 2016), detection analysis of individuals, clusters and transmissions (Wallenstein 1980), network analysis of the disease (Ali and Keil 2006, 2011), etc. In all these cases, we see examples of high level analytical and assessment approaches to detect, monitor, and control the spread of the disease. These are essential for the overall city management, the ultimate containment of the disease outbreak, and towards recovery.

Furthermore, throughout the process, it is vital to have people as the priority to reach containment and recovery. In their study on 'how to vaccinate a city against panic', Glass and Schoch-Spana (2002, p. 217) provide their suggestions in the form of five guidelines to integrate the public into planning processes, including: "*(1) treat the public as a capable ally in the response to an epidemic, (2) enlist civic organisations in practical public health activities, (3) participate the need for home-based patient care and infection control, (4) invest in public outreach and communication strategies, and (5) ensure that planning reflects the values and priorities of affected populations*". In reality, we can verify that people are part of governance. They are not only meant to receive updates and communication from the governments but are proactive actors towards containment and recovery. As discussed before, the methods we utilise to reach containment ought to develop in integrative ways that are multi-sectoral, multi-spatial, and multi-objective. These should involve a larger group of stakeholders, and particularly the general public. In summary, the step-by-step process to containment and then recovery should be taken accurately and with the highest level of attentiveness (i.e. both care and caution). In doing so, managing the city would succeed to reach containment and then recovery.

6.3 '10 Recommendations' for the City in Need

In this section, we reflect on all the lessons learned from current practices, literature review, and the experience of the recent COVID-19 outbreak. In doing so, we summarise a list of 10 recommendations (Fig. 6.1), which could be utilised as the general guide for saving the city in need during the future outbreak events. Intentionally, these recommendations are proposed as general points so that we can easily utilise and implement them for any city that suffers from a disruptive outbreak event at any time. This list serves as the summary of practical guidelines, which also resembles the main aim of the book; i.e. to enhance the city's resilience, and to optimise the city's management. These recommendations are not meant to be paradigms, but they should get regularly updated to remain adaptive and reflective of the realities of the outbreak and the context. In the least effective way, if not anything more, the following recommendations ought to support the city's immediate action plan against the adversities of the outbreak.

(1) *Temporary utilisation of places and facilities*

In order to minimise the impacts on public health services and facilities, especially if they are limited in numbers and/or with limited workforces, it is recommended to separate the treatment operations from other operations. A possible approach is to temporary utilise places and facilities that are either redundant, or are not linked to primary parts of the city. The examples of this recommendation include assembly of temporary structures and units in designated areas, utilisation of large scale open areas with access to medical and emergency services, utilisation of redundant but safe areas/buildings that include basic facilities, and utilisation of buildings with large open spaces for treatment (Cheshmehzangi 2020b). In all cases, any use of such facilities should be carefully monitored and controlled in terms of access, usage,

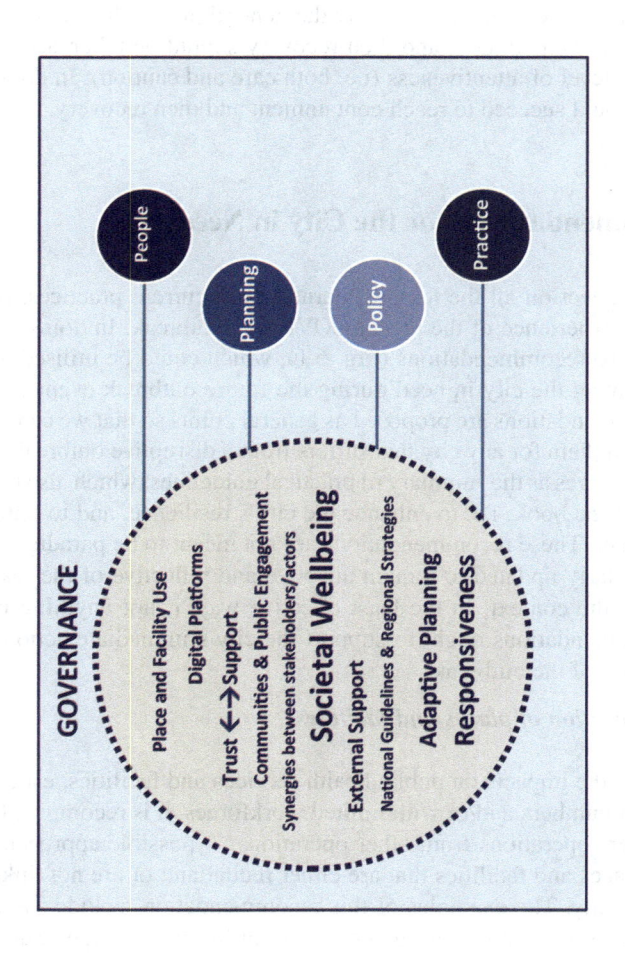

Fig. 6.1 Summary of 10 recommendations supporting governance from four perspectives of people, planning, policy and practice. *Source* Author's own

allocation of zones, operations, deliveries, necessary supply provisions, and regular disinfection procedures. The safety of workforces must remain as the priority, and the overflow of patients (infected cases and potential cases) should be monitored throughout all operations. Partly associated with this recommendation, it is also suggested to have designated spaces for emergency equipment and storage in all healthcare services and facilities. This provision is aimed to help early preparedness and maintain the safety of healthcare workforces and units.

(2) *Digital Platforms for multiple uses*

In the age of digital technologies, it is simply wrong if we are unable to use our available and non-imaginary digital platforms in the time of need. Digital platforms can come in different forms and can offer a variety of support. They can be recognised as innovative assets for us to reduce the impact of disruptions, to maintain the healthiness of city operations, and to implement adaptive measures for any planning and management. Some examples of digital techniques are, the use of artificial intelligence in the detection and monitory procedures, data-driven approaches to health monitory assessment and risk assessment, integrated information-based models for risk management, etc. Moreover, we can use digital tools extensively for various uses. Some are already commonly used in various locations, such as digital payment methods, online or distance learning (DL) education systems, online shopping, online postal services, online banking, home-based work, online meetings, online trade, e-commerce, etc. Also, the example of official multi-functional digital platforms (as shown in Chap. 5) would be beneficial to ease many operations such as dissemination of updates, provision of guidelines, education, training, etc. Altogether, digital platforms are useful to build up the resilience of certain operations and to help with better practices of control and monitory. The use of digital technologies is certainly widespread and also more effective in medical and treatment procedures. Also, at multiple levels and as part of the operations of many urban systems, such technologies can be utilised to achieve smart-resilient approaches for city management.

(3) *Maintaining trust to generate support*

Throughout the book, the involvement of people in various procedures is highlighted as effective methods for the development of all-inclusive approaches and the ultimate development of trust. The role of people needs to be sustained as the priority and actors of decision-making processes. This would help to maintain trust between the government and people, and generate support from the communities. The possible methods of achieving this trust and ensuing support are through transparency in data and information share, official means of communication, clear and supportive announcements and updates, the involvement of community representatives in task forces or specific groups involved in contingency plans, action plans or strategic plans, empowerment of community-level support, and attention to vulnerable groups and communities (including elderly, jobless and low-income groups, single parents, people with mobility difficulties, people with high-risk health conditions, etc.). Alongside these, we recommend having a more responsive approach to a variety of factors, such as community-level safety and security, provision of support to local businesses as well as small and medium enterprises/companies, maintenance of primary services and amenities, monitory of potential disorders or associated crime, and societal well-being. The latter, if prioritised, then has to be reflected on public health services, provision of social services, and the support that will be required to address any vulnerabilities of the society. Lastly, the reliability and responsibility of the government should be maintained and respected through a healthy mutual process of support.

(4) *Empowering the communities through public engagement*

Under the overarching factors of the previous recommendation, the empowerment of communities through healthy public engagement approaches is meant to enhance the power and competence of the society (Cheshmehzangi and Dawodu 2018). If regarded as the priority for societal wellbeing matters, then community empowerment should be reflected in effective participatory approaches. According to Cheshmehzangi and Dawodu (ibid, pp. 173–177), the best practices for public engagement are expected to provide or consider the followings: power and competence, the by-products of partnership, early participation, education and empowerment, stakeholder analysis, perspectives and context, integration of top-down and bottom-up approaches, institutional support, and participatory plan. While the provision of all may not be achievable in the case of the outbreak, it is recommended to prioritise participatory plans of various kinds, such as the involvement of community representatives, enhancement of community-level safety and security, and the provision of community-level amenities/services. It is also highly recommended to enhance community education and provide opportunities for necessary training. Of all, institutional support is extremely essential to develop an effective public engagement in various ways, including also media and news dissemination.

(5) *Breaking the barriers between multiple actors and stakeholders*

The city includes many urban systems, many sectors, and many actors and stakeholders. In the whole process of outbreak progression, we see the multiplicity of impacts and hence, we suggest a multi-sectoral approach that brings together a larger group of stakeholders. The many examples of country-level and city-level task force groups or control and prevention teams represent a team of diverse stakeholders. They usually operate to address the assessment and management of risks and threats and work in line with governmental strategies and context-specific requirements. Their primary role is also to feed directly to decision-making procedures made at a higher level of governance. This set-up is mainly developed to ensure the inclusiveness of decision-making procedures and to maintain healthy communications with multiple sectors or actors (inclusive of communities or community representatives). During the outbreak, the city management should create synergies between its operational and administrative units, increase its accountability, enhance public-private partnerships, and develop a chance for more interactive opportunities of management and support. In doing so, the management should remain top-down but with the continuous objective-driven involvement of key stakeholders and actors. Therefore, all duties should be clear and any overlapping duties should be made redundant. By breaking the barriers through such an approach, we can simply strengthen the overall preparedness plan of resilience thinking.

(6) *Reducing fear, anxiety, and panic in the society*

It is essential to maintain the socio-economic stability of the city. The roles of government and governance should be clear as they are the representatives of both preparedness and responsiveness. The dissemination of information or updates should be formal yet reassuring, reflecting on educating society and making them more resilient. To reduce the fear in society, the city has to maintain its necessary support through our four dimensions of operations, institutions, services, and supplies. The city should address any vulnerabilities through a robust prioritisation plan, and reduce the possible uncertainties that may create more anxiety. In this regard, we have to take an educational approach to avoid any misinformation, and we have to sustain communication on all measures to ensure the society is aware and in full compliance with the procedures. According to the guidelines of Glass and Schoch-Spana (2002, p. 219) on vaccinating the city against panic, extracted steps for implementation should be for the general public to "*recognise that panic is rare and preventable*", which should be disseminated positively and constructively to the general public. The role of public policy is ever important to ensure public information is released timely and is accurate with adequate instructions of protective measures. The approach should also include methods that go "*beyond the hospital for mass-casualty care*" such as plans for home-based patient care and infection control, development of community-wide response to deal with mass casualties, involve alternative care providers, and use micro-level measures to detect and identify patients, provide information and therapies, and provide access to treatment for all (ibid). Lastly, it is recommended to provide accurate information, which is recognised to be as important as providing medicine (ibid), including: "*(1) plan a health communication strategy that empowers the general public, (2) produce multi-lingual and culturally relevant health information, (3) educate the educators; make use of local spokespersons to disseminate information, and (4) be timely and forthcoming with information about the limited of what is known*". In sum, to support the city overcoming its immediate shocks, we have to create participatory decision making processes to ensure the general public is fully on board, fully informed, and fully prepared against any adversities.

(7) *Welcome external support with careful monitory*

This recommendation may seem unacceptable for many city authorities, but it is an essential matter that should be taken into consideration in the time of need. In the current epoch of extended trades and communications, the city may struggle to survive on its own for a long time. Therefore, it has to work closely with externalities that are essential to its operations, institutions, services, and supplies. Hence, it is important to maintain the supply and demand of the society and keep up the relations to external bodies outside the city boundaries. If necessary, external support should be demanded as early as possible to avoid any eventual shortages or deficiencies in the containment and recovery processes. Through possible external support, the city has to address the needs of the society, such as supplies and services for medical needs (medicine, medical products/facilities, medical units, medical guides, etc.), supply chain needs, emergency units and workforces (i.e. in the case of high-level emergency or crisis), primary assets for services and operations, economic support to cover some of the losses, and the provision of any other primary products. This support, if needed and if taken, should be monitored carefully for safety checks, careful allocation, and proper use. The city authorities should realise the fact that the cityness of the city only exists in the larger network of cities and communities.

(8) *Maintain support from national guidelines and regional strategies*

Associated with the previous recommendation, it is recommended for the city to maintain support from national guidelines and regional strategies. At the regional level, the city needs to strategically maintain the safety and delivery of some of its primary amenities and supplies, such as food, water, energy, as well as regional productions that are important for the city and its communities. At the national level, the city has to follow the new and provisional guidelines—and some in the form of new regulations, new policies, and action plans, and then adapt and localise them to the needs of the city. This should be done in a reflective approach to ensure mandatory measures are in place and advisory notifications/suggestions are considered through a high level of multiple assessments. Through the regional strategies (i.e. at the possible level of province, state, county, or equivalent), the city should develop a set of emergency planning as well as prevention and safety measures. Through the national guidelines, the city has to evaluate and adapt general recommendations (or regulations) in the local context. The procedures should be kept formal and effective, and if needed, they should be taken forward at the multi-level governance. It is highly recommended to stop the potential competing effects of urban competitiveness with neighbouring rivals, and instead, focus more on regional-level support, regional production chains, and regionally implemented control measures. More importantly, it is at the local and regional level that essential strategies could be successfully implemented.

(9) *Consideration of adaptive planning*

By taking into account certain adaptive measures, it is recommended for the city to develop adaptive planning from the inception of the outbreak event, or as early as possible. This means that the relevant strategies developed for the action plan should be adaptable. Hence, they should follow the progress of the outbreak in a reflective approach. The conditions and characteristics of the outbreak in each phase should be assessed carefully as they should continuously be used for any potential alterations, amendments, and readjustments. It is critical that throughout the outbreak progression, we continue in a procedural way to ensure adaptability is part of the strategic planning and decision-making processes. In doing so, the city authorities could reflect on the realities of the city and its communities, such as the actual resilience of the community, recognition of vulnerable groups/sectors/communities, conditions of the health system, socio-economic factors of the city, etc. We also need to do an adequate investigation of the disease itself, and how it may affect the daily operations of the city, how it influences the societal wellbeing, and ultimately what support or measures are then required to reduce those impacts and vulnerabilities. We have to evaluate the speed and mode of the disease transmission and adapt or develop new measures for further implementation. These should be phased properly, addressed reflectively, and developed comprehensively.

(10) *Avoid prolonged stages of the outbreak progression*

As we recommend to respond to the outbreak progression procedurally, it is also recommended to avoid prolonged stages all through. Based on our earlier explanations, we identified six distinct phases of the outbreak progression, namely (1) identification, (2) response and containment, (3) transmission, (4) transition, (5) recovery, and (6) post-recovery (see Chap. 2, Sect. 2.2). During the course of outbreak progression, time is very crucial. In each phase, we recommend having a set of action plans, which could accurately address the specific issues of each stage. Any unintended delays could have adverse impacts on the effectiveness of our response to the outbreak conditions and progression. In the identification phase, we suggest to speed up investigation and research through the right channels. In the case of an outbreak, the government should be informed in the earliest possible time, and with the considerate use of scientific research and the experts' guidelines. In the response and containment phase, actions should be reflective and the relevant city authorities must provide immediate guidelines and precautionary measures. The society should be informed clearly and detailed guidelines should be provided. In the transmission phase, detection of cases, sustained transmissions, pattern(s) of spread, infected clusters, and affected communities should all be the priority. In this phase, mobility should be minimised and primary services/amenities should be protected. In the transition phase, potential risks and threats should be identified and managed with high-level control and monitory measures. Any reversing patterns should be identified and addressed immediately. In the recovery phase, the city may require more time to first contain the outbreak and then have gradual progress to recovery. It is also feasible to skip transmission and transition phases, which is recognised as a successful approach directly from containment to full recovery. In the final phase of post-recovery, the city requires time to heal and the society needs more support to recover. In all phases, we stress the importance of the position of people while we highly recommend avoiding prolonged stages. In doing so, the city authorities can prove to be responsive, reflective, and all-inclusive in saving the city in need. The exception cases are for intentional and well-planned delays in containment processes of response and containment phase and recovery phase. In both phases, we anticipate more planning and attempts to contain the disease outbreak.

6.4   For the City in Need

Apart from a set of practical guidelines on urban resilience and city management for the future outbreak events and their inevitable disruptions, this book also aimed to shed light on some of the realities. It highlighted the many deficiencies and flaws of our institutions, as well as the issues of many governments and societies. It also questioned the governance of multiple levels, and it interrogated some of our wrongly developed structures of political, social, and economic systems. The post-pandemic era of this recent COVID-19 outbreak must bring substantial readjustments, if not changes, to our systems. We have to reflect more cautiously and more responsibly on the realities and address them with better preparedness and responsiveness plans. We have to rethink our never-ending patterns of economic growth, and address the societal needs that are more essential for much healthier societies and more resilient communities. After all, as the highly-respected Mahatma Gandhi puts it well in his famous quote: *"The world has enough for everyone's need, but not enough for everyone's greed"*. Hence, we have to truly understand how we have to restructure our societies more comprehensively, more inclusively, and more competently.

With the globe at a temporary halt, we have to realise how fragmented we have become with so many disputes and inequalities between nations, regions, races, communities, neighbours, etc. With so many scenes of ghost towns across the globe, we have to question the over-commercialisation of our cities. We have to question the specific globalisation trends that made us relying on false media and disconnect us from the realities, science, and excellent education. With many effective global initiatives and agenda in place, such as the sustainable development goals (SDGs), we have to understand that implementation alone is not enough, and we have to achieve what is the best for our societies. With so many entertainment industries facing temporary closures, this should become a moment to realise how much of our so-called normal lives have turned into some sort of entertainment and consumer-based activities, and perhaps often more than what we really need. With so many social disorders, we have to reflect on our education systems and realise what has gone wrong that has weakened the general public knowledge and attitudes of our societies. With so many inactions, we have to realise that when a man's act of worship is more important than the care he must have for his fellow citizens, then we may just be at the fall of humanity itself. With so many fictions against facts, we have to realise how vulnerable we can be that one disease outbreak could even bring the biggest economies to their knees. It could then put us all in a global economic crisis. With so many deficiencies in our capabilities and capacities in global health, we have to see huge paradigm shifts in our healthcare services and systems.

Furthermore, we have to learn from those successful examples, such as those persistent management models that were led by China, Singapore, South Korea, etc. (e.g. through the examples of early webinars and events that were specifically arranged to create a knowledge-share platform between countries/regions, and more). We have to realise that the inactions are not necessarily based on the failures of democracies, but are based on how effective the society plays their part in a collective structure and how robust the governments play their parts in taking the lead. Then we have to realise the issues of misconduct and those that could have made us much stronger in our eventual preparedness and responsiveness plans. We have to remember that for some countries, the deployment of armies across the globe to make wars and kill other people appeared to be much easier than saving their own people inside their country boundaries during the disease outbreak. These failures have to be remembered, and we have to reflect on these realities with an open mind. This should help us to increase our awareness, willingness, and resilience. Finally, how could it be that in the name of humanity, many could not implement stronger measures? But then how could they, in the name of humanity, prioritise the economy against people's health and wellbeing? Undeniably, wellbeing comes with the economic values, but shouldn't they be dealt with through our stronger institutions to start with? Hence, the post-pandemic era ought to include many changes, and it will be more thought-provoking.

To sum, what we have left is a range of questions regarding the multiplicity of actions and inactions. What pragmatic or innovative reactions are the ones identified to be the most impactful in achieving early recovery of full containment? At the city level, how safety and security can be maintained during the outbreak progression? And to what extent of multiple sectors, can safety and security be maintained? In facing the longer degree of lost time and lost activities, the city's recovery phase may get postponed. It then becomes a major challenge to restart the operations, to get back to normalisation, and to minimise the impacts. The city and its communities may suffer from decline and lack of businesses and the afterward economic crisis. It will become a challenge to bring back the businesses on their feet and have the workforces back to gradual operations. While the city struggles to maintain the supply side of the economy, it has to keep the security and safety of its multiple stakeholders. But how exactly can the city cope with the extending list of measures that are put in place during the pandemic event? And how will it cope in a prolonged event that puts pressure on the resilience of communities, public health, economy, businesses, services, industries, etc.? With the many failures we experienced this time, will we be able to restructure our social and political systems? Will we be able to define and refine our economic systems? And will be able to strengthen the sense of global governance in the time of need? And if we are able to achieve any of these matters, or at least, reflect on them more prudently and more thoughtfully, what would be the new normals? And how would they play out to replace some of our outdated and perhaps unsustainable business-as-usual practices? In the continuing events, there are many uneventful instants, many dynamic moments, and many who are no longer with us. We witnessed many devoid of actions and compliance, we noticed growing insecurity and unfairness, and we came to the recognition of many attempts towards recovery and stability. Some temporary provisions may become permanent, and some may simply cut the automation → unemployment → endless economic growth → global unsustainability knot!

If there is anything that could be done to help the city and make it any safer, nobody should hesitate by all means. If we all believe in this principle, then we could have a much stronger society. In facing the disease outbreak, the city needs the support of all stakeholders, particularly the government and society. Hence, we certainly need to acquire resilience both individually and collectively. It is evident that through the experiences of the recent outbreak, the government and society are simply the driving forces of resilience enhancement and city management. The recent COVID-19 outbreak scrutinises many issues related to our health systems, our global economy, our people, our resilience, our management, our mistrust, our international relations, and our behaviours. It also sheds light on many negative factors, such as our acts of racism, our neglect, our irresponsible behaviours, our acts of misinformation, our misjudgement, our lack of knowledge, our vulnerable societies, our weak infrastructures, our delicate cities, and our fragile relations. All of sudden, all these together felt like a major test of humanity. It then brought some key questions that how resilient are we? How prepared are we? How could we respond more promptly? And will this be the right time to revisit our beliefs?

At first, China was an easy target to receive the blames, but as time progressed, it was evident that China responded more effectively than ever; and also perhaps better than the others. The disease could have started anywhere else, as in many previous events; but the way we ought to respond to the situation should not make any difference to what is required in the best possible way. Many communities suffered throughout these few months, and many will continue to deal with the adversities for some time. China and many other countries went through some unprecedented situations, but none may be similar to what Wuhan and its people experienced.

For the majority, life became so static all the sudden and time became meaningless. But for those who suffered directly, lost their loved ones, or fought the disease at the frontline, life was never more dynamic. We learned how to remain individual, we learned how to care about the others, and we learned the ups and downs of disruptions. All these changed our lifestyles, and we had to adapt ourselves to the ever-changing conditions on a daily basis. For many, this became a living nightmare, for many it was time for anxiety and fear, and for the majority, it was just an unprecedented experience. In this process, we lost many of our social values, and then we remembered that we are more individual than social. This outbreak changed our perceptions and made us realise how vulnerable and how fragile everything is; and how weak(er) we could become if the outbreak was more severe than what we experienced this time. Despite all these, we have to reflect on the realities around us, and then evolve intellectually to become more resilient.

At the time when social distancing became a norm, we realised the lack of adaptability in our presumed cultural and social needs. We realised what has become a normal lifestyle is simply not sustainable. We realised what we often can do is overlooked by our actions and inactions. We also realised that organising our social and political systems are ever important. And in these realisations, we hopefully became aware of our unsustainable patterns of production and consumption. After all, with or without us, life will go on!

At the time this book is completing, China has passed through its peak of the outbreak under the shadow of future waves of the outbreak, several other countries are upsettingly struggling in different regions, the number of cases is increasing rapidly in Europe, South America, Southeast Asia, North America, and the Middle

East, there are major worries about poorer or more vulnerable nations of the global south, the spread is continuing globally with many travel bans and restrictions, many cities and regions are under full lockdown, many borders are closed, there are signs of longer-term economic crisis, and the race for vaccine development is ongoing. It is important to note that many of these issues could have been averted or curtailed if we acted more promptly and much robustly. Also, these could have been evaded with better resilience and management. Clearly, there were also signs of many earlier neglect and inactions. By the time this book is published, we hope we have passed through some of these adversities and remember how to be well prepared and how to respond more effectively next time. This book should be available as a general guide, and should reflect on some of the realities we experienced this time. In the future, it is very likely for us to have more outbreak events and it is likely that we have to reflect on these days more thoroughly and thoughtfully. The city will survive and we have to remain as resilient as we can. We may need to face similar situations in the future, and once again, we have to find the right ways to save the city in need.

References

Ali, H. S., & Keil, R. (2006). Global cities and spread of infectious disease: The case of severe acute respiratory syndrome (SARS) in Toronto, Canada. *Urban Studies, 43*(3), 492–509.

Ali, H. S., & Keil, R. (Eds.). (2011). *Networked disease: Emerging infections in the global city.* Part of IJURR Studies in Urban and Social Change Book Series 45, Chichester: Blackwell Publishing.

Alterman, R. (1988). Adaptive planning. *Cognitive Science, 12,* 393–421.

Bacallao, J., Schneider, M. C., Najera, P., Aldighieri, S., Soto, A., Marquino, W., et al. (2014). Socioeconomic factors and vulnerability to outbreak of leptospirosis in Nicaragua. *International Journal of Environmental Research and Public Health, 11*(8), 8301–8318.

Banach, D. B., Johnston, B. L., Al-Zebeidi, D., Bartlett, A. H., et al. (2017). Outbreak Response and Incident Management: SHEA Guidance and Resources for Healthcare Epidemiologists in United States Acute-Care Hospitals. *Infection Control and Hospital Epidemiology, 38*(12), 1393–1419.

Batica, J., Gourbesville, P., & Hu, F.-Y. (2013). *Methodology for flood resilience index.* In *International conference on flood resilience experiences in Asia and Europe–ICFR*, Exeter, United Kingdom.

Brogt, E., Grimshaw, M., & Baird, N. (2015). Clergy views on their role in city resilience: Lessons from Canterbury earthquakes. *Kotuitui: New Zealand Journal of Social Sciences Online, 10*(2), 83–90.

Bronen, R., & Chapin, F. S. (2013). Adaptive governance and institutional strategies, for climate-induced community relocations in Alaska. *Proceedings of the National Academy of Sciences, 110*(23), 9320–9325.

Brunner, R., & Lynch, A. (2010). *Adaptive governance and climate change.* Boston: The American Meteorological Society.

California Hospital Association (CHA). (2015). *What is the relationship between an Emergency Management Program (EMP) and an Emergency Operations Plan (EOP)?* Retrieved March 12, 2020, from https://www.calhospitalprepare.org/post/what-relationship-between-emergency-man agement-program-emp-and-emergency-operations-plan-eop-0.

Chamberlain, A. T., Lehnert, J. D., & Berkelman, R. L. (2017). The 2015 New York City Legionnaires' disease outbreak: A case study on a history-making outbreak. *Journal of Public Health Management and Practice, 23*(4), Article no. 410.

Chelleri, L. (2012). From the resilient city to urban resilience: A review essay on understanding and integrating the resilience perspectives for urban systems. *Documents d'analisi geografica, 58*(2), 287–306.

Chelleri, L., Schuetze, T., & Salvati, L. (2015). Integrating resilience with urban sustainability in neglected neighbourhoods: Challenges and opportunities of transitioning to decentralizsed water management in Mexico City. *Habitat International, 48,* 122–130.

Cheshmehzangi, A. (2016). City Enhancement beyond the Notion of "Sustainable City": Introduction to Integrated Assessment for City Enhancement (iACE) Toolkit. *Energy Procedia, 104,* 153–158.

Cheshmehzangi, A. (2020a). Comprehensive Urban Resilience for the City of Ningbo (in Chinese: 宁波市城市综合抗灾弹性框架). Report submitted to local government units in February 2020, Ningbo, China.

Cheshmehzangi, A. (2020b). *10 Adaptive Measures for Public Places to face the COVID-19 Pandemic Outbreak,* City & Society, Article ID: CISO_12282, https://doi.org/10.1111/CISO. 12282.

Cheshmehzangi, A., & Dawodu, A. (2018). *Sustainable urban development in the age of climate change–people: The cure or curse.* Singapore: Palgrave Macmillan.

Coafee, J. (2016). *Terrorism, risk, and the global city: Towards Urban Resilience.* New York: Routledge.

de Roo, G. (2015). Going for Plan B-Conditioning adaptive planning: About urban planning and institutional design in a non-linear, complex world. In R. Geyer & P. Cairney (Eds.), *Handbook on complexity and public policy.* Cheltenham: Edward Elgar Publishers.

Emergency Medical Services Authority (EMSA). (2014). *Hospital incident command system (HICS) guidebook* (5th ed.). Retrieved March 12, 2020, from https://www.uchcoalition.org/wp-content/uploads/HICS-Guidebook-2014.pdf.

Errigo, M. F. (2018). The adapting city: Resilience through water design in Rotterdam. *TeMA Journal of Land Use, Mobility and Environment, 11*(1), 51–64.

Folke, C., Hahn, T., Olsson, P., & Norberg, J. (2005). Adaptive governance for social-ecological systems. *Annual Review of Environment and Resources, 30,* 441–473.

Fraser, E. D. G. (2003). Social vulnerability and ecological fragility: Building bridges between social and natural sciences using the Irish Potato Famine as a case study. *Conservation Ecology, 7*(2).

Fraser, E. D. G., Mabee, W., & Figge, F. (2005). A framework for assessing the vulnerability of food systems to future shocks. *Futures, 37*(6), 465–479.

Garrett, V., Bornschlegel, K., Lange, D., Reddy, V., Kornstein, L., Kornblum, J., et al. (2006). A recurring outbreak of Shigella Sonnei among traditionally observant Jewish children in New York City: The risks of dyacare and household transmission. *Epidemiology and Infection, 134*(6), 1231–1236.

Gimenez, R., Labaka, L., & Hernantes, J. (2017). A maturity model for involvement of stakeholders in the city resilience building process. *Technological Forecasting and Social Change, 121,* 7–16.

Gimenez, R., Labaka, L., & Hernantes, J. (2018). Union means strength: Building city resilience through multi-stakeholder collaboration. *Journal of Contingencies and Crisis Management, 26*(3), 385–393.

Girard, L. F. (2011). Creativity and the human sustainable city: Principles and approaches for nurturing city resilience. In T. Baycan & L. F. Girard (Eds.), *Sustainable city and creativity: Promoting creative urban initiatives* (pp. 55–96). New York: Routledge.

Glass, T. A., & Schoch-Spana, M. (2002). Bioterrorism and the people: How to vaccinate a city against panic. *Clinical Infectious Diseases, 34*(2), 217–223.

Green, O. O., Garmestani, A. S., Albro, S., Ban, N. C., Berland, A., Burkman, C. E., et al. (2016). Adaptive governance to promote ecosystem services in urban green spaces. *Urban Ecosystems, 19*(1), 77–93.

Heffernan, R., Mostashari, F., Das, D., Karpati, A., Kulldorff, M., & Weiss, D. (2004) Syndromic surveillance in public health practice, New York City. *Emerging Infectious Diseases, 10*(5). Retrieved March 13, 2020, from https://wwwnc.cdc.gov/eid/article/10/5/03-0646_article.

Henderson, D. A., Inglesby T. V. Jr., O'Toole, T., Inglesby, T. V., Grossman, R., & O'Toole, T. (2001). A plague on your city: Observations from TOPOFF. *Clinical Infectious Diseases, 32*(3), 436–445. Retrieved March 13, 2020, from https://academic.oup.com/cid/article/32/3/436/283538.

Hong, S., & Lee, S. (2018). Adaptive governance and decentralization: Evidence from regulation of the sharing economy in multi-level governance. *Government Information Quarterly, 35*(2), 299–305.

Hudec, O. (2017). Cities of resilience: Integrated adaptive planning. *Quality Innovation Prosperity, 21*(1), 106–118.

Jabareen, Y. (2013). Planning the resilient city: Concepts and strategies for coping with climate change and environmental risk. *Cities, 31,* 220–229.

Jabareen, Y. (2015). *The risk city-cities countering climate change: Emerging planning theories and practices around the world, chapter on: The risk city resilience trajectory* (pp. 137–159). Singapore: Springer.

Jim, C. Y., Lo, A. Y., & Byrne, J. A. (2015). Charting the green and climate-adaptive city. *Landscape and Urban Planning, 138,* 51–53.

Karpouzoglou, T., Dewulf, A., & Clark, J. (2016). Advancing adaptive governance of social-ecological systems through theoretical multiplicity. *Environmental Science & Policy, 57,* 1–9.

Kato, S., & Ahern, J. (2008). 'Learning by doing': Adaptive planning as a strategy to address uncertainty in planning. *Journal of Environmental Planning and Management, 51*(4), 543–559.

Keil, R., & Ali, H. S. (2007). Governing the sick city: Urban governance in the age of emerging infectious disease. *Antipode, 39*(5), 846–873.

Liao, K.-H. (2012). A theory on urban resilience to floods–a basis for alternative planning practices. *Ecology and Society, 17*(4).

Marana, P., Labaka, L., & Sarriegi, J. M. (2018). A framework for public-private-people partnerships in the city resilience-building process. *Safety Science, 110,* 39–50.

McPhearson, T., Hamstead, Z. A., & Kremer, P. (2014). Urban ecosystem services for resilience planning and management in New York City. *Ambio, 43*(4), 502–515.

Ministry of Health (MoH) website. (2020). Singapore's MoH Main page on coronavirus updates. Retrieved March 11, 2020, from https://www.moh.gov.sn/highlights/details/confirmed-imported-cases-of-novel-coronavirus-infection-in-Singapore-multi-ministry-taskforce-ramps-up-precautionary-measures.

Monstadt, J., & Schmidt, M. (2019). Urban resilience in the making? The governance of critical infrastructures in German cities. *Urban Studies, 56*(11), 2353–2371.

Nash, D., Mostashari, F., Fine, A., Miller, J., O'Leary, D., et al. (2001). The outbreak of West Nile virus infection in the New York City area in 1999. *New England Journal of Medicine, 344*(24), 1807–1814.

Ng, T. S. T., & Xu, J. (2015). *An integrated framework for resilience management of inter-network city infrastructures.* In *Proceedings for 2015 International Conference on Building Resilience.*

Nieddu, G. T., Billings, L., Kaufman, J. H., Forgoston, E., & Bianco, S. (2017). Extinction pathways and outbreak vulnerability in a stochastic Ebola model. *Journal of the Royal Society, Interface, 14*(127), 20160847.

Obhubunwo, C., Ameh, C., Oduyebo, O., Ahumibe, A., Mutiu, B., et al. (2016). Clinical profile and containment of the Ebola virus disease outbreak in two large West African cities, Nigeria, July-September 2014. *International Journal of Infectious Diseases, 53,* 23–29.

Onowhakpor, A. O., Adam, V. Y., Sakpa, O. E., & Ozokwelu, L. U. (2018). Status of Ebola Virus Disease (EVD) preventive practices among health-care practice among health care workers (HCWs) in Benin City: A year after disease containment in Nigeria. *The Pan Africa Medical Journal, 30*(50), 1–8.

Parkins, J. R., & MacKendrick, N. A. (2007). Assessing community vulnerability: A study of the mountain pine beetle outbreak in British Columbia, Canada. *Global Environmental Change, 17*(3–4), 460–471.

Peck, J. (2012). Receptive city: Amsterdam, vehicular ideas and adaptive spaces of creativity policy. *International Journal of Urban and Regional Research, 36*(3), 462–485.

Pyrko, I., Howick, S., & Eden, C. (2017). Risk systemicity and city resilience. In *EURAM 2017*.

Rancourt, N. (2013). *Gender and vulnerability to cholera in sierra leone: Gender analysis of the 2012 Cholera outbreak and an assessment of Oxfam's response*. Report Document, OXFAM GB.

Rauws, W. (2017). Embracing uncertainty without abandoning planning: Exploring an adaptive planning approach for guiding urban transformations. *disP-The Planning Reviews, 53*(1), 32–45.

Rauws, W., & de Roo, G. (2016). Adaptive planning: Generating conditions for urban adaptability: Lessons from Dutch organic development strategies. *Environment and Planning B: Planning and Design, 43*(6), 1052–1074.

Reiner, M., & McElvaney, L. (2017). Foundational infrastructure framework for city resilience. *Sustainable and Resilient Infrastructures, 2*(1), 1–7.

Rosenthal, J. K., & Brechwald, M. (2013). Climate adaptive planning for preventing heat-related health impacts in New York City. In J. Knieling & W. Leal Filho (Eds.), *Climate change governance*. Part of the Climate Change Management book series (CCM) (pp. 205–225) Singapore: Springer.

Schemann, K., Firestone, S. M., Taylor, M. R., Toribio, J-A. LML, Ward, M. P., & Dhand, N. K. (2013). Perceptions of vulnerability to a future outbreak: A study of horse managers affected by the first Australian equine influenza outbreak. *BMC veterinary Research, 9*(1), Article no. 152.

Seeliger, L., & Turok, I. (2014). Avering a downward spiral: Building resilience in informal urban settlements through adaptive governance. *Environment and Urbanization, 26*(1), 184–199.

Shafqat Akanda, A., Jutla, A. S., Gute, D. M., Sack, R. B., Alam, M., Huq, A., et al. (2013). Population vulnerability to biannual cholera outbreaks and associated macro-scale drivers in the Bengal Delta. *The American Journal of Tropical Medicine and Hygiene, 80*(5), 950–959.

Spaans, M., & Waterhout, B. (2017). Building up resilience in cities worldwide–Rotterdam as participant in the 100 Resilient Cities Programme. *Cities, 61*, 109–116.

Termeer, C. J. A. M., Dewulf, A., & Van Lieshout, M. (2010). Disentangling scale approaches in governance research: Comparing monocentric multilevel and adaptive governance. *Ecology and Society, 15*(4), Article no. 29.

Trist, E. L. (1976). *Action research and adaptive planning*. In A. W. Clark (Ed.), *Experimenting with organizational life: The action research approach* (pp. 223–236). Singapore: Springer.

Tsui, K.-L., Wong, Z. S.-Y., Goldsman, D., & Edesess, M. (2013). Tracking infectious disease spread of global pandemic containment. *IEEE Intelligent Systems, 28*(6), 60–64.

Van Veelen, P. C. (2016). *Adaptive planning for resilient coastal waterfronts: Linking flood risk reduction with urban development in Rotterdam and New York City*, A+BE, Architecture and the Built Environment.

Verebes, T. (2013). *Masterplanning the adaptive city: Computational urbanism in the twenty-first century*. New York: Routledge.

Wallenstein, S. (1980). A test for detection of clustering over time. *The American Journal of Epidemiology, 111*, 367–372.

Weiss, D., Boyd, C., Rakeman, J. L., Greene, S. K., et al. (2017). A large community outbreak of Legionnaires' disease associated with a cooling tower in New York City, 2015. *Public Health Reports, 132*(2), 241–250.

Zapata Campos, M. K., & Zapata, P. (2012). Changing La Chureca: Organizing city resilience through action nets. *Journal of Change Management, 12*(3), 323–337.

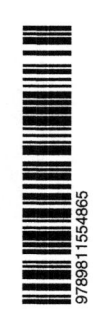

9789811554865